# WITHDRAWN

## Carnegie Mellon

# MODELING REMEDIAL ACTIONS AT UNCONTROLLED HAZARDOUS WASTE SITES

# MODELING REMEDIAL ACTIONS AT UNCONTROLLED HAZARDOUS WASTE SITES

by

S.H. Boutwell, S.M. Brown
B.R. Roberts, D.F. Atwood

Anderson-Nichols & Co., Inc.
Palo Alto, California

NOYES PUBLICATIONS
Park Ridge, New Jersey, U.S.A.

Copyright © 1986 by Noyes Publications
Library of Congress Catalog Card Number 86-5253
ISBN: 0-8155-1084-5
ISSN: 0090-516X
Printed in the United States

Published in the United States of America by
Noyes Publications
Mill Road, Park Ridge, New Jersey 07656

10 9 8 7 6 5 4 3 2 1

Library of Congress Cataloging-in-Publication Data

Modeling remedial actions at uncontrolled hazardous
    waste sites.

    (Pollution technology review, ISSN 0090-516X ;
no. 130)
    Includes bibliographies and index.
    1. Hazardous waste sites--Mathematical models--
Handbooks, manuals, etc.  2. Water--Pollution--
Mathematical models--Handbooks, manuals, etc.
3. Water, Underground--Pollution--Mathematical models--
Handbooks, manuals, etc.  I. Boutwell, S.H.
II. Series.
TD811.5.M63   1986     363.7'28    86-5253
ISBN 0-8155-1084-5

# Foreword

Guidance for the selection and use of models for evaluating the effectiveness of remedial actions at uncontrolled hazardous waste sites is presented in this book. In the past, assessment of remedial action performance at uncontrolled hazardous waste sites has largely been accomplished using best engineering judgement. This may be appropriate for many sites, but there are sites where conditions are not understood, and selection and design of appropriate remedial actions is not apparent. Models can be used to supplement best engineering judgement by providing a quantitative assessment of site conditions and remedial action performance. This could allow more accuracy and confidence in decisions concerning technical and cost-effectiveness of remedial actions. This book will give the reader pertinent information on the selection and application of predictive tools for the control of specific problems at uncontrolled waste sites. It is a valuable guide for this purpose.

The book is composed of four volumes. Volume 1 presents a model selection methodology based on flow charts and matrices. Three basic decisions form the framework: Is modeling required? If so, what level of modeling is required? What are the required model capabilities? Volumes 2, 3, and 4 describe guidance for simplified methods for evaluation of subsurface and waste control actions, numerical models for the same actions, and analytical and numerical model use for evaluation of remedial actions in surface water, respectively. Techniques and literature data useful in estimating parameter values are provided.

Model application guidance includes (1) the general capabilities of different types of models, including sources of information on specific models; (2) factors to consider when linking different models to assess complicated site and remedial action conditions; (3) the steps to follow in applying models for remedial action assessment; (4) user expertise and resource requirements; (5) alternative ways of analyzing remedial action performance; and (6) key assumptions and limitations affecting the use of specific models.

The information in the book is from *Modeling Remedial Actions at Uncontrolled Hazardous Waste Sites,* prepared by S.H. Boutwell, S.M. Brown, B.R. Roberts, and D.F. Atwood of Anderson-Nichols & Co., Inc. for the U.S. Environmental Protection Agency, April 1985.

The table of contents is organized in such a way as to serve as a subject index and provides easy access to the information contained in the book.

> Advanced composition and production methods developed by Noyes Publications are employed to bring this durably bound book to you in a minimum of time. Special techniques are used to close the gap between "manuscript" and "completed book." In order to keep the price of the book to a reasonable level, it has been partially reproduced by photo-offset directly from the original report and the cost saving passed on to the reader. Due to this method of publishing, certain portions of the book may be less legible than desired.

## NOTICE

The material in this book was prepared as an account of work sponsored by the U.S. Environmental Protection Agency. It has been subject to the Agency's peer and administrative review and has been approved for publication as an EPA document.

This document is intended to present information on the selection and application of predictive tools for the control of specific problems caused by uncontrolled waste sites. It is not intended to address every conceivable waste site problem or all possible applications. Mention of trade names or commercial products does not constitute endorsement or recommendation for use.

# Acknowledgments

This document represents the combination of efforts for the U.S. Environmental Protection Agency, Office of Research and Development (ORD), by Anderson-Nichols & Co., Inc. (ANCo) in Palo Alto, CA. Mr. Douglas Ammon of the Hazardous Waste Engineering Research Laboratory (HWERL) coordinated the development of this document. Mr. Lee A. Mulkey of the Environmental Research Laboratory (ERL) in Athens, GA, was the project officer.

The technical project monitors, authors, and reviewers are listed below.

*Volume 1:* Selection of Models for Remedial Action Assessment. Technical project monitor: Mr. Thomas O. Barnwell, Jr., ERL; Authors: Mr. Scott H. Boutwell, Mr. Stuart M. Brown, and Dr. Benjamin R. Roberts.

*Volume 2:* Simplified Methods for Subsurface and Waste Control Actions. Technical project monitor: Mr. Douglas Ammon, HWERL; Author: Mr. Stuart M. Brown.

*Volume 3:* Numerical Modeling of Surface, Subsurface and Waste Control Actions. Technical project monitor: Mr. Douglas Ammon; Authors: Mr. Stuart M. Brown, Mr. Scott H. Boutwell, Dr. Benjamin R. Roberts, and Ms. Dorothy Fisher Atwood.

*Volume 4:* Analytical and Numerical Models for the Evaluation of Remedial Actions in Surface Water. Technical project monitor: Mr. Thomas O. Barnwell, Jr.; Authors: Mr. Scott H. Boutwell and Dr. Benjamin R. Roberts.

Technical review of this document was provided by Mr. Anthony S. Donigian, Jr. who also served as the project director. Other reviewers are listed below.

Charles R. Cole, Battelle Pacific Northwest Laboratories
William Fallon, Office of Research and Development

      Wayne C. Huber, Ph.D., University of Florida
      Yasuo Onishi, Ph.D., Battelle Northwest Laboratories
      Richard Stanford, Clean Sites Inc.
      Paul K.M. Van der Heijde, International Ground Water Modeling
        Center
      David T. Williams, U.S. Army Waterways Experiment Station
      David B. Watson, Anderson-Nichols & Co., Inc.

Dr. Richard T.Y. Lo participated in the development of model application guidelines and remedial action modeling requirements in Volume 3.

Ms. Susan Reutter-Harrah supervised report production. Word processing was provided by Ms. Carol McCullough and Ms. Dorothy Inahara. Graphics were developed by Ms. Marythomas Hutchins and Ms. Virginia Rombach.

# Contents and Subject Index

INTRODUCTION . . . . . . . . . . . . . . . . . . . . . . . . . . . . . . . . . . . . . . 1

### VOLUME 1
### SELECTION OF MODELS FOR REMEDIAL ACTION ASSESSMENT

1. EXECUTIVE SUMMARY . . . . . . . . . . . . . . . . . . . . . . . . . . . . . . . 4

2. INTRODUCTION . . . . . . . . . . . . . . . . . . . . . . . . . . . . . . . . . . . 6
   2.1 Purpose of This Report . . . . . . . . . . . . . . . . . . . . . . . . . . . . 6
   2.2 Definition of Models . . . . . . . . . . . . . . . . . . . . . . . . . . . . . 7
   2.3 The Role of Remedial Action Assessment in the Remedial
       Response Process . . . . . . . . . . . . . . . . . . . . . . . . . . . . . . . 8
       2.3.1 The Role of Models in Remedial Action Assessment . . . . . . . . 8
   2.4 Framework and Organization of Report . . . . . . . . . . . . . . . . . . . 10
       2.4.1 Basic Framework for Model Selection . . . . . . . . . . . . . . . . 10
       2.4.2 Organization of Report . . . . . . . . . . . . . . . . . . . . . . . . 10
   2.5 Caveats of Use . . . . . . . . . . . . . . . . . . . . . . . . . . . . . . . . 12
       2.5.1 Assumed User Knowledge and Expertise . . . . . . . . . . . . . . . 12
       2.5.2 Understanding of the Site and Remedial Actions . . . . . . . . . . 12
       2.5.3 Limitations of the Model Selection Methodology . . . . . . . . . . 13

3. THE FIRST DECISION: IS MODELING NECESSARY? . . . . . . . . . . . . . 14
   3.1 Overview . . . . . . . . . . . . . . . . . . . . . . . . . . . . . . . . . . . . 14
   3.2 The Decision to Model: Flow Chart . . . . . . . . . . . . . . . . . . . . 14

4. METHODOLOGY FOR MODEL SELECTION FOR SOIL AND
   GROUND-WATER CONTAMINATION PROBLEMS . . . . . . . . . . . . . . 18
   4.1 Overview . . . . . . . . . . . . . . . . . . . . . . . . . . . . . . . . . . . . 18
   4.2 What Level of Modeling Is Required? . . . . . . . . . . . . . . . . . . . 19

x  Contents and Subject Index

 4.3 Required Model Capabilities for Each Level . . . . . . . . . . . . . . . . . . 23
  4.3.1 Level I Analysis. . . . . . . . . . . . . . . . . . . . . . . . . . . . . . . . . . 23
  4.3.2 Level II Analysis . . . . . . . . . . . . . . . . . . . . . . . . . . . . . . . . . 27
 4.4 Resource and Data Availability . . . . . . . . . . . . . . . . . . . . . . . . . . . 32
  4.4.1 Level I Analysis. . . . . . . . . . . . . . . . . . . . . . . . . . . . . . . . . . 32
  4.4.2 Level II Analysis. . . . . . . . . . . . . . . . . . . . . . . . . . . . . . . . . 34
 4.5 Model Selection Criteria for Soil and Ground-Water
   Contamination Problems. . . . . . . . . . . . . . . . . . . . . . . . . . . . . . . 36

5. METHODOLOGY FOR MODEL SELECTION FOR SURFACE
  WATER CONTAMINATION PROBLEMS . . . . . . . . . . . . . . . . . . . . . . 38
 5.1 Overview. . . . . . . . . . . . . . . . . . . . . . . . . . . . . . . . . . . . . . . . . . 38
 5.2 What Level of Modeling Is Required? . . . . . . . . . . . . . . . . . . . . . . 40
 5.3 Required Model Capabilities for Each Level . . . . . . . . . . . . . . . . . . 44
  5.3.1 Level I Analysis. . . . . . . . . . . . . . . . . . . . . . . . . . . . . . . . . . 44
  5.3.2 Level II Analysis. . . . . . . . . . . . . . . . . . . . . . . . . . . . . . . . . 48
 5.4 Resource and Data Availability . . . . . . . . . . . . . . . . . . . . . . . . . . . 52
 5.5 Model Selection Criteria for Surface Water Remedial Action
   Assessment. . . . . . . . . . . . . . . . . . . . . . . . . . . . . . . . . . . . . . . . 52

REFERENCES . . . . . . . . . . . . . . . . . . . . . . . . . . . . . . . . . . . . . . . . . . 56

## VOLUME 2
## SIMPLIFIED METHODS FOR SUBSURFACE AND
## WASTE CONTROL ACTIONS

1. INTRODUCTION . . . . . . . . . . . . . . . . . . . . . . . . . . . . . . . . . . . . . . 60
 1.1 Purpose of Report. . . . . . . . . . . . . . . . . . . . . . . . . . . . . . . . . . . . 60
 1.2 Report Organization . . . . . . . . . . . . . . . . . . . . . . . . . . . . . . . . . . 61

2. SUMMARY AND CONCLUSIONS. . . . . . . . . . . . . . . . . . . . . . . . . . . 62

3. REMEDIAL ACTION EVALUATION WITH SIMPLIFIED METHODS . . . 65
 3.1 Overview. . . . . . . . . . . . . . . . . . . . . . . . . . . . . . . . . . . . . . . . . . 65
 3.2 Subsurface Control Measures. . . . . . . . . . . . . . . . . . . . . . . . . . . . 70
  3.2.1 Capping and Top Liners . . . . . . . . . . . . . . . . . . . . . . . . . . . 70
  3.2.2 Seepage Basins and Ditches. . . . . . . . . . . . . . . . . . . . . . . . . 71
  3.2.3 Subsurface Drains, Ditches and Bottom Liners . . . . . . . . . . . 72
  3.2.4 Impermeable Barriers. . . . . . . . . . . . . . . . . . . . . . . . . . . . . 72
  3.2.5 Ground-Water Pumping . . . . . . . . . . . . . . . . . . . . . . . . . . . 74
  3.2.6 Interceptor Trenches . . . . . . . . . . . . . . . . . . . . . . . . . . . . . 74
 3.3 Waste Control. . . . . . . . . . . . . . . . . . . . . . . . . . . . . . . . . . . . . . 75
  3.3.1 Permeable Treatment Beds . . . . . . . . . . . . . . . . . . . . . . . . . 75
  3.3.2 Bioreclamation . . . . . . . . . . . . . . . . . . . . . . . . . . . . . . . . . 75
  3.3.3 Chemical Injection. . . . . . . . . . . . . . . . . . . . . . . . . . . . . . . 76
  3.3.4 Solution Mining (Extraction). . . . . . . . . . . . . . . . . . . . . . . . 77
  3.3.5 Excavation/Hydraulic Dredging . . . . . . . . . . . . . . . . . . . . . . 77

4. THEORY UNDERLYING AVAILABLE SIMPLIFIED METHODS......79
   4.1 Overview..............................................79
   4.2 Well Hydraulics......................................79
       4.2.1 Confined Aquifers..............................80
       4.2.2 Leaky Aquifers.................................83
       4.2.3 Water Table Aquifers..........................87
       4.2.4 Available Well Hydraulics Solutions..........91
   4.3 Drain Hydraulics....................................94
   4.4 Ground-Water Mounding Estimation Methods........101
   4.5 Seepage/Infiltration Estimation Methods.........106
   4.6 Superposition......................................108
   4.7 Transformation Methods............................119
   4.8 Conformal Mapping..................................124
   4.9 Contaminant Transport.............................128

5. AVAILABLE HAND-HELD CALCULATOR AND MICRO-
   COMPUTER PROGRAMS.....................................135
   5.1 Overview............................................135
   5.2 Available Programmable, Hand-Held Calculator Programs........136
   5.3 Available Programs for Micro-Computers...........146

6. EXAMPLE APPLICATIONS..................................156
   6.1 Overview............................................156
   6.2 Example 1: Water Table Suppression with an Interceptor Trench...156
   6.3 Example 2: Plume Capture with a Pumping/Injection Doublet....161
   6.4 Example 3: Ground-Water Pumping with and Without an
       Impermeable Barrier................................167
   6.5 Example 4: Recirculation System for Ground-Water Clean-Up....175
   6.6 Example 5: Drain Recirculation System............178

REFERENCES................................................183

## VOLUME 3
## NUMERICAL MODELING OF SURFACE, SUBSURFACE AND WASTE CONTROL ACTIONS

1. INTRODUCTION..........................................196
   1.1 Purpose of Report..................................196
   1.2 Report Organization...............................197

2. CONCLUSIONS...........................................199

3. MIGRATION AND FATE PROCESSES........................201
   3.1 Overview............................................201
   3.2 Processes Controlling Movement Within Zones.....205
       3.2.1 Advection....................................205
       3.2.2 Dispersion..................................205
       3.2.3 Erosion......................................206

        3.2.4 Sorption/Retardation............................. 206
   3.3 **Processes Controlling Transfers Between Zones**................ 207
        3.3.1 Evapotranspiration ............................. 207
        3.3.2 Infiltration..................................... 207
        3.3.3 Drainage ..................................... 207
        3.3.4 Volatilization.................................. 207
   3.4 **Processes Controlling Transformation/Degradation** ............ 208

**4. REMEDIAL ACTIONS AND AFFECTED PROCESSES**............ 209
   4.1 **Overview**........................................... 209
   4.2 **Surface Control**...................................... 211
        4.2.1 Grading...................................... 211
        4.2.2 Revegetation ................................. 211
        4.2.3 Surface Water Diversion and Collection ................ 216
   4.3 **Subsurface Control** .................................. 216
        4.3.1 Capping and Top Liners .......................... 216
        4.3.2 Subsurface Drains and Bottom Liners ................. 217
        4.3.3 Ground-Water Pumping .......................... 217
        4.3.4 Interceptor Trenches ............................ 217
        4.3.5 Seepage Basins and Ditches....................... 221
        4.3.6 Impermeable Barriers............................ 221
   4.4 **Waste Control**...................................... 221
        4.4.1 Permeable Treatment Beds ....................... 224
        4.4.2 Bioreclamation ................................ 224
        4.4.3 Chemical Injection.............................. 224
        4.4.4 Solution Mining (Extraction)........................ 227
        4.4.5 Excavation and Hydraulic Dredging .................. 227

**5. NUMERICAL MODEL APPLICATION GUIDELINES**.............. 229
   5.1 **Overview**........................................... 229
   5.2 **Numerical Model Capabilities**........................... 230
        5.2.1 Surface Zone Models ............................ 231
        5.2.2 Unsaturated Zone Models ........................ 236
        5.2.3 Saturated Zone Codes ........................... 237
   5.3 **Interactions Between Models**........................... 239
        5.3.1 Soft Linkage of Codes ........................... 240
        5.3.2 Generic Bridge Program Design ..................... 242
   5.4 **Model Application Process** ............................ 243
   5.5 **User Expertise and Resource Requirements** ................ 248
   5.6 **Analysis of Remedial Action Performance Using Numerical Models** ............................................. 250
        5.6.1 Assessment of Reductions in Future Exposure Levels....... 250
        5.6.2 Uncertainty Regarding Remedial Action Performance ...... 251
        5.6.3 Optimization of Remedial Action Design................ 254
        5.6.4 Assessment of Design Life and Impacts of Failure ......... 256

**6. REMEDIAL ACTION MODELING REQUIREMENTS**.............. 261
   6.1 **Overview**........................................... 261

   6.2 Modeling Requirements . . . . . . . . . . . . . . . . . . . . . . . . . . . . 262
       6.2.1 Capping, Grading, and Revegetation . . . . . . . . . . . . . . . . 264
       6.2.2 Surface Water Diversion and Collection . . . . . . . . . . . . . . . 269
       6.2.3 Ground-Water Pumping and Interceptor Trenches . . . . . . . . 270
       6.2.4 Impermeable Barriers. . . . . . . . . . . . . . . . . . . . . . . . . . . 271
       6.2.5 Subsurface Drains and Solution Mining . . . . . . . . . . . . . . . 276
       6.2.6 Excavation. . . . . . . . . . . . . . . . . . . . . . . . . . . . . . . . . . 277
       6.2.7 Hydraulic Dredging and Seepage Basins . . . . . . . . . . . . . . . 278
       6.2.8 Bioreclamation and Chemical Injection . . . . . . . . . . . . . . . 280
       6.2.9 Permeable Treatment Beds . . . . . . . . . . . . . . . . . . . . . . . 281
   6.3 Parameter Estimation Guidance . . . . . . . . . . . . . . . . . . . . . . . . 282
       6.3.1 Surface Zone Model Parameter Guidance. . . . . . . . . . . . . . 282
           6.3.1.1 Channel/Surface Roughness. . . . . . . . . . . . . . . . . 282
           6.3.1.2 Evapotranspiration . . . . . . . . . . . . . . . . . . . . . . 284
           6.3.1.3 Interception. . . . . . . . . . . . . . . . . . . . . . . . . . . 284
           6.3.1.4 Infiltration. . . . . . . . . . . . . . . . . . . . . . . . . . . . 287
           6.3.1.5 Soil Erodibility. . . . . . . . . . . . . . . . . . . . . . . . . 287
       6.3.2 Subsurface Modeling Parameters. . . . . . . . . . . . . . . . . . . . 289
           6.3.2.1 Flow-Related Parameters . . . . . . . . . . . . . . . . . . 290
           6.3.2.2 Transport-Related Parameters . . . . . . . . . . . . . . . 295

REFERENCES. . . . . . . . . . . . . . . . . . . . . . . . . . . . . . . . . . . . . . . . . 309

APPENDIX A: SUPPORTING INFORMATION ON HSPF, FEM-
WATER/FEMWASTE AND FE3DGW/CFEST . . . . . . . . . . . . . . . . . . . . 319
   A.1 Code Selection . . . . . . . . . . . . . . . . . . . . . . . . . . . . . . . . . . 319
   A.2 Linkage of HSPF, FEMWATER/FEMWASTE and CFEST. . . . . . . . 322
   A.3 Model Code Implementation. . . . . . . . . . . . . . . . . . . . . . . . . 324
   A.4 Sources of Code Documentation and User Assistance. . . . . . . . . . 325
   A.5 Parameter Adjustments Required for Each Remedial Measure . . . . 326

## VOLUME 4
## ANALYTICAL AND NUMERICAL MODELS FOR THE
## EVALUATION OF REMEDIAL ACTIONS IN SURFACE WATER

1. INTRODUCTION. . . . . . . . . . . . . . . . . . . . . . . . . . . . . . . . . . . . . 344
   1.1 Background . . . . . . . . . . . . . . . . . . . . . . . . . . . . . . . . . . . . 344
   1.2 Purpose of Report. . . . . . . . . . . . . . . . . . . . . . . . . . . . . . . . 345
   1.3 Report Content. . . . . . . . . . . . . . . . . . . . . . . . . . . . . . . . . . 346

2. MIGRATION AND FATE. . . . . . . . . . . . . . . . . . . . . . . . . . . . . . . . 347
   2.1 Overview. . . . . . . . . . . . . . . . . . . . . . . . . . . . . . . . . . . . . . 347
   2.2 Physical Processes . . . . . . . . . . . . . . . . . . . . . . . . . . . . . . . . 347
       2.2.1 Overview. . . . . . . . . . . . . . . . . . . . . . . . . . . . . . . . . . . 347
       2.2.2 Rivers/Streams . . . . . . . . . . . . . . . . . . . . . . . . . . . . . . . 351
       2.2.3 Impoundments. . . . . . . . . . . . . . . . . . . . . . . . . . . . . . . 351
       2.2.4 Estuaries . . . . . . . . . . . . . . . . . . . . . . . . . . . . . . . . . . . 352
   2.3 Chemical Processes . . . . . . . . . . . . . . . . . . . . . . . . . . . . . . . 353

2.3.1 Hydrolysis . . . . . . . . . . . . . . . . . . . . . . . . . . . . . . . . . 353
   2.3.2 Oxidation . . . . . . . . . . . . . . . . . . . . . . . . . . . . . . . . . 359
   2.3.3 Photolysis . . . . . . . . . . . . . . . . . . . . . . . . . . . . . . . . 359
   2.3.4 Volatilization . . . . . . . . . . . . . . . . . . . . . . . . . . . . . . 359
   2.3.5 Adsorption. . . . . . . . . . . . . . . . . . . . . . . . . . . . . . . . 360
   2.3.6 Bio-Degradation . . . . . . . . . . . . . . . . . . . . . . . . . . . . 360
   2.3.7 Bio-Accumulation or Bio-Magnification. . . . . . . . . . . . . . 361
   2.3.8 Precipitation/Dissolution . . . . . . . . . . . . . . . . . . . . . . 361

3. REMEDIAL ACTIONS AND AFFECTED CRITICAL PROCESSES . . . . 363
   3.1 Overview . . . . . . . . . . . . . . . . . . . . . . . . . . . . . . . . . . . . 363
   3.2 Dilution . . . . . . . . . . . . . . . . . . . . . . . . . . . . . . . . . . . . . 366
   3.3 Containment Actions . . . . . . . . . . . . . . . . . . . . . . . . . . . . 366
      3.3.1 Booms . . . . . . . . . . . . . . . . . . . . . . . . . . . . . . . . . . . 366
      3.3.2 Silt Curtains . . . . . . . . . . . . . . . . . . . . . . . . . . . . . . 367
      3.3.3 Cofferdams . . . . . . . . . . . . . . . . . . . . . . . . . . . . . . . 367
      3.3.4 Barriers/Diversions . . . . . . . . . . . . . . . . . . . . . . . . . 367
      3.3.5 Capping . . . . . . . . . . . . . . . . . . . . . . . . . . . . . . . . . 371
   3.4 Removal Measures . . . . . . . . . . . . . . . . . . . . . . . . . . . . . . 371
      3.4.1 Skimming . . . . . . . . . . . . . . . . . . . . . . . . . . . . . . . . 371
      3.4.2 Hydraulic Dredging . . . . . . . . . . . . . . . . . . . . . . . . . 371
      3.4.3 Mechanical Dredging . . . . . . . . . . . . . . . . . . . . . . . . 372
      3.4.4 Excavation . . . . . . . . . . . . . . . . . . . . . . . . . . . . . . . 372
   3.5 Treatment Measures . . . . . . . . . . . . . . . . . . . . . . . . . . . . . 372

4. USE OF REMEDIAL ACTIONS AND MODELING: CASE
   HISTORIES . . . . . . . . . . . . . . . . . . . . . . . . . . . . . . . . . . . . . . 374
   4.1 Overview . . . . . . . . . . . . . . . . . . . . . . . . . . . . . . . . . . . . 374
   4.2 Case Histories . . . . . . . . . . . . . . . . . . . . . . . . . . . . . . . . . 374
      4.2.1 Hudson River PCB Spill . . . . . . . . . . . . . . . . . . . . . . . 374
      4.2.2 Waukegan Harbor PCB Spill . . . . . . . . . . . . . . . . . . . . 377
      4.2.3 Iron Mountain Mine Site . . . . . . . . . . . . . . . . . . . . . . 377
      4.2.4 Kepone Contamination in the James River . . . . . . . . . . . 378
      4.2.5 Formalin Spill on the Russian River . . . . . . . . . . . . . . . 378
      4.2.6 Triana DDT Site . . . . . . . . . . . . . . . . . . . . . . . . . . . . 379
      4.2.7 Marathon Oil Spill . . . . . . . . . . . . . . . . . . . . . . . . . . 379
      4.2.8 Chlorine Barge Spill . . . . . . . . . . . . . . . . . . . . . . . . . 379
   4.3 Summary . . . . . . . . . . . . . . . . . . . . . . . . . . . . . . . . . . . . 379

5. USE OF ANALYTICAL AND SIMPLIFIED ASSESSMENT
   TECHNIQUES FOR REMEDIAL ACTION SCREENING AND
   ASSESSMENT . . . . . . . . . . . . . . . . . . . . . . . . . . . . . . . . . . . . 381
   5.1 Overview . . . . . . . . . . . . . . . . . . . . . . . . . . . . . . . . . . . . 381
   5.2 Uses of Simplified Assessment Techniques . . . . . . . . . . . . . . . 382
   5.3 Classification of Simplified Assessment Techniques . . . . . . . . . 384
      5.3.1 Near-Field Analyses . . . . . . . . . . . . . . . . . . . . . . . . . 384
      5.3.2 Far-Field Analyses . . . . . . . . . . . . . . . . . . . . . . . . . . 386
      5.3.3 Transformation Equations . . . . . . . . . . . . . . . . . . . . . 388

  5.3.4 Sediment-Water Interactions . . . . . . . . . . . . . . . . . . . . . . . 388
 5.4 Analytical Models . . . . . . . . . . . . . . . . . . . . . . . . . . . . . . . . . . 389
  5.4.1 Overview . . . . . . . . . . . . . . . . . . . . . . . . . . . . . . . . . . . . . 389
  5.4.2 Selected Analytical Models . . . . . . . . . . . . . . . . . . . . . . . 390

6. USE OF NUMERICAL MODELS FOR REMEDIAL ACTION
 ASSESSMENT . . . . . . . . . . . . . . . . . . . . . . . . . . . . . . . . . . . . . . . 393
 6.1 Overview . . . . . . . . . . . . . . . . . . . . . . . . . . . . . . . . . . . . . . . . . 393
 6.2 Capabilities of Available Codes . . . . . . . . . . . . . . . . . . . . . . . . . 394
 6.3 The Model Development and Application Process . . . . . . . . . . . . 402

7. MODEL REQUIREMENTS FOR SURFACE WATER REMEDIAL
 ACTIONS . . . . . . . . . . . . . . . . . . . . . . . . . . . . . . . . . . . . . . . . . . 405
 7.1 Overview . . . . . . . . . . . . . . . . . . . . . . . . . . . . . . . . . . . . . . . . . 405
 7.2 Modeling Requirements . . . . . . . . . . . . . . . . . . . . . . . . . . . . . . 408
  7.2.1 Dilution . . . . . . . . . . . . . . . . . . . . . . . . . . . . . . . . . . . . . . 408
  7.2.2 Containment: Booms and Partial Barriers . . . . . . . . . . . . . . 410
  7.2.3 Containment: Cofferdams and Full Barriers . . . . . . . . . . . . . 411
  7.2.4 Containment: Silt Curtains . . . . . . . . . . . . . . . . . . . . . . . . 412
  7.2.5 Containment: Capping . . . . . . . . . . . . . . . . . . . . . . . . . . . 413
  7.2.6 Removal: Hydraulic and Mechanical Dredging . . . . . . . . . . 414
  7.2.7 Removal: Excavation . . . . . . . . . . . . . . . . . . . . . . . . . . . . 417
  7.2.8 Treatment . . . . . . . . . . . . . . . . . . . . . . . . . . . . . . . . . . . . 417
 7.3 Parameter Estimation Guidance . . . . . . . . . . . . . . . . . . . . . . . . . 418
  7.3.1 Source Term Parameters . . . . . . . . . . . . . . . . . . . . . . . . . . 418
  7.3.2 Sediment Parameters . . . . . . . . . . . . . . . . . . . . . . . . . . . . 419
   7.3.2.1 Sediment Transport Parameters . . . . . . . . . . . . . . . 420
   7.3.2.2 Contaminated Sediment Bed Parameters . . . . . . . . . 425
  7.3.3 Boundary Condition Parameters . . . . . . . . . . . . . . . . . . . . . 428
  7.3.4 Dispersion Parameters . . . . . . . . . . . . . . . . . . . . . . . . . . . 429
   7.3.4.1 Longitudinal Dispersion Coefficient ($K_x$) . . . . . . . . . 429
   7.3.4.2 Transverse Mixing Coefficient . . . . . . . . . . . . . . . . 431
   7.3.4.3 Vertical Mixing Coefficient . . . . . . . . . . . . . . . . . . 431

REFERENCES . . . . . . . . . . . . . . . . . . . . . . . . . . . . . . . . . . . . . . . . . . 433

# Introduction

The National Contingency Plan (NCP) sets forth a process for the evaluation and selection of remedial actions at uncontrolled hazardous waste sites. One element in this process is the Engineering Feasibility Study. This study is itself a staged process that involves the screening of remedial action technologies, the detailed analysis of potentially feasible alternatives and the conceptual design of the most cost-effective alternative.

Throughout the feasibility study process a number of factors are considered when evaluating remedial actions. These factors include technical feasibility, costs, institutional constraints, and potential environmental and public health impacts. The level of attention given to each of these factors depends upon which stage in the process is being performed. During the screening stage, the intent is to reduce the large number of potential technologies to a workable number by identifying those that are clearly infeasible or inappropriate. Best engineering judgement supplemented by order-of-magnitude estimates of remedial action performance are usually sufficient during this stage.

Once a set of potentially feasible alternatives has been identified, each one has to be evaluated in detail. Again, best engineering judgement supplemented by more quantitative estimates of performance largely provide the basis for the identification of one or more cost-effective actions.

The final step is to develop a conceptual design for one or more alternatives. This involves identifying the performance expectations for the alternative, design criteria, preliminary layout and process diagrams, operation and maintenance requirements, monitoring requirements, and costs. This step requires an even more quantitative analysis of performance.

Modeling, whether it be through the use of relatively simple analytical solutions or more sophisticated numerical codes, is beginning to be used more and more throughout the feasibility study process. This four volume series is intended to provide guidance on both the selection and use of a range of modeling techniques applicable to the evaluation of remedial actions for ground-water and surface water contamination problems.

Volume 1, Selection of Models for Remedial Action Assessment, provides a methodology for the selection of models. The methodology addresses three key decisions: 1) whether modeling should be considered; 2) if so, what level of model sophistication is appropriate; and 3) what capabilities should the model or models have. The first decision is critical because modeling is appropriate for only certain sites. The second decision is important because the level of model sophistication will determine the level of resources that must be allocated. The final decision ensures that the selected model will be appropriate for the site conditions and remedial actions that need to be assessed. Once a selection has been made, the remaining volumes can be consulted to obtain guidance on model use.

Volume 2, Simplified Methods for Subsurface and Waste Control Actions, provides a compilation of simplified methods, or analytical and semi-analytical solutions, applicable to the evaluation of subsurface and waste control actions. The primary emphasis of this volume is on identifying the methods that can be used to evaluate specific actions and the assumptions and limitations affecting their use. A compilation of available hand-held calculator and micro-computer programs for different types of methods is also provided. The simplified methods contained in Volume 2 are useful for screening remedial actions and, in some cases, detail analysis and conceptual design.

Volume 3, Numerical Modeling of Surface, Subsurface and Waste Control Actions, provides guidance on the use of numerical models for sites where more detailed analyses are required and where sufficient resources are available. The volume focusses on the use of surface, unsaturated and saturated zone models. Important considerations related to the use of numerical models are discussed, as are modeling requirements for specific surface, subsurface and waste control actions. Modeling requirements include: 1) the type of model required to analyze an action; 2) the dimensionality and grid configuration required to represent an action; and 3) model parameter adjustments required to simulate the effects of implementing an action. Guidance on the estimation of model parameters is presented for situations where site characterization data are unavailable.

Volume 4, Analytical and Numerical Models for the Evaluation of Surface Water Remedial Actions, provides remedial action modeling guidance for surface water contamination problems. Simplified methods and analytical and numerical models applicable to the analysis of specific actions are discussed. Considerations related to the application of both types of models are presented along with modeling requirements for different actions.

# VOLUME 1

Selection of Models
for
Remedial Action Assessment

# 1. Executive Summary

This volume provides general guidance in the selection of models for remedial assessment at uncontrolled hazardous waste sites. Guidance is provided in the form of a series of flow charts and matrices leading to model selection. The purpose of this format is to make the model selection methodology as utilitarian and "user-friendly" as possible. As the methodology is used at different sites, user confidence and expertise will increase. With successful application experience, the model selection guide may become an integral tool in the Remedial Investigation/Feasibility Study process.

The selection of models is a function of the objectives of the modeling study, complexity of the site, and type of remedial actions being considered. These areas are represented in the methodology by: the required level of modeling flow chart; matrices of remedial actions vs. processes and required dimensionality; and discussions of time frame criteria. Other criteria that are important include model performance and data/resource availability.

The same model selection methodology can be applied to the selection of models for site characterization and exposure assessment. The user again must identify the purpose of the modeling study in order to determine the level of modeling required, and evaluate required model capabilities based on the complexity of the site, including significant environmental pathways of exposure, and potential receptors. In short, many of the model selection criteria for remedial action assessment are applicable to exposure assessment and site characterization as well.

This volume is designed to provide guidance for model selection and use in the Remedial Investigation/Feasibility Study process. The document as a whole should provide a comprehensive set of guidelines to Federal and State officals for the incorporation of models into the remedial action planning process at state and federal superfund sites.

# 2. Introduction

## 2.1 PURPOSE OF THIS REPORT

Existing state and federal funds for the clean-up of uncontrolled hazardous waste sites are limited. Consequently, proper selection of remedial actions is critical to ensure that effective measures are implemented at as many sites as possible, and that future costs as a result of inadequate actions are minimal.

Mathematical models can be used to assess the performance of remedial actions, and thus complement the analyst's expertise and judgement for selection and design of these actions. Although the concept for using models for remedial action assessment is relatively new, successful demonstrations are evident. The recent application of models for remedial action evaluation at the La Bounty landfill site in Charles City, Iowa, the Gilson Rd. site in Nashua, New Hampshire, and the Love Canal site in Niagara, New York has proven that they can provide information useful in selecting and designing actions.

Effective use of models for this purpose depends on the selection of models most suitable to the job. There are many models available today, varying in terms of complexity and purpose of use. Actual selection can be difficult, especially if one is not completely familiar with the important site and remedial action criteria needed to choose the appropriate model. This volume was developed for use as a model selection guide for assessing remedial action performance at uncontrolled hazardous waste sites. It is intended to assist state or regional staff in assessing the need for analytical predictive tools at these sites, for use by themselves or to evaluate site contractor proposals.

It should be emphasized that the model selection methodology was designed for remedial action assessment for surface water and subsurface contamination problems. However, many of the same models and model selection criteria are appropriate for both site characterization and exposure assessments. Reports by Adkins et al. (1983) and Freed et al. (1983) also provide methodologies for assessing exposure to chemical substances.

The model selection methodology is specifically for surface and ground-water contamination. Although some of overland and unsaturated zone models consider air contamination problems such as volatile emissions and fugitive dust release, model selection for air contamination control is not covered in this report. The user should refer to other reports such as Farino et al. (1983), Thibodeaux (1981) and Dynamac (1982) for models that simulate air emissions, fugitive dust, and their appropriate control technologies at hazardous waste sites.

## 2.2 DEFINITION OF MODELS

Before the model selection methodology is presented, definition of the terms "model" and "level" must be clarified. Throughout this volume we will refer to two general classes of models, based on level of complexity. They are analytical and numerical, designated as Levels I and II, respectively.

Analytical and numerical models incorporate equations to quantitatively predict results, with varying levels of accuracy. The major difference between the two types of models is the level of simplification. Analytical models (Level I) rely on simplifying assumptions such as isotropic (Hydraulic conductivity is equal for all directions: $x$, $y$, $z$) and homogeneous conditions, steady flow, and regular geometry. Their range of accuracy is around 1 order of magnitude (EPA, 1982). Numerical models (Level II) utilize the same equations, but can simulate varying processes, fluxes, and geometries, by nature of their solution techniques. Their range of accuracy is closer to a factor of 2-4 (EPA, 1982). However, for both types of models, accuracy will be also dependent upon application. Because of the number of calculations required, all numerical models and some analytical models require a computer to solve equations that calculate exposure concentrations.

2.3 THE ROLE OF REMEDIAL ACTION ASSESSMENT IN THE REMEDIAL RESPONSE PROCESS

2.3.1 The Role of Models in Remedial Action Assessment

In the past, selection and design of remedial actions has largely been accomplished through field data collection and best engineering judgement. These approaches may be sufficient for sites where environmental pathways and potential receptors are clearly defined, and where previous installations of a given remedial action have proved successful. This past experience of both identifying pathways and applying specific remedial measures may also be sufficient for sound selection at new sites with similar characteristics. However, there are a number of relatively complex sites where best engineering judgement may not provide enough guidance to allow for the proper selection and design of a remedial action. For those sites, the use of analytical and numerical predictive tools may be appropriate to obtain a _quantitative_ assessment of remedial action performance. Best engineering judgement, then, may be supplemented with quantitative results. This will potentially lead to more accurate and confident decisions.

Models have potential use in the screening of alternatives, analysis of alternatives, and conceptual design tasks. Descriptions of model use for each task are provided below. Actual selection of the _level_ of model required is dependent upon site and remedial action criteria, and to a lesser extent, the current phase of the remedial investigation/ feasibility study. Other criteria of importance can include resource and data availability and previous model performance. This information will be covered in detail in Sections 4 and 5.

_Screening of alternatives_ is performed to eliminate those actions deemed unfeasible due to technical, public health/institutional, and cost reasons. Models may be used at this stage to determine the general technical feasibility, and any potential environmental impacts arising from implementation of different remedial actions. Since the screening analysis is essentially the first iteration of subsequent, more detailed analyses, a ballpark or order-of-magnitude estimate of effectiveness is usually all that is required. For this reason, Level I models are often sufficient.

Those remedial actions that pass the initial screening effort will then be subject to <u>analysis of alternatives</u>. Models may be used to obtain information on the effectiveness, durability, and expected exposure levels as a result of the implementation of different actions. The effectiveness of an action is the extent to which it meets a design objective or site clean-up goal. The durability of an action is the length of time it is effective. Durability can be assessed by incorporating design life and risk of failure considerations into modeling of the selected actions. Such considerations may allow the prediction of exposure levels in the event of progressive or catastrophic failure of an action. Exposure levels of contaminants expected with implementation of a remedial action can be determined so compliance with regulatory criteria can be ascertained. Models may be used in the detailed analysis phase for the purposes mentioned above. Level I (Analytical) models are limited to well characterized sites and selected remedial actions. Level II (Numerical) models are more appropriate in cases where site conditions or remedial actions of interest require that changes in material properties and multiple dimensions be considered.

The stage subsequent to the feasibility study is to develop a <u>conceptual design</u> for the most cost-effective action. Again, models can assist in this process by simulating different configurations of the selected action. For example, a ground-water pumping action may be conceptually designed by evaluating pumping rates, number and placement of wells, and location of screened intervals. Again, while analytical or Level I models may be sufficient for some sites and actions, the use of numerical models may be more appropriate for complex configurations.

Because of the unique characteristics of each site/remedial action scenario, the determination, selection and use of modeling have to be addressed on a site-specific basis. However, with sufficient user expertise, field data, and guidance, models may become common tools for the Remedial Investigation/Feasibility Study process.

2.4 FRAMEWORK AND ORGANIZATION OF REPORT

2.4.1 Basic Framework for Model Selection

There are three basic decision points in the methodology discussed in this volume. Figure 2.1 is a flow chart that illustrates the framework. The decisions are:

1. Is modeling necessary

2. If so, what level of modeling is required and

3. What are the required model capabilities of that level

Flow charts and matrices are used to facilitate the model selection process. The sections of the volume that correlate to these decision points are identified below.

2.4.2 Organization of Report

Section 3 discusses the first major decision: "Is modeling necessary?" A flow chart is used to illustrate the hierarchy of questions that must be asked in making this basic decision.

Sections 4 and 5 comprise the major portions of the methodology; Section 4 deals with model selection for remedial action assessment for soil or ground-water contamination problems, and Section 5 covers the same issues for surface water problems. Each section contains the two decision points:

o  What level of modeling is required and

o  What are the required model capabilities

A summary of general model selection criteria is provided in each section to help the user identify his/her specific model requirements and to direct the user to the appropriate model use volume(s) for more information. Data and resource considerations when applying the methodology are also discussed. Section 6 contains references.

Introduction  11

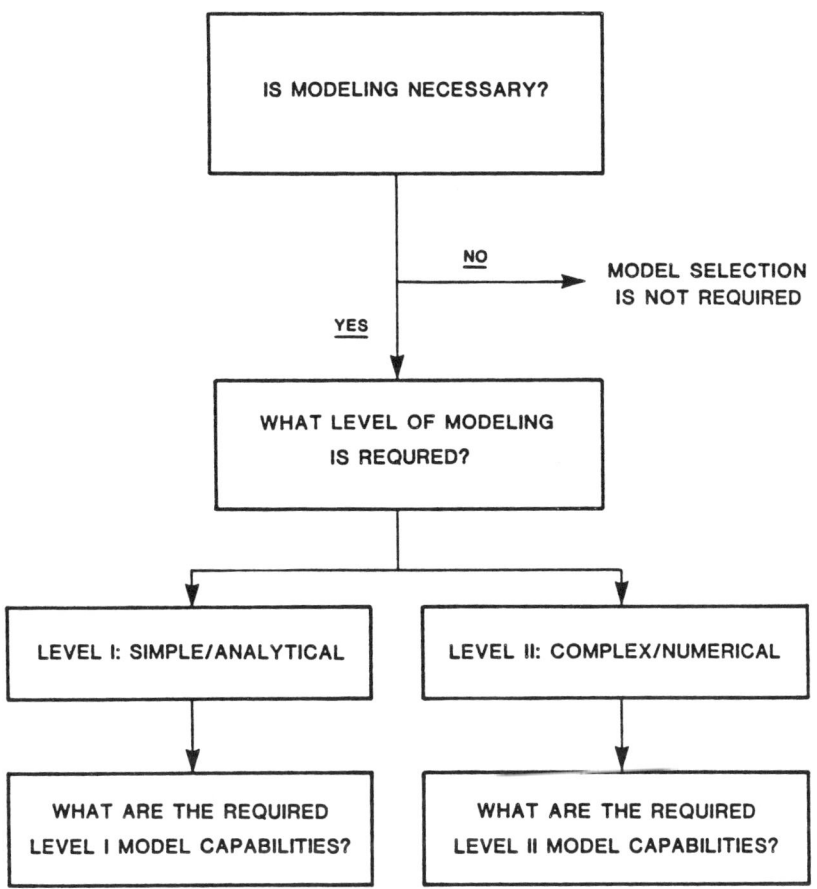

Figure 2.1  Three basic decisions in model selection.

2.5 CAVEATS OF USE

There are a number of assumptions made in this volume regarding user expertise and knowledge of the site, and limitations of the model selection methodology. The user is advised to assess his/her expertise and anticipated support in light of the caveats of use.

### 2.5.1 Assumed User Knowledge and Expertise

The primary group of users is expected to be Federal (EPA) regional and state environmental officials and staff. These people often must evaluate field inspection reports, direct efforts for data acquisition, and evaluate site contractor proposals for remedial action, including recommended models for exposure and remedial action assessment. To make sound decisions using this methodology, the user should have at the minimum a general understanding of a mix of disciplines, such as Hydrology, Civil Engineering, Soil Science, and Environmental Chemistry and an understanding of the basic concepts of chemical transport and fate modeling, such as levels of model complexity and expected accuracy, processes that can be simulated, and parameter estimation techniques. Ideally, academic background in any of the above disciplines supplemented with experience, job training, and/or exposure (e.g., short course attendance) in the other disciplines provides a profile of the recommended background of a user.

### 2.5.2 Understanding of the Site and Remedial Actions

In addition to required expertise, the user should be able to characterize site conditions. This means that major environmental pathways and potential receptors must be identified. Such pathways and receptors can include contaminated runoff into surface waters, leachate migration into ground water, direct contamination of a drinking water aquifer, and release of toxic volatile emissions. This qualitative assessment of the site is a function of the users expertise and the available data, such as observed concentrations in surface water and ground water, and knowledge of the history of the site, including type of contaminants, methods of disposal, and release rates, if such information is available.

The user should be able to categorize groups of remedial actions and correlate them to the major pathways and potential receptors. As models will be evaluated according to their ability to simulate various remedial actions, it is important that the user be familiar with the purpose and basic engineering design of the available remedial action technologies.

### 2.5.3 Limitations of the Model Selection Methodology

This methodology was designed to provide <u>guidance</u> on the selection of models for remedial action assessment in surface and ground waters. Given the myriad possibilities of site conditions, the methodology is directed towards model selection at a <u>generic</u> level. That is, there may be decision points in the methodology where more than one answer exists. It is at these points that sufficient user expertise and knowledge of the site is most critical. Therefore, the information derived from the methodology should supplement existing knowledge and expertise for application to site-specific conditions.

The user should also refer to the subsequent volumes for additional information about specific models, such as user manuals and test case applications, before actual model selection is determined.

It should be emphasized that this manual alone will not be sufficient for actual model selection; the references cited above should be examined, and, if necessary, outside guidance should be obtained in order to facilitate the selection process.

# 3. The First Decision: Is Modeling Necessary?

3.1 OVERVIEW

The decision to use models for remedial action assessment is the first and perhaps most critical in the model selection methodology. This section will help the user to answer the question "Will the resources expended and results obtained be worth the modeling effort?". If so, models can supplement best engineering judgement for remedial action assessment. If not, the user should explore other methods of remedial action assessment, such as collection and analysis of more field data.

The decision to use models is a function of the nature and complexity of the site being considered, as well as the extent of contamination and range of potential remedial actions. As mentioned in the introduction, site characteristics and remedial action criteria form the basis of model selection throughout this report. Examination of these issues allow the user to make the decision to model, and if models are deemed necessary, they set the stage for selection of required model capabilities for subsurface and surface water problems in Sections 4 and 5, respectively.

3.2 THE DECISION TO MODEL: FLOW CHART

Figure 3.1 is a flow chart that can be used to determine if modeling is required. The major issues or decision points are represented using a flow chart in order to facilitate ease of use. Each decision point is discussed below.

The first step is to develop a conceptual understanding of the site. To do this, the user should make assumptions as to the type and degree of hazards at the site, based on

The First Decision: Is Modeling Necessary? 15

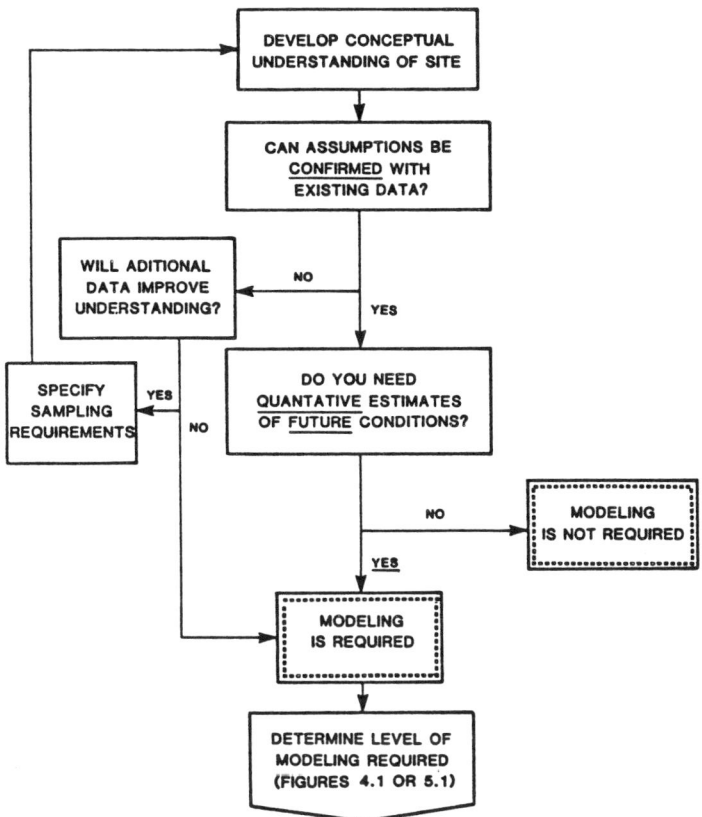

Figure 3.1  Flow chart to determine if modeling is required.

existing data and best engineering judgement. Such assumptions may include: the location of potential sources; the relative importance of different contaminant migration and fate pathways (e.g., air, surface water and ground water); the relative importance of different transport, transformation, and inter-media transfer processes (e.g., contaminated runoff from land into surface water, and volatilization of a pollutant from water to air); and the general type(s) of remedial measures that may be applicable, such as waste isolation, removal, or treatment, leachate or runoff control, or water diversion.

Once an understanding is developed, the user should ask the following question: "<u>Can assumptions be confirmed with existing data?</u>" Such assumptions can include the extent or plume of contaminants and rate of spreading. For example, can the contamination and fate pathways be identified by examining contamination levels in different media?

If the assumption concerning important pathways and receptors cannot be confirmed with existing data, the following question should be asked: "<u>Will additional data improve understanding?</u>" Quite often additional data gathered from sampling programs in the Remedial Investigation/Feasibility Study will be sufficient to confirm one's understanding of the site and help identify appropriate remedial measures. However, in some instances existing data may be confusing or contradict the user's assumptions and understanding of the site. For example, pump tests may reveal that a wide range of hydraulic conductivities are present. Or, estimates and location of the source mass cannot be verified. In these cases, models may serve to <u>interpret</u> and <u>interpolate</u> site conditions to provide a better understanding.

If site pathways are well characterized (i.e., contaminated runoff into a simple water body) and conditions are fairly homogeneous (i.e., one general soil type or single layer aquifer), additional data will probably suffice to confirm assumptions. The user should then <u>specify sampling requirements</u>, obtain more data, and re-iterate the decision process by developing a <u>new</u> understanding of the site based on the new data, and proceed from there.

If additional data will <u>not</u> serve to improve the understanding of the site, <u>modeling is required</u>. The decision to model in these cases is usually dependent primarily on site complexity, as opposed to remedial action criteria.

If the question "Can assumptions be confirmed with existing data?" was answered in the affirmative, the site itself is well enough characterized not to warrant additional data or modeling. The next step is to examine remedial action criteria.

The question "Do you need quantitative estimates of future conditions?" addresses the need to predict potential contamination levels and the effectiveness of remedial actions in reducing those contamination levels. Qualitative estimates of future conditions can be made based on the conceptual understanding of the site and judgement. When contamination pathways are well characterized and past experience indicates that appropriate remedial actions will work, a qualitative assessment of future conditions may be sufficient. However, if multiple or complex pathways are present, selection of the appropriate remedial actions and their configurations are not apparent, and specific regulatory criteria for contamination levels must be met, a quantitative assessment of future conditions may be required. For these cases an affirmative response to the question posed above means that modeling is required. A negative response indicates that modeling is not required. In this latter case, the remainder of the model selection methodology presented herein is not applicable, and data and experience are sufficient for remedial action assessment.

As evidenced by Figure 3.1, the decision to model may be a result of site complexity (i.e., additional data will not improve understanding) and the need for a quantitative assessment of remedial action performance over time.

As the user proceeds through the Remedial Investigation/ Feasibility Study, he/she may arrive at different answers for the need for modeling, depending on the current task at that time. In light of this, the user should consider any future modeling decisions (e.g., in the detailed analysis or conceptual design stages) to be made. This can effect future resource allocation and data collection. As mentioned before, the methodology to decide to model does involve iteration, and the user should expect to reassess the need to model throughout the Remedial Investigation/Feasibility Study.

If modeling is required, the next step in the model selection methodology is to determine the level of modeling required. Sections 4 and 5 include flow charts for this purpose, for both ground water and surface water problems, respectively. These sections will assist the user in the third step of the decision framework by identifying the required model capabilities.

# 4. Methodology for Model Selection for Soil and Ground-Water Contamination Problems

## 4.1 OVERVIEW

At this point, the user has ascertained the need for modeling in remedial action assessment, and has identified the specific media (e.g., air, subsurface, and surface water) that are affected and are subject to contamination control. This section helps to answer the second two questions for subsurface contamination problems in the modeling decision framework: 1) What <u>level</u> of modeling is required?, and 2) What are the required <u>model capabilities</u> of that level? In conjunction with these decisions data and resource availability issues for each level of model are also discussed. Section 5 covers these same issues for surface water contamination problems. The formats of both sections are similar. Flow charts and matrices are used to guide the user towards model selection. The matrices will introduce the user to the interacting relationships of the remedial actions, environmental processes, and flow fields (for required model dimensionality). Information describing the remedial actions and environmental processes of concern in the soil and ground-water systems is provided in Volume 3: Numerical Modeling of Surface, Subsurface, and Waste Control Actions. Other sources of information on remedial actions include: JRB (1982), Ehrenfield and Bass (1983) and SCS (1981). A list of soil and ground-water remedial actions considered for simulation is provided in Table 4.1.

This methodology can also be used to select models for site characterization and exposure assessment. Selection criteria will be based on site complexity and modeling objectives, and may be less stringent than criteria for remedial action assessment. In many of these cases, analytical models may suffice.

## 4.2 WHAT LEVEL OF MODELING IS REQUIRED?

There are seven questions to be answered when determining the required level of model. Figure 4.1 is a flow chart that illustrates the hierarchy of decisions to be made. Each question or decision must be answered in the affirmative for analytical (Level I) modeling to be chosen. A "no" answer at any decision point pushes the user towards the use of a numerical (Level II) model, whereupon data and resource availability should be examined. This hierarchy was developed to define the strict and limited conditions of analytical model use in remedial action assessment. These decisions are described below.

The first decision or question is:

"Are order-of-magnitude <u>predictions</u> acceptable?"

This is primarily a function of the current task of the remedial investigation/feasibility study. In the screening of alternatives, remedial actions are ranked for potential use according to their general technical feasibility. Therefore, order-of-magnitude assessments are usually acceptable. In the analysis of alternatives and conceptual design tasks, the selected remedial actions are subjected to a more rigorous analysis. Quantitative assessments of remedial action performance are needed at these stages so that the most effective action or combination of actions in terms of reducing concentration levels is chosen. Therefore, it is possible that order-of-magnitude estimates may not be acceptable, and Level II modeling is required.

The next three questions to be asked concern the degree of variability or heterogeneity in site conditions. The first is:

"Is it reasonable to assume that <u>media properties</u> are uniform, and do not vary spatially?"

In actuality, site conditions or media properties are never truly homogeneous; different levels of heterogeneity exist, depending on site complexity and the size of area being considered. However, in terms of modeling requirements, assumptions can often be made to simplify site conditions. If a high degree of accuracy is not critical, and properties are <u>relatively</u> uniform (i.e., one soil type or similar soil characteristics, single layer aquifer), Level I

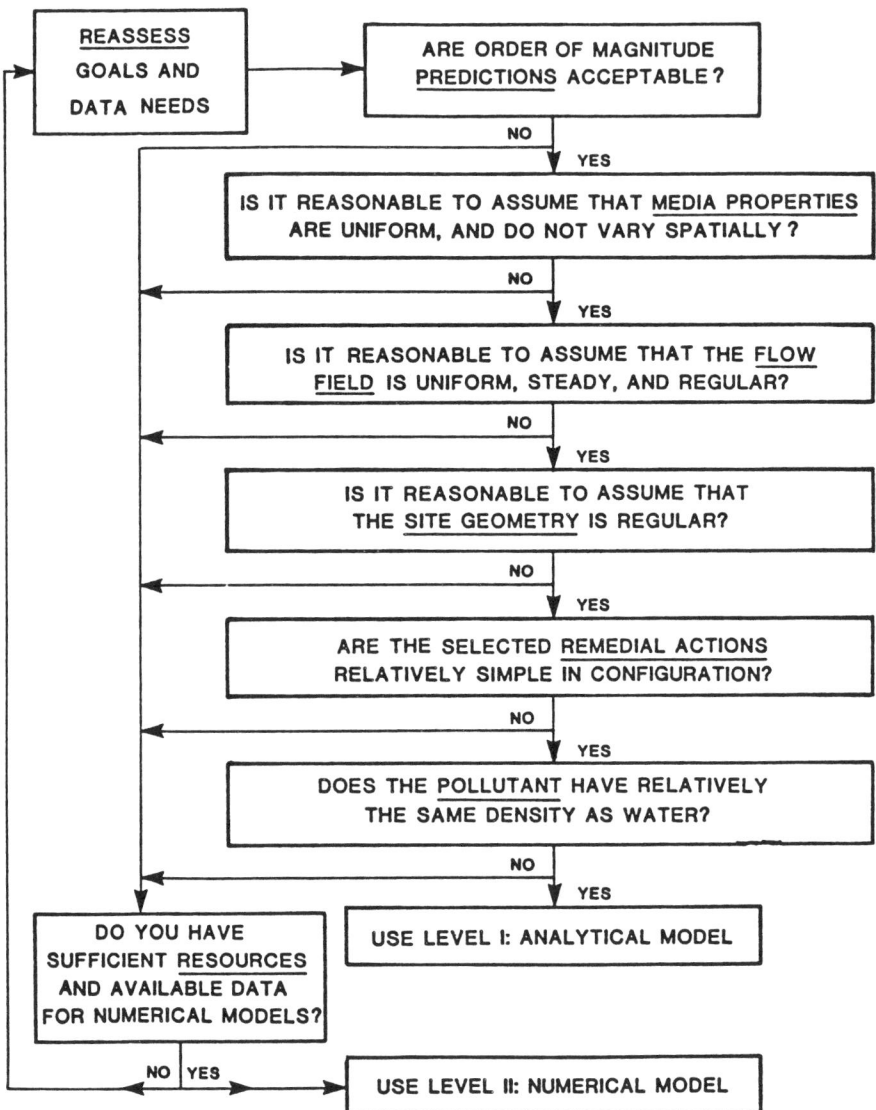

Figure 4.1 Flow chart to determine the level of modeling required for soil and ground-water systems.

modeling may be appropriate. If media properties <u>cannot</u> be simplified, a Level II model is required. This question is posed because only numerical models can explicitly represent variability in media properties, such as porosity and hydraulic conductivity. The number of simplifying assumptions made is dependent upon both the user's expertise and knowledge of the site. Therefore it is critical that the user be able to characterize media properties, and be cognizant of the limitations associated with these assumptions.

The second of these questions concerning site conditions is:

"Is it reasonable to assume that the <u>flow field</u> is uniform and steady?"

Unsaturated zone flow is most often unsteady and irregular, except in those cases where infiltration is relatively constant, as occuring under seepage/recharge basins. In many cases, flow and transport processes in the unsaturated zone may be neglected so that simulation of saturated zone processes may be sufficient. Flow conditions must be reasonably uniform, steady, and regular for analytical (level I) models to be applied. Uniform flow refers to flow that is in one direction (e.g., radial flow), and does not vary across the width of the flow field. Steady flow does not change over time. This occurs where boundary conditions such as pumping/injection and recharge from rainfall or a stream are constant over time.

The third site condition question to be asked is:

"Is it reasonable to assume that the <u>site geometry</u> is regular?"

Examples of regular site geometry include constant aquifer thickness and rectangular, circular or square shaped site on a plan view. As with media property variability, no site is totally rectangular, square, or conical, nor are aquifers equally thick everywhere. However, some hazardous waste sites can be approximated in this manner. Those with rectangular surface impoundments and single or double layer shallow aquifers are an example. Some examples of where these simplifications cannot be made include aquifers with fractured bedrock or aquifers with highly irregular boundaries. If this question can be answered in the affirmative, go on to the next question.

The next question deals with remedial action requirements in terms of model selection:

> "Are the selected <u>remedial actions</u> relatively simple in configuration?"

As noted in Section 2, remedial action criteria are an integral part of model selection, and thus a central issue in the selection methodology in this manual. Some remedial actions and specific configurations cannot be simulated by analytical models because they must be represented using variable media properties, or they cause perturbations in the flow field so that flow is not uniform or steady. Remedial actions and configurations that fit this category include: permeable treatment beds and partially-penetrating wells and drains. If these actions are <u>not</u> to be selected for detailed analysis and conceptual design, then proceed to the next question.

Pollutant characteristics also affect the level of model required. The primary question to be asked here is:

> "Does the <u>pollutant</u> have relatively the same density as water?"

If the pollutant has relatively the same density as water it will be advected by the water although it can also be retarded, or slowed, if it sorbs to the media. Pollutants that are much heavier or lighter than water will not mix or be advected entirely by the water; the result being two-phase flow. In these cases the pollutant mass, either as a liquid or a gas, exhibits its own flow with that of water. This phenomena is extremely difficult to represent and only a select group of complex numerical models are capable of adequately representing pollutant movement under these conditions.

If all of the above questions were answered in the affirmative, a <u>Level I: Analytical Model</u> is appropriate for use. However, the user may have to use this flow chart iteratively when working through the Remedial Investigation/Feasibility Study, as objectives according to each stage vary.

If <u>any</u> of the above questions were answered in the negative, the user must ask the following question:

> "Do you have sufficient <u>resources</u> and available <u>data</u> for numerical models?"

If the user has on the order of four to eight man-months of time and has the requisite data sets (see Section 4.4.2) for calibration and validation, the answer is affirmative, and a Level II: Numerical Model is appropriate for use. If, however, resources (including computer facilities, expertise and time) and data are not adequate, the user must reassess the goals and data needs for the current stage, and re-iterate the level of modeling decision process again.

## 4.3 REQUIRED MODEL CAPABILITIES FOR EACH LEVEL

Once the appropriate level of model is chosen, the user should identify the required model capabilities for the site, based on flow field, critical processes, and remedial action criteria. Section 4.3.1 covers the required model capabilities for Level I models, Section 4.3.2 covers the same for Level II models.

Matrices are used at this stage to facilitate the identification of required model capabilities based upon a wide range of potential scenarios. Figure 4.2 is a flow chart that illustrates the framework for identification of model capabilities and model selection. After the required model capabilities are identified, a discussion on general model selection criteria, and data and resource availability for both levels of models is provided.

### 4.3.1 Level I Analysis

A Level I analysis is appropriate for remedial action assessment if the user has answered "yes" to the first six questions in Figure 4.1: Flow Chart To Determine the Level of Modeling Required.

The available Level I methods for subsurface remedial action assessment are fairly specific in terms of the type of remedial actions that can be evaluated, and have limitations for use that should be observed. For example, the conformal mapping method is appropriate for simulation of fully-penetrating and hanging impermeable barriers. Table 4.2 is a matrix of remedial action configurations vs. simplified and analytical methods. The remedial action configurations listed on the "Y" axis are the same as those listed in Table 4.1.

TABLE 4.1   SOIL AND GROUND-WATER REMEDIAL ACTIONS

Grading

Revegetation

Surface Water Diversion

Capping

Seepage Basins and Ditches

Interception Trenches:
- Fully-penetrating
- Partially-penetrating

Ground-Water Pumping:
- Fully-penetrating levels
- Partially-penetrating wells

Impermeable Barriers
- Fully-penetrating
- Hanging

Drains

Permeable Treatment Beds

Bioreclamation/Chemical Injection

Excavation/Hydraulic Dredging

Solution Mining/Extraction

Soil and Ground-Water Contamination Problems 25

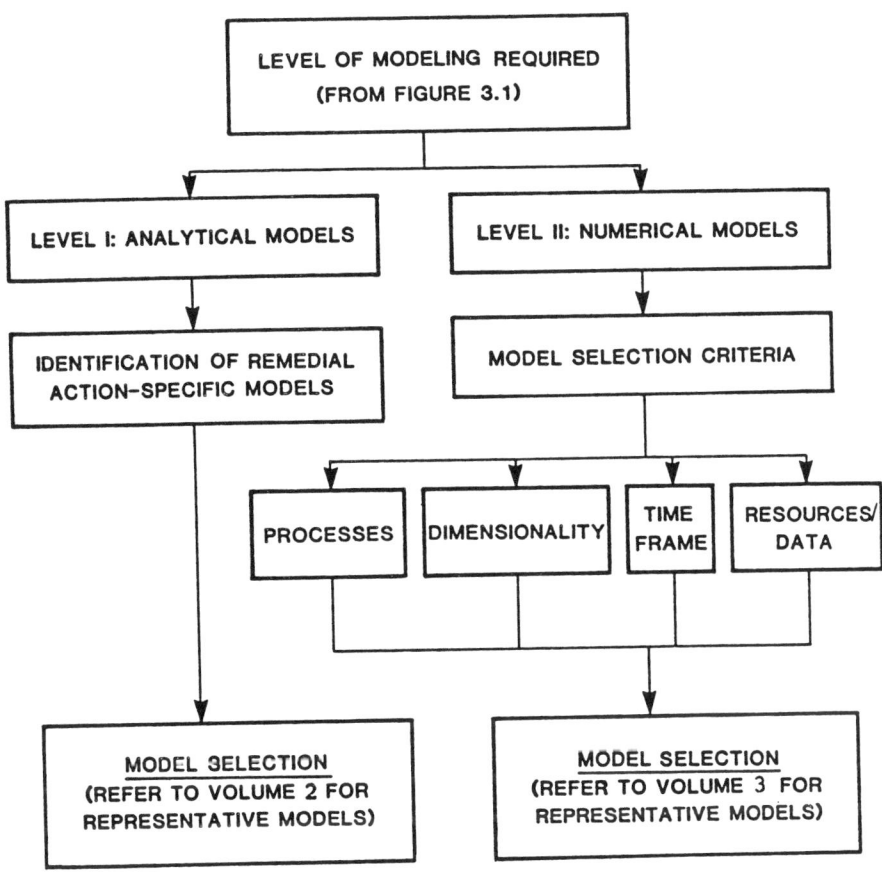

Figure 4.2  Flow chart for required model capabilities for soil and ground-water systems.

TABLE 4.2  REMEDIAL ACTIONS VS. SIMPLIFIED AND ANALYTICAL METHODS MATRIX

| REMEDIAL ACTIONS | Runoff-Estimation Methods | Sediment Yield Methods | Well Hydraulics | Drain Hydraulics | Ground-water Mounding | Superposition | Pond Seepage Estimation Methods | Infiltration Estimation Methods | Transformation Methods | Contaminant Transport | Conformal Mapping |
|---|---|---|---|---|---|---|---|---|---|---|---|
| • Grading | ● | ● | | | | | | | | | |
| • Revegetation | ● | ● | | | | | | | | | |
| • Surface Water Diversion | ● | | | | | | | | | | |
| • Capping | | | | | | | | ● | ● | ● | |
| • Seepage Basins and Ditches | | | | | ● | ● | ● | | ● | ● | |
| • Interceptor Trenches | | | | | | | | | | | |
| • Fully-penetrating | | | | ● | | ● | | | ● | ● | |
| • Partially-penetrating | | | | ● | | ● | | | ● | ● | |
| • Ground-water Pumping: | | | | | | | | | | | |
| • Fully-penetrating Wells | | | ● | | | | | | ● | ● | |
| • Partially-penetrating Wells | | | ● | | | | | | ● | ● | |
| • Impermeable Barriers | | | | | | | | | | | |
| • Fully-penetrating | | | | | | | | | ● | ● | ● |
| • Hanging | | | | | | | | | ● | ● | |
| • Drains | | | | ● | | | | | ● | ● | |
| • Permeable Treatment Beds | | | | | | | | | ● | ● | |
| • Bioreclamation/Chemical Injection | | | ● | | ● | ● | | ● | ● | ● | |
| • Excavation/Hydraulic Dredging | | | | | | | | | ● | ● | |
| • Solution Mining/Extraction | | | ● | | | ● | | ● | ● | ● | |

There are 11 simplified and analytical methods available for Level I remedial action assessment. These methods include: runoff estimation; sediment yield; well hydraulics; drain hydraulics, ground-water mounding; superposition; pond seepage estimation; infiltration estimation; transformation; conformal mapping; and contaminant transport. Some of these methods encompass the theory used to develop different types of solutions (e.g., well hydraulics, mounding, conformal mapping and contaminant transport), whereas others encompass the theory behind the use of these solutions to evaluate relatively complex geohydrological conditions (e.g, superposition and transformation). Volume 2: Simplified Methods for the Evaluation of Subsurface Remedial Actions, provides a good summary of selected analytical methods. The user should refer to this volume for descriptions of the methods, examples of applications, and appropriate references. The user should identify all the methods that are applicable to the remedial actions being screened and consult the appropriate references in order to choose the specific techniques that are applicable to the specific site.

### 4.3.2 Level II Analysis

If the user answered 'no' to any of the first six questions in Figure 4.1, Level II (Numerical) methods should be considered for remedial action assessment at this site. Resources and data availability must also be answered in the affirmative in Figure 4.1 to allow numerical model use.

Identification of required model capabilities for Level II analysis is more complex than for Level I analysis, due to the large number of models available with variable capabilities. It is accomplished by evaluating three primary groups of remedial action criteria: environmental processes that are affected by remedial actions and thus should be represented in a model; the minimum required dimensionality; and time frame. The first two areas are represented by matrices. The time frame requirements are discussed in the text. Time frame requirements are extremely site-specific, thus general guidance is given. These areas are described below.

A measure of remedial action effectiveness is how well the action controls or affects specific environmental processes that are responsible for off-site contamination problems. For example, impermeable barriers control the movement of ground water and contaminants. Remedial actions are most often simulated by adjusting the parameters of specific environmental process equations in the model. Therefore, affected processes for each remedial action should be identified in order to ensure that the selected model considers them.

Table 4.3 is a matrix of remedial actions vs. required processes. The processes are divided into 3 zones: overland (or surface), unsaturated, and saturated. These zones also correspond to the general types of models available. The processes include: runoff, evapotranspiration, erosion, and infiltration in the overland zone; percolation/leaching, dispersion, retardation, degradation, and drainage in the unsaturated zone; and ground-water movement, dispersion, retardation, and degradation in the saturated zone. The processes of infiltration and drainage can be considered as inter-zone, as they describe water movement between the overland-unsaturated zones and unsaturated-saturated zones, respectively. A '◆' in the blocks indicates that the selected model should simulate that specific process in order to represent the effects of a given type of remedial action.

Once the user has identified processes to be represented in the selected model based upon the remedial actions subject to analysis, the required model dimensionality should be identified. Table 4.4 is a matrix of required dimensionalities for remedial actions as a function of zone. The zones specify the number of land segments (parcels of the surface zone separated into areas of uniform properties) required for the overland zone, and the number of dimensions and direction(s) for both unsaturated and saturated zones. By reviewing this matrix, the user can identify the required spatial domain or dimensionality of the selected model.

The third area of remedial action criteria is the time frame requirements. Numerical models can simulate chemical transport and fate in two modes: using a <u>steady-state</u> mode, where fluxes such as water velocity and <u>pollutant</u> loading are constant or time invariant; or using a <u>transient</u> mode, where flow and/or contaminant transport may <u>vary over</u> time. Most models may run in either mode, depending on the input data and specifications by the user. The flow field throughout the system is usually established or calculated first, then the transport part of the model utilizes the flow field velocities to calculate pollutant transport. In this way, time frame may be specified separately for both

Soil and Ground-Water Contamination Problems 29

TABLE 4.3 REMEDIAL ACTIONS VS. REQUIRED PROCESSES MATRIX

| REMEDIAL ACTIONS | OVERLAND ZONE | | | | UNSATURATED ZONE | | | | | SATURATED ZONE | | | | |
|---|---|---|---|---|---|---|---|---|---|---|---|---|---|---|
| | RUNOFF | EVAPOTRANS-PIRATION | EROSION | INFILTRATION | PERCOLATION/LEACHING | DISPERSION | RETARDATION | DEGRADATION | DRAINAGE | GROUND-WATER MOVEMENT | DISPERSION | RETARDATION | DEGRADATION | |
| • Grading | ♦ | | ♦ | ♦ | | | | | | | | | | |
| • Revegetation | ♦ | ♦ | ♦ | ♦ | | | | | | | | | | |
| • Surface Water Diversion | ♦ | | ♦ | ♦ | | | | | | | | | | |
| • Capping | ♦ | | ♦ | ♦ | ♦ | ♦ | | | | | | | | |
| • Seepage Basins | | | | ♦ | ♦ | | | | | | | | | |
| • Interceptor Trenches | | | | | | | | | | | | | | |
| • Fully-penetrating | | | | | | | | | | ♦ | ♦ | | | |
| • Partially-penetrating | | | | | | | | | | ♦ | ♦ | | | |
| • Ground-water Pumping | | | | | | | | | | | | | | |
| • Fully-penetrating wells | | | | | | | | | | ♦ | ♦ | | | |
| • Partially-penetrating wells | | | | | | | | | | ♦ | ♦ | | | |
| • Impermeable Barriers: | | | | | | | | | | | | | | |
| • Fully-penetrating | | | | | | | | | | ♦ | ♦ | | | |
| • Hanging | | | | | | | | | | ♦ | ♦ | | | |
| • Drains | | | | | ♦ | ♦ | | | ♦ | | | | | |
| • Permeable Treatment Beds | | | | | | | | | | ♦ | ♦ | ♦ | ♦ | |
| • Bioreclamation/Chemical Injection | | | | | | | | | | ♦ | ♦ | ♦ | ♦ | |
| • Excavation/Hydraulic Dredging | | | | ♦ | ♦ | ♦ | ♦ | | ♦ | | | | | |
| • Solution Mining/Extraction | | | | | ♦ | ♦ | ♦ | | ♦ | | | | | |

♦ = Indicates that selected model should simulate these processes in order to represent the remedial action.

TABLE 4.4  REMEDIAL ACTIONS VS. REQUIRED MODEL DIMENSIONALITY MATRIX

| REMEDIAL ACTION | Overland | Zone Unsaturated | | Saturated | |
|---|---|---|---|---|---|
| | No. of Segments | No. of Dimensions | Direction(s) | No. of Dimensions | Direction(s) |
| Grading | S | | | | |
| Revegetation | S | | | | |
| Surface Water Diversion | M | | | | |
| Capping | | 1 | $Z^a$ | | |
| Seepage Basins | | 1 | Z | $2^b$ | $X,Y^b$ |
| Drains | | | | | |
|   Fully-penetrating wells | | | | 2 | X,Y |
|   Partially-penetrating wells | | | | 3 | X,Y,Z |
| Ground-water Pumping | | | | | |
|   Fully penetrating wells | | | | 2 | X,Y |
|   Partially-penetrating wells | | | | 3 | X,Y,Z |
| Impermeable Barriers | | | | | |
|   Fully-penetrating | | | | 2 | X,Y |
|   Hanging | | | | 3 | X,Y,Z |
| Interceptor Trenches | | | | 2 | X,Y |

(continued)

TABLE 4.4 (continued)

| REMEDIAL ACTION | Overland | Unsaturated | | Saturated | |
|---|---|---|---|---|---|
| | No. of Segments | No. of Dimensions | Direction(s) | No. of Dimensions | Direction(s) |
| Permeable Treatment Beds | | | | $2^d$ <sup>c</sup> | $x,y^d$ <sup>c</sup> |
| Bioreclamation/ Chemical Injection | | | | | |
| Excavation/Hydraulic Dredging | | 1 | z | | |
| Solution Mining/ Extraction | | 1 | z | | |

[a] Denotes vertical direction
[b] Only if Ground-water mounding from the seepage basin is significant
[c] Assumes treatment bed is constructed so as not to modify the flow field
[d] Assumes injection/extraction wells are fully penetrating

S = single
M = multiple

the flow and transport 'modules'. The time frames required to properly represent the effects of a remedial measure depend on the hydrologic zone, the important processes, and the remedial measure itself. For example, remedial actions designed for control of erosion and runoff such as grading and surface water diversion could require a transient (or dynamic) simulation with short time steps because rainfall and runoff processes fluctuate rapidly. In the saturated zone, the flow field is usually more steady, and fluctuations occur on a scale of months and years. Thus, a steady-state simulation is usually applicable. However, a transient simulation may be required if recharge and discharge from pumping/injection or hydraulically-connected streams fluctuate over the simulation period, causing the flow to be unsteady. Contaminant transport usually requires a transient mode when simulating remedial actions, as model results of interest include the <u>variation</u> in concentrations in pre-and post-restorative periods.

## 4.4 RESOURCE AND DATA AVAILABILITY

As mentioned in Section 2, resource and data needs must be examined when the level of modeling is determined. This question is particularly important when it is apparent that a Level II numerical model is required for simulation. A brief overview of these issues is provided below; a detailed examination of procedures for model use including estimation of parameters for subsurface remedial actions are provided in Volume 3.

### 4.4.1 Level I Analysis

Resource and data availability is not as critical to performing a Level I analysis as is the user's expertise and understanding of the site. Analytical and simplified methods require very little data, can often be solved by hand or with personal computers, and do not require a large amount of time for implementation. However, use of these methods does require an understanding of the assumptions and limitations behind their development. A summary of basic Level I (Analytical) model data needs is presented in Table 4.5. Sampling programs in the technology screening of alternatives stage of the Remedial Investigation/ Feasibility Study should attempt to satisfy the data needs of at least a Level I analysis.

TABLE 4.5  DATA NEEDS FOR LEVEL I (ANALYTICAL) METHODS FOR SUBSURFACE PROBLEMS (after Javandel et al., 1983)

I. Geometry of System

- Average thickness and depth of aquifer
- Positions of significant features:
  - Source(s) of contamination
  - Discharge and recharge areas

II. Fluid (Water) Velocity

- Direction and magnitude of average regional velocity in vicinity of study area

III. Concentration of Pollutant

- Source release rate
- History of operation of facility

IV. Dispersivity of Media

- Representative value of longitudinal dispersivity for one-dimensional problems
- Representative values for both longitudinal and transverse dispersivities for two-dimensional problems

V. Pollutant Characteristics

- Retardation factors or distribution coefficients for solutes that sorb to media
- Decay rate for solutes that are non-conservative

### 4.4.2 Level II Analysis

Resource and data availability issues are more critical to performing a Level II analysis compared to a Level I analysis. Data sets that represent the range of values over time and space are required for calibration and testing (performing simulations to ensure "agreement" between observed and predicted data), and remedial action simulation, where parameters and site configurations are adjusted to represent the selected measures. The flexibility provided by numerical models to represent spatial variability can lead to expanded data needs from the increased number of elements or compartments discretized in the model. Table 4.6 provides a list of required data for Level II methods. Since numerical models are more applicable to detailed analysis and conceptual design tasks in the Remedial Investigation/ Feasibility Study, corresponding sampling programs in this stage should be directed at meeting the needs for model simulations, provided the need to model for a specific site is warranted.

Similarly, resource needs are more intensive than for Level I. Demands will vary according to the complexity of the modeling study, but generally, <u>four to eight</u> man-months of time for an <u>experienced</u> user should be allocated. However, time requirements can vary greatly, and will depend on the application, also. This manpower requirement should be used as a rough estimate, and is dependent on a number of factors, including: expertise and experience of the user with the selected model; availability of computer facilities and software support; sufficient time to conduct the study; and sufficient money to train the user, pay for his/her time, and pay for the model, if it is proprietary and can be obtained.

Resource needs are also affected by model performance related criteria. The utility of a model depends not only on its ability to represent site and remedial action conditions, but also on model design, implementation, testing, and documentation. These factors influence the accuracy of model predictions, ease of use, data requirements, and computer run costs. An additional, less tangible, attribute of a model is its perceived reliability, which is dependent on the number of times it has been successfully implemented, the verification or testing of model results against field measurements, and the technical

TABLE 4.6   DATA NEEDS FOR LEVEL II (NUMERICAL) METHODS FOR SUBSURFACE PROBLEMS (after Javandel et al., 1983)

I. Geometry of System

- Real extent of aquifer
- Location of natural or mathematical boundaries
- Thickness of aquifer and its variation with the space
- Location and rates of discharge and recharge areas

II. Fluid (Water) Velocity

- Distribution throughout the system

III. Concentration of pollutant

- History of operation of facility
- Present and future source rates of pollutant and chemical composition
- Positions of sources relative to aquifer

IV. Dispersivity of Media

- Representative value of longitudinal dispersivity for one-dimensional problems
- Representative values for both longitudinal, transverse, and vertical dispersivities for two- and three-dimensional problems

V. Pollutant Characteristics

- Retardation factors or distribution coefficients for solutes that sorb to media
- Decay rate for solutes that are non-conservative

(theoretical) basis for model calculations. These criteria can be used to distinguish between models which satisfy all of the site and remedial action criteria discussed earlier. The most desirable models have extensive documentation, have been applied to a number of diverse situations, have been tested against several comprehensive data sets, and are relatively efficient in terms of data preparation, requirements and computer time. Models that exhibit these characteristics will not require as much time and effort compared to ones that do not exhibit the same characteristics.

## 4.5 MODEL SELECTION CRITERIA FOR SOIL AND GROUND-WATER CONTAMINATION PROBLEMS

By evaluating the matrices for Level I and II models the user will be able to correlate model capabilities with his/her site characteristics and selected remedial actions. Some general trends of model selection criteria are apparent in the matrices. These trends are described below to clarify any confusion on the appropriate model capabilities required.

In a Level I analyses, the selection of a specific method or group of methods is primarily a function of the selected remedial actions. For example, drain hydraulic methods are applicable to drains, well hydraulic methods are applicable to ground-water pumping/injection. The configuration of the remedial action measure will also affect the type of techniques that are applicable. For example, partially-penetrating interceptor trenches and wells create more complex flow patterns than do fully-penetrating drains and wells; thus they require different techniques to account for the more complex flow field.

For Level II analyses, there are general trends in each major group of model selection criteria: processes, required dimensionality, and time frame. These criteria are discussed below.

The processes required for remedial action simulation are often a function of the specific zone that is affected and the pollutant characteristics. For example, remedial actions such as grading, revegetation, and surface water diversion require representation of the runoff and infiltration processes in the selected model. Similarly, if the pollutant has a high affinity for sorption, the mechanisms of retardation and soil erosion should be represented. As it is expected that most subsurface contamination problems will concern ground water, the selected saturated zone transport model should usually consider dispersion, retardation and degradation.

The minimum required dimensionality will also vary according to the complexity of the site and selected remedial actions. Some general requirements are apparent for each zone, however. In the overland zone, the actions require a single segment model or one that allows only uniform properties for slope, roughness, and soil type. The other remedial action requires a multiple segment model, or one that allows the segmentation of the drainage area. Some overland zone models allow multiple segments; such models give the user flexibility to simulate all of the above actions.

In the unsaturated zone, a one-dimensional simulation in the "Z" or vertical direction is often sufficient for the evaluation of the few remedial actions that affect that zone, such as excavation, seepage basins, and hydraulic dredging. However, if there are soil layers with varying permeability, a two dimensional horizontal-vertical (x-z) simulation is required to represent the percolation of water (Z direction) and the lateral interflow (x direction).

In the saturated zone, a two-dimensional simulation is the minimum required dimensionality for most actions. If mounding is not a problem, actions such as fully-penetrating wells and interceptor trenches, fully-penetrating cut-off walls, and drains may require a x-y flow pattern simulation. Serious mounding problems also require a minimum of two-dimensional x-y simulation, if not a three-dimensional (x,y,z). Other actions can be represented with a x-y simulation if the flow at the site has a neglible vertical component. If it doesn't, a three dimensional (x,y,z) simulation is required.

In terms of time frame, overland and unsaturated zone measures such as grading, revegetation, surface water diversion, and capping will require a dynamic simulation. Remedial measures used in the saturated zone can often be simulated using a steady-state mode for flow simulation, although boundary conditions and monitoring data should be evaluated to determine if a transient simulation is necessary. Contaminant transport will almost always require a transient simulation to examine reductions in concentrations due to implementation of specific remedial actions.

# 5. Methodology for Model Selection for Surface Water Contamination Problems

5.1 OVERVIEW

This section will assist the user in answering the second two questions of the modeling decision framework for surface water:

1. What level of modeling is required and

2. What are the required model capabilities for that level

Flow charts and matrices form the basis of the methodology. The matrices will introduce the user to relationships of remedial actions to model selection criteria as processes, dimensionality, and time frame. This section parallels Section 4 for soil and ground-water contamination problems. General model selection criteria and resource and data availability issues are discussed in the latter part of this section. Descriptions of surface water remedial actions including required dimensionality and affected processes, are provided in Volume 4: Simplified Methods and Numerical Models for the Evaluation of Surface Water Remedial Actions. If additional information is required, the user should refer to available remedial action technology guides, such as JRB (1982), Ehrenfield and Bass(1983), and SCS (1981). A list of surface water remedial actions considered for simulation is provided in Table 5.1. Environmental process descriptions, along with the significant parameters, environmental conditions of concern, and relation to other processes are also provided in Volume 4. Additional sources of information for these processes include Mills et al. (1982) and Callahan et al. (1979).

TABLE 5.1 SURFACE WATER REMEDIAL ACTIONS

I. No Action

II. Physical/Mechanical Measures

- Mechanical dredging
- Hydraulic dredging
- Excavation
- Dilution
- Barriers/diversions
- Skimming
- Cofferdams
- Booms
- Silt curtains
- Capping

III. Treatment

- In-situ
- On-site

The state-of-the-art for remedial action assessment modeling in surface water is not as advanced as modeling for ground water. The available Level I and Level II methods can adequately represent the wide range of waterbody conditions, but have not been as extensively tested for remedial action assessment as have ground-water methods. A key problem is estimating parameter values for specific remedial actions. Quite often, best engineering judgement will have to suffice for making the appropriate parameter adjustments for remedial action simulation. Therefore, the user should consider the uncertainty inherent in the representation of remedial actions when selecting a model.

## 5.2 WHAT LEVEL OF MODELING IS REQUIRED?

There are six basic questions to be answered when determining the level of model required for surface water contamination problems. Figure 5.1 is a flow chart that illustrates the hierarchy of decisions to be made. The first five questions should be answered in the affirmative for analytical or simplified methods (Level I) to be chosen. A "no" answer for any of these questions forces the user to consider the use of a numerical model (Level II), whereupon a resource and data availability decision must be made. If both resources and data are available, the numerical model should be chosen. If either resources or data are not available, the user will be directed to reassess project goals and/or data needs, and re-iterate the decision process flow chart. This hierarchy was developed as such in order to define the strict and limited conditions of analytical and simplified methods used in remedial action assessment.

Environmental conditions such as unsteady flow regimes, non-uniform geometry, and complex sediment-water interactions, cannot be accurately represented by Level I methods. However, the use of Level I methods for remedial action assessment in surface water, while more limited than analogous use for soil and ground-water problems, may be more appropriate given the limited testing and parameter estimation available.

As with subsurface problems, the flow chart questions may also be posed when selecting models for exposure assessment. However, model requirements for conducting exposure assessments will most likely be less stringent, due to the fact that only the complexity of site conditions and modeling objectives determine model selection. The

### Surface Water Contamination Problems 41

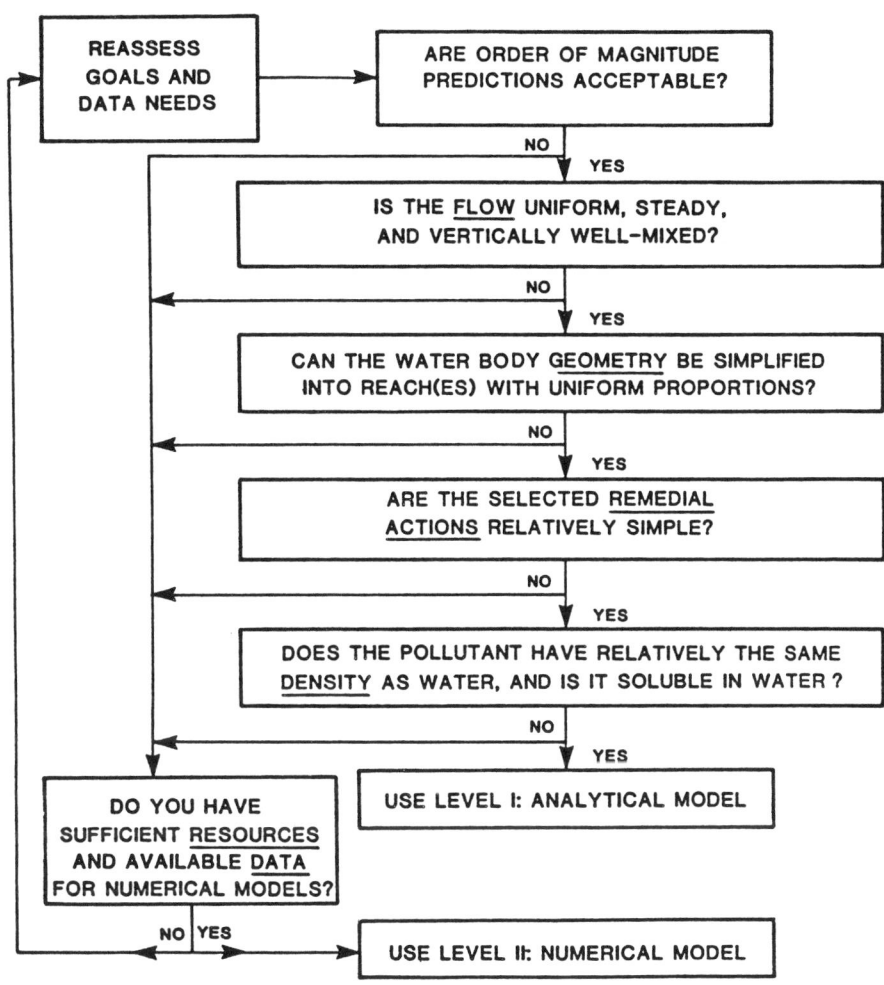

Figure 5.1  Flow chart to determine the level of modeling required for surface water systems.

questions to determine the level of modeling required for remedial action assessment are described below.

The first question to be asked is universal to both surface water and subsurface contamination problems:

"Are order-of-magnitude predictions acceptable?"

This decision is a function of the current task at hand of the Remedial Investigation/Feasibility Study, and the relative complexity of the site. For screening of alternatives, order-of-magnitude estimates are often acceptable. When detailed analysis and conceptual design is initiated, there is a need for higher accuracy in quantitative assessments of remedial action performance. Numerical models are much more accurate than analytical models when site conditions cannot be well-characterized, and are usually more appropriate for use. There are many cases however, where accuracy needs are independent of the current stage of the Remedial Investigation/Feasibility Study. These cases are dependent on the complexity and number of remedial actions being evaluated. If order-of-magnitude estimates are acceptable, proceed to the next question.

The next two questions concern the degree of heterogeneity in site conditions. The first question is:

"Is the flow uniform, steady, and vertically well-mixed?"

As with soil and ground-water problems, the flow field must be relatively simple if analytical models are to be used. Again uniform flow refers to flow that does not vary spatially i.e., along the length of the waterbody. Steady flow refers to flow that does not vary over time. The degree of vertical mixing is important in model level selection because un-mixed, stratified waterbodies, such as impoundments and estuaries, have complex and sometimes bi-directional flow and thus require numerical methods. A vertically well-mixed waterbody, then, has essentially uniform flow in the "Z" or vertical dimension.

The second of the two questions concerning site heterogeneity is:

"Can the waterbody geometry be simplified into reach(es) with uniform proportions?"

One of the limitations of analytical models is that the system being represented must be simple or regularly shaped. In surface waters, this means the waterbody has to be segmented into reaches or lengths of uniform size. Obviously, some waterbodies, such as estuaries and delta-like rivers, do not lend themselves to such simplification. The user must examine the shape and configuration to determine if an average width, length, and depth can be utilized.

The next question is:

"Are the selected remedial actions relatively simple?"

The use of analytical or simplified methods for remedial action assessment in surface waters is limited to very few measures. Some surface water measures usually cause perturbations in the flow, or involve complex sediment processes such as sediment transport and exchange from bed sediments, both of which often cannot be accurately represented by analytical or simplified methods. Such methods include: hydraulic dredging, barriers, skimming, and booms. Other measures such as mechanical dredging can be represented with Level I methods, but only with gross simplifications as to their effect in the waterbody.

The most important question to ask concerning pollutant criteria is:

"Does the pollutant have relatively the same density as water, and is it soluble in water?"

If the pollutant is soluble and has the same density as water, it will be transported along with the water, and may be accurately represented by a Level I method. If the pollutant floats on the surface or sinks to the bottom, or is insoluble, two different types of transport processes must be coupled (or interact with each other): one for water movement or advection, the other for the pollutant movement. This phenomenon can be simulated only by a select group of numerical models, and is similar to the two-phase flow phenomenon mentioned in Section 4.2.

If all the above questions can be answered in the affirmative, a Level I Method is appropriate for use. If any of the above questions is answered in the negative, the user must consider numerical models for use. Before selection of such methods however, the following question should be asked:

"Do you have sufficient resources and data for numerical models?"

Numerical models are both cost and data intensive. Therefore a minimum of four to eight man-months of time, and requisite data (see Section 5.3.2) for testing and prediction purposes should be available. If the answer is affirmative, Level II (Numerical) methods should be selected. If, however, the resources and/or data are inadequate, the user must reassess the goals and data, and re-iterate the level of modeling decision process again.

## 5.3 REQUIRED MODEL CAPABILITIES FOR EACH LEVEL

Matrices are used at this stage to correlate remedial actions with their required model capabilities for simulation. Figure 5.2 is a flow chart that illustrates the sequence of events that lead to model selection. After the level of modeling is determined, the user should refer to the specific matrices or text to identify required model capabilities. In a Level II analysis three main groups of model selection criteria are examined: environmental processes, waterbody/dimensionality, and time frame. After model capabilities are identified, the user is referred to Volume 4, which contain model profiles and sources of information on models.

### 5.3.1 Level I Analysis

A Level I analysis is appropriate for remedial action assessment if the user has answered "yes" to the first five questions in Figure 5.1. It is likely that Level I methods will be sufficient in the screening of alternatives task of the Remedial Investigation/ Feasibility Study; order of magnitude estimates of technical performance are desired. However, Level I methods are also appropriate for the analysis of alternatives task if site conditions and remedial actions can be simplified (i.e., steady, uniform flow and selection of such actions as excavation, dilution, or on-site treatment). Unlike Level I methods available for soil and ground-water, surface water methods are not specific in terms of remedial action simulation. Like numerical methods, they have various capabilities and hence, various uses. Table 5.2 is a matrix of simplified

Surface Water Contamination Problems 45

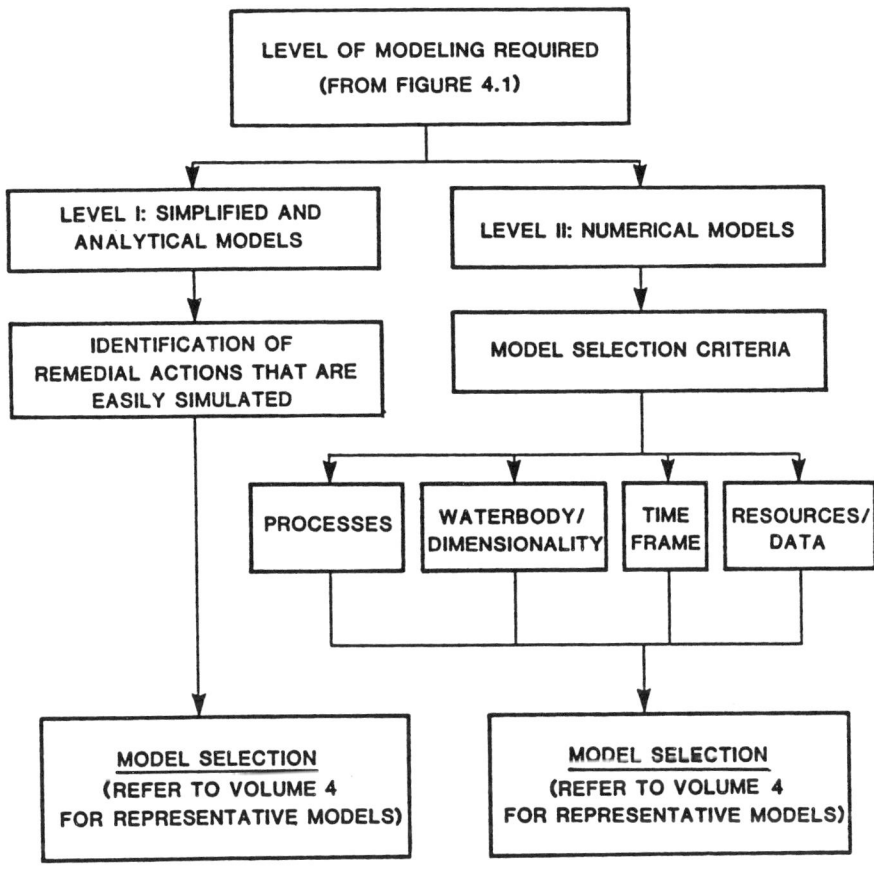

Figure 5.2  Flow chart for required model capabilities for surface water systems.

TABLE 5.2   SIMPLIFIED AND ANALYTICAL SURFACE WATER
            MODELS VS. REMEDIAL ACTIONS MATRIX

Simplified Analytical Methods

| Remedial Actions | Contaminant Transport | Sediment Processes | Estimation of Loading | Transformation Processes |
|---|---|---|---|---|
| Mechanical[1] dredging | | | | |
| Hydraulic[1] dredging | | | | |
| Excavation | | T,P | B,V | |
| Barriers[1] | | | | |
| Skimming[1] | | | | |
| Dilution | D,A | | M | 2 |
| Cofferdams | D,A | | | |
| Booms[1] | | | | |
| Silt curtains[1] | | | | |
| Capping | | T,P | B,V | |
| On-site[1] treatment | | | | |
| In-situ[1] treatment | D,A | P | | 2 |

[1] This action cannot be represented accurately using simplified or analytical methods.

[2] The transformation process required will be a function of the specific pollutant characteristics.

Legend:
A = Advection-Dispersion Equation
B = Bed Exchange Analysis
D = Dilution Analysis
M = Mixing Zone Analyses
P = Partitioning (Sorption)
T = Sediment Transport Analysis
V = Vertical Distribution of Sediments Analysis

techniques and analytical methods vs. remedial actions. Four major groups of model capabilities are listed across the 'X' axis: contaminant transport, sediment processes, estimations of loading, and transformation processes. In the boxes of the matrix the specific methods of these major groups are identified for each remedial action. Level I methods include simplified assessment techniques and analytical models.

There are a number of simplified assessment techniques that can be used to represent contaminant transport, sediment processes, estimation of loading and transformation processes. Volume 4: Simplified Methods and Numerical Models for the Evaluation of Surface Water Remedial Actions provides descriptions of these techniques and appropriate references.

Like the simplified assessment techniques, analytical models require steady-state flow conditions and uniform geometry. Their applicability can be limited, given the unsteady flow regimes, non-uniform geometry, and complex sediment-water interactions that characterize the environmental conditions when remedial actions are implemented. Within this general group of models, differences can include: complexity of geometry allowed, mode of pollutant loading (instantaneous or continuous), degree of mixing and dispersion (if any), ability to calculate transfer of mass between the sediment bed and the water column, methods of estimating sediment transport (user input suspended sediment concentrations, or concentrations calculated for each reach separately), lumped or specific first-order decay reactions, and the range of default values available for model parameters.

As was mentioned earlier in this section, Level I methods have very limited roles for remedial action assessment in surface waters. Of the thirteen measures listed in Table 5.2, five cannot be accurately represented using simplified or analytical methods. Some require simulation of complex geometry configurations (i.e., barriers), complex sediment processes (dredging), and low/high density pollutant transport and fate (skimming, booms). However, many of these complex remedial actions, such as skimming and booms, are usually employed under emergency response conditions, and this would not allow the time or resources for numerical simulation. Volume 4 contains a matrix of representative methods vs. model capabilities, and descriptions of representative models. Some other good sources for compilations of available methods include Mills et al.

(1982) for simplified methods, and Codell et al. (1982) and Onishi et al. (1982) for analytical models.

### 5.3.2 Level II Analysis

If the user answered 'no' to any of the first five questions in Figure 5.1, and determined that sufficient data and resources were available, Level II methods should be considered for remedial action assessment. This subsection will help identify the required model capabilities according to site and remedial action factors. The selection of numerical models is appropriate when site/remedial action conditions cannot be reasonably simplified, or when specific configurations of a remedial action, such as a barrier, must be evaluated for the best conceptual design.

Identification of required model capabilities for Level II analysis is more complex than for Level I, due to the large number of models available with variable capabilities. Similar to identification of required model capabilities for soil and ground-water problems, it is accomplished by evaluating three groups of remedial action criteria: _environmental processes_ that are affected by remedial actions and thus should be represented in a model; _dimensionality_ according to the remedial action and waterbody: and the _time frame_. These groups are described below.

A measure of remedial action effectiveness is how well the action controls or affects specific environmental processes that are responsible for off-site contamination problems. For example, barriers control the advection and dispersion of surface water and contaminants. Remedial actions are most often simulated by adjusting the parameters of specific environmental process equations in the model. Therefore, affected processes for each remedial action must be identified in order to ensure that the selected model simulates them.

Table 5.3 is a matrix of remedial actions vs. environmental processes. The processes are divided into two major groups across the 'x' axis: transformation and physical. Transformation processes include: hydrolysis, oxidation, photolysis, volatilization, bio-degradation, and bio-accumulation. Physical processes include adsorption, sediment (transport and bed-exchange of contaminated sediments), advection, and dispersion. Short descriptions

TABLE 5.3 REMEDIAL ACTIONS VS. PROCESSES MATRIX

| ACTIONS | \ PROCESSES | HYDROLYSIS | OXIDATION | PHOTOLYSIS | VOLATILIZ. | BIO-DEG. | BIO-ACC. | ADSORPTION | SEDIMENT | ADVECTION | DISPERSION |
|---|---|---|---|---|---|---|---|---|---|---|---|
| DILUTION | | 0 | 0 | 0 | 0 | 0 | 0 | 0 | 0 | + | + |
| **REMOVAL** | | | | | | | | | | | |
| MECHANICAL DREDGING | | 0 | 0 | + | + | - | 0 | - | + | 0 | + |
| EXCAVATION | | 0 | 0 | 0 | 0 | - | 0 | - | + | + | + |
| HYDRAULIC DREDGING | | 0 | 0 | 0 | 0 | - | 0 | - | + | + | + |
| BARRIERS/ DIVERSIONS | | 0 | 0 | 0 | 0 | 0 | 0 | 0 | 0 | + | + |
| SKIMMING | | 0 | 0 | - | - | 0 | 0 | 0 | 0 | - | - |
| **CONTAINMENT** | | | | | | | | | | | |
| COFFERDAMS | | 0 | 0 | 0 | 0 | 0 | 0 | 0 | + | + | + |
| BOOMS | | 0 | 0 | - | + | 0 | 0 | 0 | 0 | - | - |
| SILT CURTAINS | | 0 | 0 | 0 | 0 | 0 | 0 | - | - | + | + |
| CAPPING | | 0 | 0 | 0 | 0 | - | 0 | - | - | + | + |
| **TREATMENT** | | | | | | | | | | | |
| IN-SITU | | 0 | 0 | 0 | 0 | 0 | 0 | - | - | 0 | 0 |
| ON-SITE | | 0 | 0 | 0 | + | 0 | 0 | + | + | + | + |

Legend:
+ = Enhances the process in relation to no action
- = Mitigates the process in relation to no action
0 = Does not affect the process

of these processes are provided in Volume 4. A '+' in the matrix boxes indicates that the remedial action enhances the process in relation to no action. A '-' indicates that the remedial action mitigates that process in relation to no action. If either of these effects for remedial action are identified, the selected model should be able to simulate those processes. A '0' indicates that the remedial action has no effect on that process; hence representation of that process is not critical in the selected model.

The second major criteria group is the dimensionality requirements. Table 5.4 is a matrix of remedial actions vs. waterbody. The minimum model dimensionality is a function primarily of the waterbody type. The waterbodies across the 'X' axis are grouped as estuary, lake, or river, with subgrouping within each according to system geometry and degree of mixing. Numbers and letters in the matrix text denote the type of simulation needed for that remedial action in the specific waterbody. For example, "2L" denotes that a two-dimensional (lateral/longitudinal) simulation is required for that remedial action/waterbody scenario. A "0" indicates that the remedial action is not suited for use under the specific waterbody conditions. Some remedial actions, such as dilution and barriers or diversions, often may be simulated by adjusting the model boundary conditions and system geometry. Most of the remedial actions require a two-Dimensional (longitudinal/lateral) simulation. However, as the mixing becomes more turbulent or complex (as in estuaries and large lakes), a "pseudo" two-Dimensional simulation (longitudinal/ vertical) with coefficients for the horizontal or lateral dimension, or even three-dimensional simulation, may be required.

The third area of remedial action criteria is the time frame requirements. Like ground-water models, numerical models for surface water can simulate chemical transport and fate in two modes: using a <u>steady-state</u> mode, where fluxes such as water velocity and pollutant loading are constant or time invariant; or using a <u>transient</u> mode, where flow and/or contaminant transport may vary over time. Most models may run in either mode, depending on the input data and specifications by the user. The flow field throughout the system is usually established or calculated first, then the transport part of the model utilizes the flow field velocities to move contaminant particles. In this way, time frame may be specified for both the flow and transport 'modules'. The time frames required to properly represent the affects of a remedial measure depends on the flow regime or type of waterbody, the important processes, and the

TABLE 5.4  REMEDIAL ACTIONS VS. WATER BODY MATRIX

| REMEDIAL ACTIONS | ESTUARIES | | | | | LAKES | | | RIVERS | |
|---|---|---|---|---|---|---|---|---|---|---|
| | NARROW, WELL-MIXED, SHALLOW | NARROW, STRATIFIED | BAY, WELL-MIXED, SHALLOW | BAY, STRATIFIED | MONO. OR DIMICTIC | WELL-MIXED | RESERVOIRS | SINGLE STEM, WIDE, UNEVEN | SINGLE STEM, NARROW | BRANCHING/ DELTA |
| NO ACTION | | | | | | | | | | |
| **REMOVAL** | | | | | | | | | | |
| MECHANICAL DREDGING | 1 | 2V | 0. | 0 | 3 | 2L | 2P | 0 | 1 | 2L |
| EXCAVATION | 1 | 2V | 0 | 0 | 2L | 2L | 2L | 0 | 1B | 2L |
| HYDRAULIC DREDGING | 1 | 2V | 2L | 3 | 3 | 2L | 2L | 2L | 1B | 2L |
| BARRIERS/ DIVERSIONS | 2L | 3 | 2L | 3 | 3 | 2L | 3 | 2L | 2L | 2L |
| SKIMMING | 0 | 0 | 0 | 2V | 2V | 2V | 2V | 0 | 0 | 0 |
| DILUTION | 1 | 2V | 2L | 3 | 0 | 0 | 2V | 1B | 1 | 1B |
| **CONTAINMENT** | | | | | | | | | | |
| COFFERDAMS | 2L | 3 | 2L | 3 | 3 | 2L | 3 | 2L | 2L | 2L |
| BOOMS | 3 | 3 | 0 | 0 | 3 | | 3 | 0 | 0 | 0 |
| SILT CURTAINS | 2V | 2V | 0 | 0 | 2V | 2V | 2V | 0 | 2V | 0 |
| CAPPING | 0 | 2L | 0 | 0 | 0 | 0 | 2L | 0 | 2L | 0 |
| **TREATMENT** | | | | | | | | | | |
| IN-SITU | 0 | 0 | 0 | 0 | 0 | 0 | 0 | 2L | 2L | 2L |
| ON-SITE | | * | DEPENDANT ON REMOVAL ACTION USED IN CONJUNCTION | | | | | | | * |

LEGEND:
1 = 1-DIMENSIONAL
2 = 2-DIMENSIONAL
3 = 3-DIMENSIONAL
L = LATERALLY AVERAGED
V = VERTICALLY AVERAGED
0 = ACTION IS NOT APPLICABLE TO THIS WATERBODY
B = BRANCHING OR NETWORK

measure itself. For example, remedial actions that alter the flow regime, such as hydraulic dredging and barriers may require a transient or dynamic simulation of the flow system. Similarly, some waterbodies, such as branching estuaries and snowmelt-fed rivers, require a dynamic simulation when the flows fluctuate within the simulation period, which could be a number of days (for an estuary), or over the course of a year (for a river). As with ground-water models, the contaminant transport simulation must always be in a transient or dynamic mode. When simulating remedial actions, the interest is in predicting variations in concentrations from baseline conditions to different configurations of selected remedial actions.

5.4 RESOURCE AND DATA AVAILABILITY

The same issues of resource and data availability for subsurface methods apply to surface water methods. Tables 5.5 and 5.6 list the data needs for surface water assessment for simplified and analytical, and numerical methods, respectively. The user should refer to Section 4.4 for a generic overview of resource and data needs.

5.5 MODEL SELECTION CRITERIA FOR SURFACE WATER REMEDIAL ACTION ASSESSMENT

At this point, the reader has become familiar with remedial actions in terms of affected processes, minimum dimensionality required, and time frame requirements.

In light of this information derived from the matrices, some trends become apparent:

- o  Most removal measures affect sediment-water interactions, particularly adsorption and sediment deposition, erosion, and transport

- o  Physical processes such as longitudinal dispersion and advection are more greatly affected by remedial actions than are chemical/biological processes

- o  Remedial actions are specific to the type of waterbody as well as to the type of discharge or chemical

TABLE 5.5   DATA NEEDS FOR LEVEL I (ANALYTICAL) METHODS FOR SURFACE WATER PROBLEMS

I. Geometry of System

　　o Uniform reach or waterbody size: length, width, depth

II. Flow

　　o Average representative flow or velocity

III. Source of Pollutant

　　o Representative continuous rate, or specific pulse, or
　　o Initial concentration from near field analysis

IV. Pollutant Characteristics

　　o Lumped decay or specific transformation rates if pollutant is non-conservative
　　o Sediment concentrations and sorption coefficient, or
　　o Sediment size, diameter, sorption coefficient, and channel slope (if pollutant is hydrophobic)

TABLE 5.6   DATA NEEDS FOR LEVEL II (NUMERICAL) METHODS FOR SURFACE WATER PROBLEMS

I. Geometry of System

- o Size of specific reaches: length, width, depth

II. Flow

- o Distribution of flow or velocity (or depth and width) throughout the system

III. Source of Pollutant

- o Present and future source rates
- o Locations of sources

IV. Dispersion

- o Average representative longitudinal dispersion coefficient for one-dimensional problems, both longitudinal and transverse dispersion coefficients for two-dimensional problems
- o Time-varying coefficients for estuarial simulation

V. Pollutant Characteristics

- o Lumped decay rate or specific transformation rates if pollutant is non-conservative
- o Sorption coefficients for each sediment type; sediment density and diameter, channel slopes, and bed exchange rate if pollutant is hydrophobic

Model selection criteria for a Level I analysis is a
function of the pollutant characteristics (i.e., whether it
sorbs and can be transported by sediments) and whether the
remedial action can be accurately represented by a
simplified or analytical method. Quite often sediment
processes will have to be represented on a gross level to
represent many of the remedial actions. However, in many
cases, parameter estimation for most remedial actions will
be very difficult. Numerical simulation with default or
uncertain parameter values will lead to potentially spurious
results. In these cases it is advisable to utilize
simplified or analytical techniques. The user should refer
to Volume 4 for backup information and references for Level
I methods.

Model selection criteria for Level II analysis are more
specific. Some general guidelines include:

o   The simulation should usually be dynamic (time
    varying) in order to simulate uneven flows (as in an
    estuary) and pulse (spill) inputs of pollutants

o   The spatial domain (dimensionality) will vary
    according to the remedial action, but most actions
    require a two-dimensional, vertical/longitudinal
    simulation

o   Many pollutants are hydrophobic; thus, the ability
    to simulate sorption and sediment transport is
    critical. Suspended sediments are a heterogeneous
    mixture, requiring a model that can simulate the
    various types, including organic matter which is
    very important for sorbing organic pollutants. The
    simulation of deposition and resuspension is
    important also for the above mentioned reasons

o   The ability to simulate degradation processes is
    very important for those pollutants that readily
    dissolve or are susceptible to volatilization and
    photolysis. This is apparent when performing a
    baseline assessment to determine the persistence of
    the pollutant in the system. In addition, toxic
    daughter products (degraded forms of the pollutant)
    may be subject to specific sorption and degradation
    effects. Therefore, degradation kinetics should
    not be discounted in long-range fate analyses.

# References

Adkins, L.M., J.J. Doria and M.T. Christopher. 1983. Methods for Assessing Exposure to Chemical Substances - Vol. 3: Methods for Assessing Exposure From Disposal Of Chemical Substances, EPA 560/5-83-016, U.S. Environmental Protection Agency, Office of Pesticides and Toxic Substances, Washington, D.C.

Callahan, M.C., M.W. Slimak, N.H. Gabel, J.P. Map, C.F. Fowler, J.R. Freed, P., Jennings, R.L. Durfee, F.C. Whitmore, B. Maestri, W.R. Mabey, B.R. Holt and C. Gould. 1979. Water-Related Environmental Fate of 129 Priority Pollutants, EPA 440/4-79-029, Vol. 1,2, U.S. Environmental Protection Agency, Washington, D.C.

Codell, R. B., K. T. Key and G. Whelan. 1982. A Collection of Mathematical Models for Dispersion in Surface Water and Groundwater, NUREG-0868, U.S. Nuclear Regulatory Commission, Washington, D.C.

Dynamac Corporation. 1982. Methods for Assessing Exposure to Windblown Particulates, U.S. Environmental Protection Agency, Office of Health and Environmental Assessment, Washington, D.C.

Ehrenfield, J.R. and J.M. Bass. 1983. Handbook for Evaluating Remedial Action Technology Plans, EPA 600/2-83-76, U.S. Environmental Protection Agency, Municipal Environmental Research Laboratory, Cincinnati, OH.

Environmental Protection Agency. 1982. Workshop Summary, Level II Predictive Exposure Assessment. April 27-29, 1982, Atlanta, Georgia. U.S. Environmental Protection Agency, Athens, GA.

Farino, W., P. Spawn, M. Jasinski and B. Murphy. 1983. Evaluation and Selection of Models for Estimating Air Emissions from Hazardous Waste Treatment, Storage and Disposal Facilities. Revised Draft Final Report for the U.S. Environmental Protection Agency, Office of Solid Waste, Washington, D.C.

Freed, J.R., T. Chambers, W.N. Christie and C.E. Carpenter. 1983. Methods for Assessing Exposure to Chemical Substances - Vol. 2: Methods for Assessing Exposure to Chemical Substances in the Ambient Environment, EPA 560/15-83-015, U.S. Environmental Protection Agency, Office of Pesticides and Other Toxic Substances, Washington, D.C.

Javandel, I., C. Doughty and C.F. Tsang. 1984. Groundwater Transport: Handbook of Mathematical Models. American Geophysical Union Water Resources Monograph, Washington, D.C.

JRB Associates. 1982. Handbook - Remedial Actions at Waste Disposal Sites, EPA-625/6-82-006, U.S. Environmental Protection Agency, Cincinnati, OH.

Mills, W., J. Dean, D. Porcella, S. Gherini, R. Hudson, W. Frick, G. Rupp and G. Bowie. 1982. Water Quality Assessment: A Screening Procedure for Toxic and Conventional Pollutants, EPA 600/6-82-004abc, Vol. 1,2, U.S. Environmental Protection Agency, Athens, GA.

Onishi, Y., G. Whelan and R.L. Skaggs. 1982. Development of a Multimedia Radionuclide Exposure Assessment Methodology for Low-Level Waste Management, PNL-3370, Pacific Northwest Laboratory, Richland, WA.

SCS Engineers. 1982. Costs of Remedial Response Actions at Uncontrolled Hazardous Waste Sites, EPA 600/2-82-035, U.S. Environmental Protection Agency, Environmental Research Laboratory, Cincinnati, OH.

Thibodeaux, L. 1981. Estimating the Air Emissions of Chemical From Hazardous Waste Landfills. Journal of Hazardous Materials.

# VOLUME 2

Simplified Methods for Subsurface and Waste Control Actions

# 1. Introduction

1.1 PURPOSE OF REPORT

During the 1950's, the development of analytical and semi-analytical solutions for flow in ground-water systems dominated the literature. Even though attention shifted to numerical modeling in the 60's, progress continued in the development of analytical methods (Walton, 1979). In particular, a number of solutions for contaminant transport were developed during this time.

Many of these solutions, and analysis methods involving the use of these solutions, are applicable to the evaluation of subsurface control (e.g., ground-water pumping and impermeable barriers) and waste control (i.e., in-situ treatment) remedial action technologies. The purpose of this volume is to provide general guidance on the use of these "simplified methods." More specifically, the volume seeks to:

1. Identify the specific simplified methods applicable to the evaluation of each subsurface and waste control action;

2. Identify key assumptions and limitations affecting the use of specific methods;

3. Provide a compilation of methods that have been programmed for use on hand-held calculators and micro-computers; and

4. Demonstrate the use of selected methods through example evaluations of different remedial actions.

It is commonly assumed that the methods discussed herein are "easy to use" because they require limited data, manpower, time and computer resources. <u>This is a dangerous assumption</u>. The proper application of these methods requires considerable <u>judgement</u> and <u>experience</u>, at times as much as would be required to use a more sophisticated numerical model.

## 1.2 REPORT ORGANIZATION

A summary and brief set of conclusions are provided in the next section.

Section 3 identifies the specific simplified methods that are applicable to the evaluation of different subsurface and waste control remedial action technologies. This section also discusses, in general, how each method can be used.

Section 4 discusses the basic theory underlying different groups of available analytical and semi-analytical methods. This section does not attempt to provide complete derivations for different methods. A number of excellent textbooks and other publications cover their derivation; where appropriate, these publications are identified for those readers interested in more background. The focus of Section 4 is on the assumptions and limitations associated with different groups of methods, and how they affect the usefulness of these methods for remedial action evaluation.

Section 5 is a compilation of the methods that have been programmed for use with either hand-held calculators or micro-computers. Tables showing many of the available programs and sources for the programs are provided.

Section 6 provides a series of example applications that serve to demonstrate how different methods can be used to evaluate selected remedial action alternatives. The example applications are largely for hypothetical sites, some of which have been patterned after existing uncontrolled hazardous waste sites.

## 2. Summary and Conclusions

A large number of the existing analytical and semi-analytical solutions for ground-water flow and transport, and associated simplified methods, are applicable to the evaluation of subsurface and waste control remedial actions. The limited data and resource requirements (i.e., time, manpower and computer facilities) associated with the use of these methods make them ideally suited to the screening of remedial action performance and, in some cases, to the detailed analysis and conceptual design of remedial actions.

A number of the more commonly used methods have been compiled in several publications that would be of use to state and Federal Superfund staff and site contractors. A relatively complete set of well hydraulics solutions, including tables and graphs of well functions, can be found in a handbook by Walton (1984a). A large number of drain hydraulic solutions have been compiled by Cohen and Miller (1983). Finally, van Genuchten and Alves (1982), Javandel et al. (1984) and Walton (1984a) have compiled a number of contaminant transport solutions.

Hand-held calculator and micro-computer programs have been written for a subset of the more commonly used methods. These programs greatly reduce the amount of work involved in making numerous repetitive calculations when using these methods. They also eliminate the need for tables and graphs of well functions, and expand the capabilities of some methods by incorporating simple numerical techniques that would be difficult to solve by hand. Some of the programs have been published in the open literature, while others can be obtained directly from their developers. The International Center for Ground Water Modeling at Holcomb Research Institute, Butler University, provides a clearinghouse service for available hand-held calculator and micro-computer programs.

Despite benefits associated with these simplified methods, there are a number of important limitations and key assumptions that must be considered when using them in a practical evaluation or remedial action performance. Many of

## Summary and Conclusions

the analytical and semi-analytical solutions for flow and transport were derived for specific types of aquifers (e.g., confined, leaky or water table) with highly idealized characteristics. Typically, the aquifers are assumed to be horizontal, infinite in extent, constant in thickness, and composed of homogeneous and isotropic properties. Since few, if any, aquifers can fully satisfy these assumptions, even on a local scale, some degree of simplification or correction is often required. Transformation methods like equivalent sections and incremental methods and corrections for anisotropy are commonly used to construct aquifers with hydraulically equivalent characteristics. The method of images is commonly used to construct aquifers that are bounded. In using the method of images, however, it is only possible to construct aquifers with highly idealized geometries like wedges, strips and rectangles.

Many of the solutions were also derived for highly idealized ground-water flow patterns. Typically, solutions are available for radial or uniform, one-dimensional (horizontal) flow patterns. Fortunately, through the use of superposition, these idealized flow patterns can be combined so that more complex flow patterns can be evaluated. The superposition of solutions has its limitations, however, particularly for water table aquifers. Superposition in water table aquifers is only appropriate when changes in water table elevations are small compared to the saturated thickness.

The other major limitation is that many of the solutions were derived for specific well or drain configurations. Typically, wells or drains are assumed to be fully penetrating. This assumption makes it possible to neglect vertical flow components. When evaluating wells or drains that are not fully penetrating, solutions derived specifically for the configuration of interest or appropriate corrections should be used. This also holds for wells with finite diameters and for flowing wells.

These and other limitations preclude the complete, detailed analysis of all remedial action design objectives and configurations with the simplifed methods discussed herein. Changes in water table elevations or piezometric heads associated with the implementation of most subsurface and waste control remedial actions can generally be evaluated. The major exception includes certain drain and impermeable barrier configurations, particularly near the ends of partially penetrating drains or barriers of finite length. The other exception is one side of a fully penetrating, impermeable barrier when the method of images is used.

Changes in ground-water flow patterns can also be evaluated for most remedial actions, especially those that involve wells or drains. The one major exception is for remedial actions implemented in a water table aquifer. If the remedial action produces large changes in head relative to the saturated thickness, it may not be possible to evaluate changes in flow patterns with these methods. Ground-water flow around the ends of impermeable barriers of finite length is another major exception. All of the available simplified methods require that impermeable barriers are assumed to be infinite in length, keyed-in at the ends, or completely surrounding.

Changes in contaminant transport cannot be fully evaluated for many remedial actions. Most of the solutions were derived for radial or one-dimensional flow patterns. Thus, their use is largely limited to remedial actions that can be treated as point sources or sinks (e.g., recovery wells and injection wells). They were also derived based on the assumption that the properties affecting contaminant retardation and degradation are homogeneous and isotropic. Therefore, the spatial changes in these properties produced by many of the waste control actions (e.g., bioreclamation and chemical injection) cannot be evaluated.

Analytical solutions for contaminant transport typically consider all of the key processes of importance (i.e., advection, dispersion, retardation and degradation), whereas semi-analytical solutions typically only consider advection and, in some cases, retardation. Semi-analytical methods, however, offer great flexibility in terms of the complexity of flow patterns that can be analyzed.

Despite their apparent "ease of use," considerable judgement and experience are required to evaluate remedial action performance with simplifed methods. In applying these methods it is important to recognize the tradeoffs that are being made between the ease of application and the accuracy with which these methods can simulate the effects of implementing different remedial actions. The reader is referred to Volume 1 of this series for guidance on how to determine whether to select simplified methods or more detailed, numerical models. Volume 3 provides those who chose numerical models with guidance on their use in remedial action evaluation.

# 3. Remedial Action Evaluation with Simplified Methods

3.1 OVERVIEW

There are a large number of remedial action technologies that can be implemented at uncontrolled hazardous waste sites. These actions can be classified as either surface, subsurface or waste control technologies; control can either be by waste removal, containment or treatment. Many of the available technolgies are described in remedial action handbooks like those by JRB Associates (1982) and SCS Engineers (1982). In Volume 3, the large number of available technologies are condensed into fourteen "remedial measures." Essentially, technologies with similar design objectives were grouped together as remedial measures. Table 3.1 shows the measures that were classified as either surface, subsurface or waste control measures.

The analytical and semi-analytical methods discussed in the next section can be used to evaluate many of the remedial measures shown in Table 3.1. Since these methods are applicable only to flow and contaminant transport in groundwater systems, only subsurface and waste control measures can be evaluated. This section will discuss which of these measures can be evaluated and which specific simplified methods can be used.

In reading this section it is important to recognize that each of the subsurface and waste control measures listed in Table 3.1 can have different configurations and design objectives. Impermeable barriers, for instance, can be installed upgradient, downgradient and completely around a site. They can be partially penetrating (i.e., hanging) or fully penetrating (i.e., keyed-in). They can be used to lower the water table, divert uncontaminated ground water around a site, or preclude further migration of contaminated ground water. A complete detailed analysis of every configuration and design objective is not possible, however, because of the assumptions and limitations inherent in most simplified methods. Tables 3.2 and 3.3 list typical configurations and design objectives

TABLE 3.1   GROUPING OF REMEDIAL MEASURES

I. SURFACE CONTROL

- o   Grading
- o   Revegetation
- o   Surface Water Diversion

II. SUBSURFACE CONTROL

- o   Capping and Top Liners
- o   Seepage Basins and Drains
- o   Subsurface Drains, Ditches and Bottom Liners
- o   Impermeable Barriers
- o   Ground-Water Pumping
- o   Interceptor Trenches

III. WASTE CONTROL

- o   Permeable Treatment Beds
- o   Bioreclamation
- o   Chemical Injection
- o   Solution Mining (Extraction)
- o   Excavation/Hydraulic Dredging

TABLE 3.2  APPLICABILITY OF SIMPLIFIED METHODS TO THE EVALUATION OF SUBSURFACE CONTROL ACTIONS

| Remedial Action | Design Objective | Applicable Simplified Method(s) | Comments |
|---|---|---|---|
| Capping and Top Liners | Reduce Infiltration | SI, GM, TM | - |
| | Reduce ground-water contamination | CT, TM | Solutions for injection wells may have to be used if mounding is significant |
| Seepage Basins and Ditches | Recharge water and modify flow patterns | SI, GM, TM (Basins) SI, DH, S, TM (Ditches) | Note limitations on superposition in water table aquifers |
| Subsurface Drains, Ditches and Bottom Liners | Capture leachate | SI (Drains and Bottom Liners) NA (Ditches) | - |
| | Reduce ground-water contamination | CT, TM (Drains and Bottom Liners) NA (Ditches) | Solutions for injection wells may have to be used if mounding is significant |
| Impermeable Barriers | Divert ground water | S (Fully-penetrating barrier) | Barrier created with method of images is assumed to be infinitely long or keyed-in at ends; flow around ends of barrier cannot be considered. |
| | | GM, TM (Partially-penetrating barrier) | Solutions available only for several idealized aquifer geometries and barrier is assumed to be infinitely long or keyed-in at the ends. |
| | Capture contaminated ground water | CT, S, TM (Fully-penetrating barrier) | Barrier created with method of images is assumed to be infinitely long or keyed-in at ends; contaminant migration around ends of barrier cannot be considered. |
| | | NA (Partially-penetrating barrier) | - |

(continued)

TABLE 3.2 (continued)

| Remedial Action | Design Objective | Applicable Simplified Method(s) | Comments |
|---|---|---|---|
| Ground-water Pumping | Divert ground water | WH, S, TM | Corrections may be necessary for partially-penetrating wells; note limitations on superposition in water table aquifers |
| | Capture contaminated ground water | CT, S, TM | — |
| Interceptor Trenches | Divert ground water | DH, S, TM | Note limitations on superposition in water table aquifers |
| | Capture contaminated ground water | CT, S, TM | Drain will have to be represented as a line of closely spaced wells; note limitations on superposition in water table aquifers |

LEGEND:  WH - Well Hydraulics
DH - Drain Hydraulics
GM - Groundwater Mounding
SI - Seepage/Infiltration
S - Superposition
TM - Transformation Methods
CM - Conformal Mapping
CT - Contaminant Transport
NA - No method applicable

TABLE 3.3  APPLICABILITY OF SIMPLIFIED METHODS TO THE EVALUATION OF WASTE CONTROL ACTIONS

| Remedial Action | Design Objective | Applicable Simplified Method(s) | Comments |
|---|---|---|---|
| Permeable Treatment Beds | In-situ treatment of ground water | CT, TM | Contaminant transport solution must be applied in a step-wise fashion to treatment bed and up gradient and down gradient portions of aquifer. |
| Bioreclamation | In-situ treatment of ground water | WH, S, CT, TM | Reductions in contaminant concentrations cannot be analyzed since applicable solutions typically neglect degradation. |
| Chemical Injection | In-situ treatment of ground water | WH, S, CT, TM | Reductions in contaminant concentrations cannot be analyzed since applicable solutions typically neglect degradation. |
| Solution Mining | Mobilize contaminants | WH, S, GM, CT, TM | Selected solution must consider retardation |
| Excavation/ Hydraulic Dredging | Improve leachate quality | SI, GM, CT, TM | Contaminant transport solutions for injection wells may have to be used if mounding is significant. |

LEGEND:  GM - Ground-Water Mounding
SI - Seepage/Infiltration
TM - Transformation Methods
CT - Contaminant Transport
WH - Well Hydraulics
S - Superposition

for each measure, as well as the applicable simplified method(s). Important limitations and considerations associated with the use of different simplified methods are also listed.

## 3.2 SUBSURFACE CONTROL MEASURES

The primary goals of subsurface control measures are to prevent leachate migration and reduce ground-water contamination by diversion, containment or plume capture. Subsurface control measures include capping and top liners; seepage basins and ditches; subsurface drains, ditches and bottom liners; impermeable barriers; ground-water pumping; and interceptor trenches.

### 3.2.1 Capping and Top Liners

As Table 3.2 shows, caps and top liners are generally implemented to reduce infiltration into a waste site, thereby reducing the quantity of leachate that is generated. In evaluating the performance of capping and top liner systems, two design objectives are of concern: 1) the reduction in the quantity of leachate that is generated and 2) the associated reduction in ground-water contamination. Different methods are required to evaluate each objective.

Methods applicable to the estimation of seepage rates for landfills (see Subsection 4.5) can be used to evaluate the effect of a cap or top liner on leachate generation. Infiltration rates can be determined for both pre- and post-restoration conditions. The ground-water mounding estimation techniques discussed in Subsection 4.4 can also be used to determine whether the reduction in leachate quantity will have any effect on the degree of mounding, if any, below the site.

Associated reductions in ground-water contamination can be evaluated with simplified methods for contaminant transport (see Subsection 4.9). The choice of which type of contaminant transport method to use will depend upon whether or not the quantity of leachate generated by the site is sufficient to cause mounding. If mounding is not significant either before or after capping, almost any of the analytical or semi-analytical solutions can be used.

If mounding is significant even after capping, only those analyical or semi-analytical methods that consider radial flow can be used. Analytical methods of this type are for injection wells; none exist for ground-water mounds. Therefore, analytical methods can only be used if a flow pattern equivalent to that of a mound can be simulated with one or more injection wells. The same limitation holds for those semi-analytical methods based on the complex velocity potential concept (see Subsection 4.9). Except, these methods often also provide a way of representing a circular source of finite radius. In some cases, a flow pattern equivalent to that around a mound can be represented with such a source. The semi-analytical solution based on a simple numerical technique discussed in Subsection 4.9 can also be used. Again, an injection well or group of wells must be used to create a flow pattern equivalent to that created by the mound.

Example Application 4 in Section 6 demonstrates the use of the simple numerical technique to evaluate contaminant transport from a site where mounding is significant.

As Table 3.2 shows, transformation methods are also useful in evaluating capping and top liner actions. Transformation methods are used to transform real world aquifers with heterogeneous and isotropic conditions into equivalent, idealized aquifers with homogeneous and isotropic conditions (see Subsection 4.7). These methods are used in conjunction with almost all analytical and semi-analytical solutions for flow and contaminant transport. Thus, transformation methods are useful in the evaluation of virtually all subsurface and waste control actions.

### 3.2.2 Seepage Basins and Ditches

The primary objective for using seepage basins and ditches is to recharge site runoff or water withdrawn by wells or drains. A second objective is to improve the efficiency of plume capture by modifying ground-water flow patterns. Thus, both seepage (i.e., recharge) rates and the extent of ground-water mounding are important when evaluating seepage basins and ditches.

Subsections 4.5 and 4.4 discuss methods for estimating seepage rates for ponded facilities and changes in water table elevations (i.e., mounding), respectively. If mounding occurs, its effect on the drawdowns at nearby wells or drains can be evaluated by using the principle of superposition. The use of superposition makes it possible to evaluate a number of alternative locations for a seepage basin. It is important to remember, however, the limitations associated with using superposition in water table aquifers (see Subsection 4.3).

The effect of seepage from ditches can be evaluated in the

same manner. Instead of using ground-water mounding estimation methods, however, the drain hydraulics methods discussed in Subsection 4.3 can be used. In using these methods, the ditch is treated as a line source of finite length.

If changes in contaminant migration patterns, as a result of recharge from a basin or ditch, are also of interest, the choice of which contaminant transport method to use will again depend on the extent of mounding and whether one or more injection wells or a circular source of finite radius can be used to represent the mound created by the seepage basin or ditch. Example Application 4 in Section 6 shows one approach for analyzing a recharge basin.

### 3.2.3 Subsurface Drains, Ditches and Bottom Liners

Subsurface drains, ditches and bottom liners are usually installed in the unsaturated zone to capture leachate before it reaches the saturated zone. The infiltration estimation techniques discussed in Subsection 4.5, in particular the HELP model (Schroeder et al., 1984a and 1984b), could be used to estimate reductions in leachate quantity associated with subsurface drains or bottom lining. Given this change in leachate quantity, changes in ground-water contamination levels could be assessed with the contaminant transport methods in Subsection 4.9. The same considerations as for capping and top lining (see Subsection 3.2.1) would apply to the selection of what type of method to use. None of the simplified methods are applicable to the evaluation of ditches.

### 3.2.4 Impermeable Barriers

Impermeable barriers are grout curtains, slurry walls and sheet piling installed in the saturated zone to divert uncontaminated ground-water around a site or limit the migration of contaminated ground-water. Barriers can be placed in a number of locations relative to a disposal site: upgradient, downgradient or completely around. Barriers designed to divert ground-water by lowering water levels can either partially or fully penetrate the saturated zone. The former must be keyed into impermeable strata to preclude water movement around the ends. Barriers designed to contain ground water are normally fully penetrating.

The analytical methods discussed in Section 4 are only applicable to a few of the many possible barrier configurations. The amount of seepage likely to occur under a partially penetrating barrier can be analyzed using the conformal mapping methods described in Subsection 4.8. These methods are for two-dimensional flow in the horizontal and vertical directions only. Thus, it is assumed that the barrier is infinitely long, keyed-in at the ends, or completely surrounds the site. If the barrier does not have one of these configurations, the conformal mapping methods will only apply over those portions of the barrier where horizontal flow components are essentially perpendicular to the barrier.

The conformal mapping methods are limited to either single-layered or two-layered saturated systems; in the latter case, the layers must be of equal thickness. Therefore, it is important to carefully consider site conditions before using these methods.

The method for two-layered systems can also be used to evaluate barriers that fully penetrate the saturated zone, but are keyed into a leaky bedrock layer. Again the same restrictions apply in terms of the barrier configurations and site conditions that can be considered.

In cases where the barrier can be keyed into an impermeable bedrock layer, the principle of superposition can be used. Specifically, the method of images can be used to obtain an impermeable boundary of infinite length by using real and imaginary pumping wells (see Subsection 4.6). Different barrier configurations, including upgradient, downgradient and completely surrounding, can be analyzed through the proper use of real and image wells. An impermeable barrier surrounding a site is analyzed with the method of images in Example Application 3 in Section 6.

Despite its flexibility, the method of images has two distinct disadvantages. First, it requires that the barrier be assumed to be infinite in length, keyed-in at the ends, or completely surrounding the site. Flow conditions and heads for barriers with other configurations cannot be considered except near the center of relatively long barriers where flow directions are essentially perpendicular to the barrier.

The second disadvantage is that only ground-water flow patterns and heads on the side of the barrier with the real wells (i.e., the real region) can be analyzed. The other side (i.e., the image region) is of no value. Thus, the real well(s) must be located on the same side of the barrier as the disposal site if flow patterns and heads around the site are of concern.

### 3.2.5 Ground-Water Pumping

Ground-water pumping actions can have a number of configurations and design objectives. Single pumping wells or a line of well points can be used to capture a plume. Single or multiple wells can be installed to divert ground water by lowering the water table. They can also be used to prevent unconfined aquifers from contaminating lower aquifers separated by leaky formations. The water withdrawn by pumping may be treated and subsequently reinjected through one or more wells. The reinjection wells may be used to flush contaminants toward the pumping wells or to create a hydraulic barrier to preclude further plume migration.

The well hydraulics, superposition, transformation and contaminant transport methods discussed in Subsections 4.2, 4.6, 4.7 and 4.9, respectively offer a relatively complete set of methods for evaluating virtually all possible configurations for ground-water pumping remedial actions. They can be used to evaluate changes in ground-water flow patterns, heads and contaminant movement. All of these are important factors when evaluating ground-water pumping schemes. In using the available analytical methods it is important to recognize the key underlying assumptions and limitations (see Table 3.2). Since pumping is often used conjunctively with impermeable barriers it is also important to recognize the limitations associated with the method of images. Example Applications 2,3 and 4 in Section 6 demonstrate approaches for evaluating ground-water pumping actions.

### 3.2.6 Interceptor Trenches

Interceptor trenches are drain systems that are installed in the saturated zone. They can be used to: 1) divert ground water by lowering the water table or 2) capture a plume.

The first design objective can be evaluated using the drain hydraulics, superposition, and transformation methods discussed in Subsections 4.3, 4.6 and 4.7, respectively. A wide range of site conditions and drain configurations can be considered with these methods. Example Applications 1 and 5 in Section 6 demonstrate an approach for evaluating how water table elevations will change following the installation of a drain.

The second objective can be evaluated using the contaminant transport methods discussed in Subsection 4.9. Since none of these methods explicitly consider drain systems, a drain must be represented by a line of closely spaced wells.

## 3.3 WASTE CONTROL

Waste control measures involve the removal or treatment of hazardous wastes or contaminated water and sediments. Removal can be accomplished through excavation or hydraulic dredging. Treatment methods include permeable treatment beds, bioreclamation, chemical injection, and solution mining (extraction). Those treatment methods in the waste control category are in-situ methods. That is, treatment is accomplished in-place. On-site treatment methods like carbon adsorption, precipitation, sedimentation, and activated sludge are considered under the subsurface control category since they are typically used in conjunction with ground-water pumping systems, subsurface drains or interceptor drains.

### 3.3.1 Permeable Treatment Beds

Permeable treatment beds are trenches backfilled with limestone activated carbon or another media that can physically or chemically remove contaminants from ground water. They are installed so as to penetrate into the saturated zone, and are normally used in areas where the water table is near the ground surface. Treatment occurs as contaminated ground water passes through the bed. Permeable treatment beds are typically designed to have the same hydraulic conductivity as the surrounding materials. As a result, their installation generally has little or no affect on ground-water movement.

In evaluating the effectiveness of permeable treatment beds in terms of reducing ground-water contamination it is important to recognize that all of the simplified methods for contaminant transport assume that aquifer properties are homogeneous and isotropic. Thus, it is difficult to represent the discontinuity produced by a treatment bed because it has sorption properties different from those of the surrounding aquifer materials. To analyze a treatment bed, the simplified methods must be applied in a step-wise fashion, first to the upgradient portion of the aquifer, then the treament bed itself and then the downgradient portion of the aquifer.

### 3.3.2 Bioreclamation

Bioreclamation is an in-situ treatment method involving the

injection of microbes, nutrients and oxygen into a plume to initiate or accelerate contaminant degradation. It is commonly used for hydrocarbons and other easily biodegradable pollutants. Injection is accomplished through the use of one or more wells. Pumping is also used to obtain water for the injection system and to enhance treatment. Recirculation between the injection and withdrawal wells is often an important design consideration.

Since bioreclamation is essentially a form of a ground-water pumping technique, there are a number of simplified methods available to examine ground-water flow patterns, changes in hydraulic heads, and pollutant movement between injection and recovery wells. Despite the availability of a large number of methods, they can only be used to evaluate a few of the design objectives affecting the performance of bioreclamation systems. The well hydraulics, superposition, and transformation methods can be used to evaluate changes in flow patterns and heads induced by the wells. Contaminant transport methods can be used to estimate the size of a region that will be treated by the injected mixture, the amount of recirculation that might occur, and the time it will take for the injected mixture to arrive at a recovery well. Reductions in contaminant concentrations cannot be directly estimated, however, since the applicable solutions typically neglect degradation. Example Application 4 in Section 6 demonstrates one approach for evaluating a bioreclamation action.

### 3.3.3 Chemical Injection

Chemical injection is used to treat the waste in a landfill/lagoon or in a contaminated saturated zone. It is usually applied to sites with well defined wastes, with shallow landfill or lagoon depths, and where the vertical and horizontal extent of the contamination is small (JRB Associates, 1982). The objective of chemical injection is to immobilize or destroy a pollutant. Numerous injection wells may be employed depending on the size of the disposal site. A water supply well is usually required for chemical dilution. The effect of this measure is to substantially increase retardation and degradation processes in either the unsaturated or saturated zones.

As with bioreclamation, only a few of the design objectives for chemical injection can be evaluated with available simplified methods. Changes in flow patterns and hydraulic heads can be evaluated with the well hydraulics, superposition, and transformation methods. The size of the zone treated by the injected fluid can be evaluated with contaminant transport methods. The extent to which chemical injection will reduce ground-water contamination levels at some point downgradient from a site cannot be assessed, however. Again, this is due to the the fact that degradation is often neglected in applicable transport solutions.

### 3.3.4 Solution Mining (Extraction)

Solution mining is similar to chemical injection in that both methods chemically alter the pollutant in the waste itself. However, solution mining involves the flooding of a landfill with a chemical solvent, which may desorb or free the pollutant so that it may be mobilized in a larger leachate flow (JRB Associates, 1982). The leachate can then be collected by drains and/or well points. The objective is to increase the mobility of the contaminant.

The evaluation of the performance of a solution mining action can be approached with the same types of methods used for bioreclamation and chemical injection. Since an important design objective for solution mining is the efficiency of recovery, several of the analytical transport methods can be used to identify which well configuration will provide for the most efficient recovery. The effect of the solvent on increasing contaminant mobility can be considered by simply adjusting the retardation factor used in these methods. Semi-analytical transport methods can be used in a similar way.

### 3.3.5 Excavation/Hydraulic Dredging

Excavation/hydraulic dredging involves the removal of the waste source itself, thus improving leachate quality. Excavation is used on solids, sediments, or sludge materials. Hydraulic dredging may be used to remove liquids and/or sludges from lagoons or surface impoundments. After the waste area has been excavated or dredged, it may be backfilled to limit infiltration.

The effectiveness of excavation/hydraulic dredging can be evaluated with a number of simplified methods. Seepage/infiltration estimation methods can be used to evaluate changes in leachate quantity. In the case of waste excavation, those methods applicable to the estimation of infiltration rates for landfills (e.g., HELP model) can be used to determine whether the amount of water passing through the site will change. The amount of change will, in part, depend upon the properties of the materials used for backfill relative to those of the excavated waste.

Ground-water mounding estimation methods can be used to determine whether the shape of the water table will change as a result of changes in infiltration/seepage rates.

The type of contaminant transport method used to evaluate changes in ground-water contamination levels will again be determined by the extent of mounding. If mounding is significant, a transport method that can consider radial flow must be used. One or more injection wells or a circular source of finite radius will need to be used to simulate the effects of the mound. If mounding is not important, any of the transport methods can be used.

# 4. Theory Underlying Available Simplified Methods

## 4.1 OVERVIEW

The basic theory underlying the simplified methods applicable to remedial action evaluation can be divided into the following areas: 1) well hydraulics, 2) drain hydraulics, 3) ground-water mounding, 4) seepage/infiltration, 5) superposition, 6) transformation methods, 7) conformal mapping, and 8) contaminant transport. Some of these areas encompass the theory used to develop different types of solutions (e.g., well hydraulics, drain hydraulics, ground-water mounding, conformal mapping and contaminant transport), whereas others encompass the theory behind the use of these solutions to evaluate relatively complex geohydrological conditions (e.g., superposition and transformation methods).

The applicable theory underlying each area will be summarized in this section. Comprehensive discussions of the applicable theories and derivations of analytical expressions will not be provided, since this material is presented in a number of standard references (e.g., Freeze and Cherry, 1979; Bear, 1979; Walton, 1970 and Harr, 1962) and handbooks (e.g. Walton, 1984a and Javandel et al., 1983). Rather, this section will focus on the types of methods available in each area, and the key assumptions and limitations governing their use.

## 4.2 WELL HYDRAULICS

Wells are used in many different types of remedial action technologies. They can be used alone to control plume movement, divert uncontaminated ground water or capture contaminated ground water. They can also be used in conjunction with other technologies for the same purposes, or as part of in-situ treatment technologies where both injection and extraction are required. As a result, well hydraulics analyses are likely to be conducted at many sites.

Fortunately, numerous solutions have been developed to calculate the change in piezometric head or water table elevation resulting from the introduction of a well. Some of the earliest and perhaps most fundamental work in the area of well hydraulics was conducted by Theis (1935). As Freeze and Cherry (1979) note, Theis utilized an analogy to heat-transfer theory to derive an analytical solution for flow to a well in a highly simplified aquifer. This aquifer has the following characteristics:

1. horizontal
2. confined between impermeable layers on the top and bottom
3. infinite in horizontal extent
4. constant thickness
5. homogeneous and isotropic

Transient flow in this type of aquifer system with no sources or sinks can be described by the following partial differential equation:

$$\frac{\partial^2 h}{\partial x^2} + \frac{\partial^2 h}{\partial y^2} = \frac{S}{T} \frac{\partial h}{\partial t} \qquad (4.1)$$

where
- $h$ = piezometric head, L
- $x,y$ = horizontal directions, L
- $S$ = storativity, dimensionless
- $T$ = transmissivity, $L^2/T$
- $t$ = time, T

Recognizing that changes in head around a well are radially-symmetric, Equation 4.1 can be rewritten, in the following form

$$\frac{\partial^2 h}{\partial r^2} + \frac{1}{r} \frac{\partial h}{\partial r} = \frac{S}{T} \frac{\partial h}{\partial t} \qquad (4.2)$$

where
- $r$ = radial distance from the well, L

### 4.2.1 Confined Aquifers

The work by Theis (1935) produced a solution to Equation 4.2 for the condition of a single, fully penetrating well with a constant withdrawal (pumping) rate, an infinitesimally small well diameter, and a uniform piezometric head prior to the initiation of pumping. Figure 4.1 shows the drawdown around a well with this type of configuration in a horizontal confined aquifer. Under these conditions, flow is strictly horizontal and unidirectional toward the well. There are no vertical flow components. Using a uniform piezometric head as an initial condition, the assumption of no drawdown at infinity and a constant pumping rate as boundary conditions, Theis obtained the following solution for transient flow to a well:

$$s = h_o - h(r,t) = \frac{Q}{4\pi T} \int_u^\infty \frac{e^{-u} du}{u} \qquad (4.3)$$

where
$u = r^2 S/4Tt$, L

$h_o$ = initial piezometric head, L
$Q$ = pumping rate, $L^3/T$
$s$ = drawdown, L

As Freeze and Cherry (1979) note, the integral in Equation 4.3 is known as the exponential integral. Given the above definition for u, the integral is also known as the well function, W(u). This gives the familiar Theis equation

$$s = \frac{Q}{4\pi T} W(u) \qquad (4.4)$$

Values for W(u) can be evaluated using a series expansion as noted by Bear (1979). Tabulated values for W(u) and a graphical relationship between W(u) and 1/u are provided in most ground-water textbooks.

In using the Theis equation it is important to recognize that steady-state conditions can never be reached in an aquifer of infinite extent. In the absence of any sources of recharge, water must be continuously withdrawn from storage to meet the demands of the pumping well. This requires that the cone of depression must continually expand radially outward from the well. From a practical point of view, however, peizometric heads do reach a quasi-steady-state as the rate of propagation of the cone of depression decreases. This is particularly true for the region near the well. Thus, the Theis equation can be used to obtain an _estimate_ of steady-state conditions when the time of pumping is assumed to be long.

Many of the hand calculator and micro-computer programs discussed in Section 5 are for what is known as "Theis condition aquifers." Theis condition aquifers are essentially

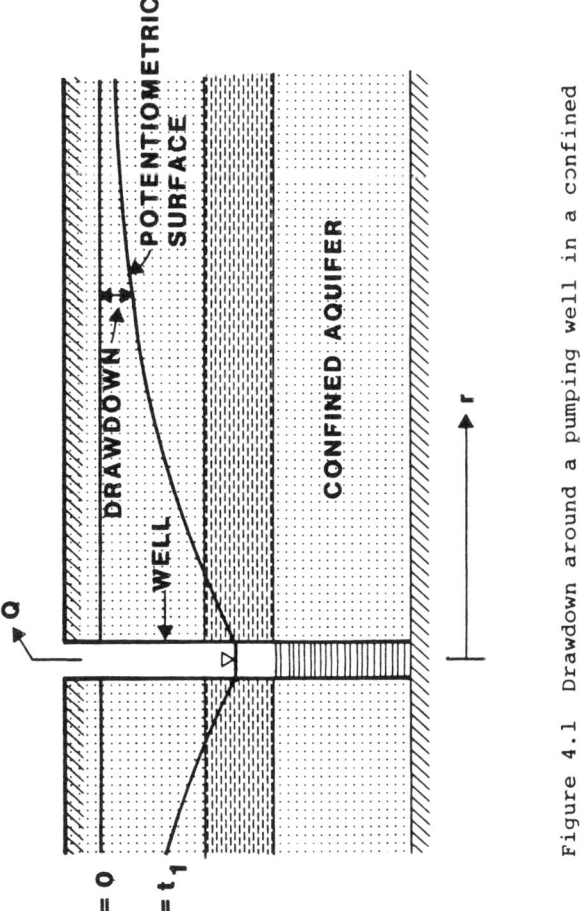

Figure 4.1  Drawdown around a pumping well in a confined aquifer.

those that have the aquifer characteristics and well conditions discussed above. Clearly few aquifers in the real word, even with reasonably acceptable simplifications, can be considered as Theis condition aquifers. Many are not confined. Some have semi-impermeable confining layers on the top and/or bottom. These are called "leaky" aquifers. Others have a free surface on top and either an impermeable or leaky layer on the bottom. These aquifers are known as "water table" aquifers.

#### 4.2.2 Leaky Aquifers

An analytical solution for leaky aquifer conditions was initially developed by Hantush and Jacob (1955) and Hantush (1956, 1960), and was later expanded by Neuman and Witherspoon (1969a, 1969b, 1972). The expression derived by Hantush and Jacob has a form similar to the Theis equation

$$s = \frac{Q}{4\pi T} W(u, r/B) \qquad (4.5)$$

where $W(u, r/B)$ is called the leaky well function and $r/B$ is a dimensionless parameter given by

$$r/B = r\sqrt{\frac{K'}{Kbb'}} = r\sqrt{\frac{K'}{Tb'}} \qquad (4.6)$$

where  $K$ = hydraulic conductivity of aquifer, L/T
$b$ = aquifer thickness, L
$K'$ = hydraulic conductivity of leaky layer, L/T
$b'$ = thickness of leaky layer, L

The assumptions and limitations underlying Equation 4.5 are essentially the same as those for the Theis equation. The aquifer is horizontal, infinite in extent, has a constant thickness, is homogeneous and isotropic, and has a uniform piezometric head prior to pumping. The well is fully penetrating with a constant pumping rate and infinitesimal diameter. Figure 4.2 shows the drawdown around such a well in a horizontal leaky aquifer. The unpumped aquifer above the leaky aquifer is often called the "supplying aquifer." The two aquifers are separated by the leaky layer.

In addition, Hantush and Jacob had to assume that: 1) the hydraulic head in the supplying aquifer remains constant; 2) the rate of leakage across the leaky layer is proportional to the difference in hydraulic heads between the pumped and

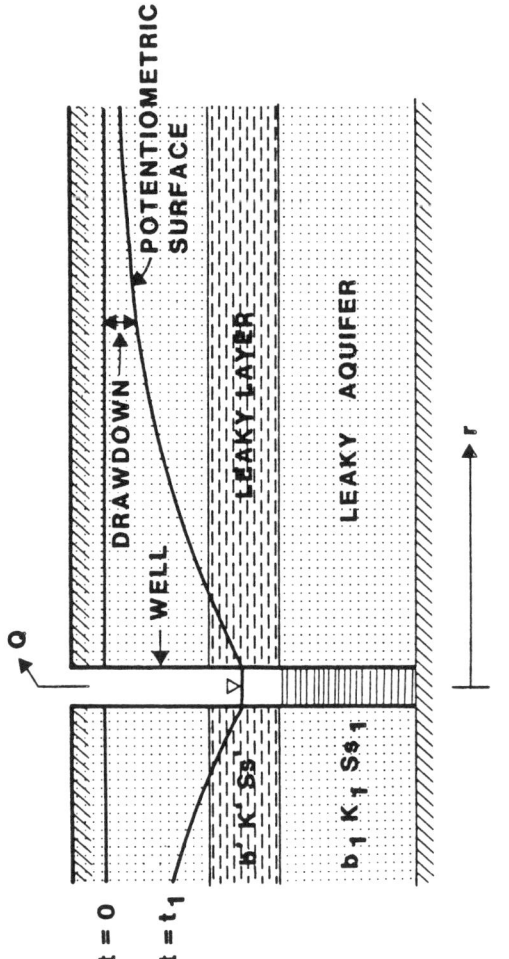

Figure 4.2  Drawdown around a pumping well in a leaky aquifer.

unpumped aquifers; and 3) flow in the pumped aquifer is strictly horizontal and unidirectional towards the well, while flow in the leaky layer is vertical. As Freeze and Cherry (1979) point out, the first assumption implies that the unpumped aquifer can provide an unlimited supply of water to the pumped aquifer. The second assumption neglects the effect of storage in the leaky layer on delaying the delivery of water. As a result, the rate of actual drawdown may be over-predicted. The third assumption neglects the potential for vertical flow components in the pumped aquifer and horizontal components in the leaky layer. Figure 4.3 shows how the assumed flow distribution (on the left side of the well) is different from the actual distribution (on the right side of the well). Huisman (1972) notes that less than 3% error will be induced if vertical flow components are neglected when $\lambda > 3H$, where $\lambda = 1/B$. Neuman and Witherspoon (1969a) note that when the hydraulic conductivity of the aquifer is at least two orders of magnitude greater than the leaky layer, neglecting vertical flow components introduces errors of no more than 5%.

Neuman and Witherspoon (1969a, 1969b) overcame the limitations imposed by the first two assumptions by generating a more rigorous leaky well function. Their expression takes the form of:

$$s = \frac{Q}{4\pi T} W(u, r/B_{11}, r/B_{21}, \beta_{11}, \beta_{21}) \quad (4.7)$$

where

$$\frac{r}{B_{11}} = r \sqrt{\frac{K'}{K_1 b_1 b'}}$$

$$\frac{r}{B_{21}} = r \sqrt{\frac{K'}{K_2 b_2 b'}}$$

$$\beta_{11} = \frac{r}{4b_1} \sqrt{\frac{K' S s'}{K_1 S_{s1}}}$$

$$\beta_{21} = \frac{r}{4b_2} \sqrt{\frac{K' S s'}{K_2 S_{s2}}} \quad (4.8)$$

$K, b$ and $S_s$ are the hydraulic conductivity, thickness and specific storage, respectively. The subscripts 1 and 2 are for the pumped and unpumped aquifers, respectively. The prime (') is for the leaky layer.

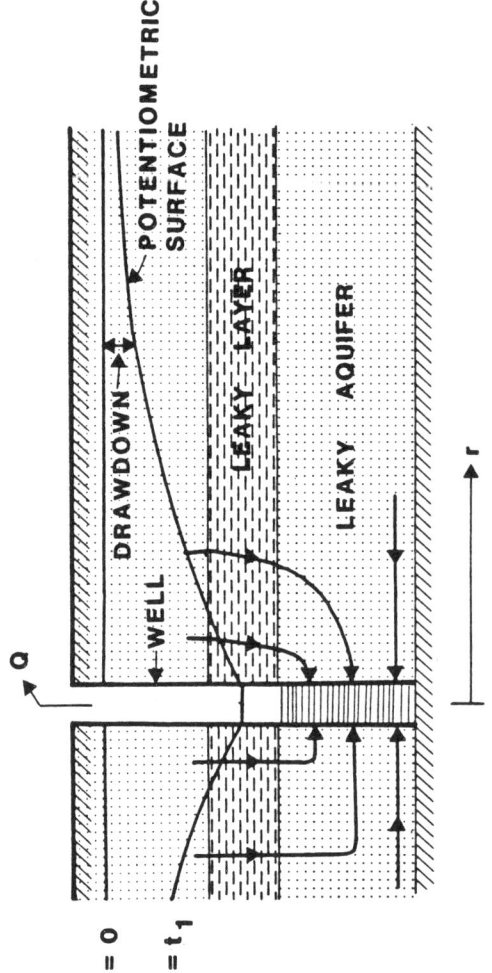

Figure 4.3 Assumed (left side of well) and actual (right side of well) flow patterns for a fully penetrating well in a leaky aquifer (adapted from Huisman, 1972).

Values for the well function in Equations 4.6 and 4.7 have been tabulated in many ground-water textbooks and the publications referenced above.

Unlike confined aquifers, water levels in leaky aquifers of infinite extent can achieve a steady-state condition. This occurs once the entire discharge of the well is derived from leakage.

Freeze and Cherry (1979) note that the simpler solution (Equation 4.5) is widely used despite its limitations. Once steady-state conditions are reached, the limitation imposed by neglecting the storage effect in the leaky layer is removed. The limitation imposed by assuming an unlimited supply of water is not removed, however. Therefore, it is important to examine the system of interest to determine whether it is reasonable to simply rely on Equation 4.6.

Freeze and Cherry (1979) also note that the Theis equation can be used in place of one of the leaky aquifer solutions, because it provides a more conservative estimate of drawdowns; drawdowns under leaky conditions will be less than those for confined conditions because of the leakage. While this may be appropriate for the analysis of wells for water supply purposes, this line of reasoning is not appropriate for remedial action evaluation. Since one of the intents of ground-water pumping as a remedial action is to lower the water table, a conservative estimate of drawdown may lead to the design of an ineffective remedial action.

### 4.2.3 Water Table Aquifers

In both confined and leaky aquifers with a fully penetrating well it is generally reasonable to assume that flow is strictly horizontal and unidirectional towards the well. In water table aquifers this assumption may or may not be reasonable. When a water table aquifer is pumped, vertical flow components are created as the water table itself changes shape around the well. Whereas in a confined aquifer, water is produced by both the compaction of the aquifer and the expansion of water, in a water table aquifer it is also produced by gravity drainage. As Walton (1970) notes, the release of water due to compaction of the aquifer and expansion of water is instantaneous. The release due to gravity drainage is not. As a result, changes in hydraulic head are initially rapid. The rate of change slows, however, as the water released by gravity drainage reaches the cone of depression. Once this occurs, the rate of change increases and the cone of depression continues to expand as gravity drainage keeps pace with declining water levels. As with confined aquifers, in the absence of any source of recharge, the water table will continue to change as long as pumping continues. Again, however, a quasi-steady condition will be reached after a reasonably long time.

A number of analytical solutions have been developed for fully penetrating wells in water table aquifers; Figure 4.4 shows drawdown around a well in a horizontal water table aquifer. The most general solution is the solution developed by Boulton (1954, 1955, 1963) and later advanced by Neuman (1972, 1973, 1975). Their work produced an solution that takes the form of

$$S = \frac{Q}{4\pi T} W(u_A, u_B, \eta) \qquad (4.9)$$

where $W(u_A, u_B, \eta)$ is known as the "unconfined well function." Just following initiation of pumping, the unconfined well function is $W(u_A, \eta)$, where

$$u_A = \frac{r^2 S}{4Tt}$$

$$\eta = \frac{r^2}{b^2} \qquad (4.10)$$

and S is now the elastic storativity of the aquifer. This elastic storativity is responsible for the instantaneous release of water. At some later time, the unconfined well function is $W(u_B, \eta)$, where,

$$u_B = \frac{r^2 S_y}{4Tt} \qquad (4.11)$$

$S_y$ is the specific yield responsible for delaying the release of water. Details regarding the use of Equation 4.9 and graphs and tables providing values of the unconfined well function are given in Neuman (1975) as well as many ground-water textbooks.

The expression developed by Neuman considers the effects of both the delay caused by gravity drainage and vertical flow components. The delay effect has its greatest impact during the early stages of pumping. Bear (1979) states that the

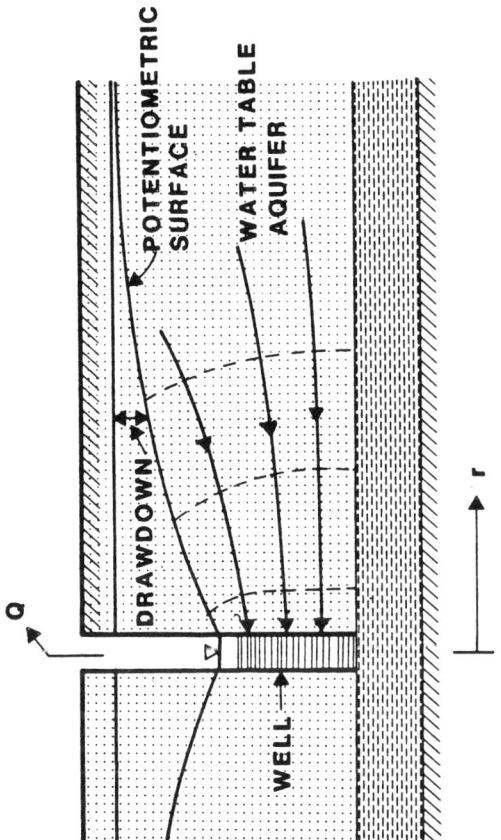

Figure 4.4 Drawdown around a pumping well in a water table aquifer.

specific yield increases at a diminishing rate with time. Therefore, if water levels following long-term pumping are of interest, the impact of the delay effect can be neglected. Bear (1979) states further, however, that lenses of silt and clay can produce significant delays.

Vertical flow components can affect water levels under certain water table conditions. Bear (1979) points out that Boulton (1954) and Hantush (1964) both suggested that vertical flow components are of importance in the region of 0<r<0.2H when

$$t > \frac{5 H_o n_e}{K} \qquad (4.12)$$

where  $H$ = hydraulic head at the well, L
       $H_o$ = initial hydraulic head, L
       $n_e$ = effective porosity, dimensionless

Stallman (1965) found vertical flow components to be important in the region

$$\frac{Tt}{n_e r^2} < 1 \qquad (4.13)$$

The importance of vertical flow components and the delay caused by drainage should be examined given specific site conditions, because the other general solutions for water table aquifers assume both can be neglected.

In cases where they can be neglected, an expression developed by Boulton (1954) in one of his earlier works can be used

$$s = \frac{Q}{2\pi T} (1 + C_f) V(\rho,\tau) \qquad (4.14)$$

where  $C_f$ = correction factor
       $V(\rho,\tau)$ = gravity well function for water table aquifers
       $\rho$ = $r/H_o$
       $\tau$ = $Kt/n_e H_o$
       $T$ = $KH_o$ = transmissivity, $L^2/T$

Values of $V(\rho, \tau)$ and $C_f$ are provided in a number of publications, including Boulton (1954), Schoeller (1959) and Hantush (1964). Bear (1979) presents several ways of approximating $V(\rho,\tau)$ for different ranges of $\tau$, and notes that an error of less than 6% is obtained if $C_f$ is assumed to be zero for $0.05<\tau<5$.

It is interesting to note for $\tau_s>5$, $V(\rho,\tau)$ is approximately equal to $1/2\ W(u)$, where $u = n_e r^2/4Tt$ (Bear 1979). Combining this result with Equation 4.14 shows that the analytical solution for confined aquifers can, in some cases, be used to estimate water levels in a water table aquifer, particularly for long time frames where conditions approach a quasi-steady state. Bear (1979) states that for a thick aquifer with small drawdowns that satisfy the following condition, a water table aquifer can be treated as a confined aquifer:

$$(H_o-h) << H_o \text{ or } H_o+h \approx 2H_o \qquad (4.15)$$

### 4.2.4 Available Well Hydraulics Solutions

The previous discussions in this section have overviewed the theory underlying some of the general analytical solutions available for flow in confined, leaky and water table aquifers. The basic assumptions upon which these solutions are derived limits their use in a number of situations. Fortunately, a number of other analytical solutions with less restrictive assumptions have been developed. Table 4.1 was adapted from tables in Walton (1984a). It provides a reasonably complete inventory of available analytical solutions for confined, leaky and water table aquifer systems, respectively. Each solution in Table 4.1 is characterized in terms of: 1) the aquifer characteristics and well configuration that can be analyzed, 2) whether the solution is for time-varying or steady-state conditions, and 3) the type of output that can be obtained.

As this table shows, solutions are available for isotropic and anisotropic conditions. Corrections for anisotropy, in addition to those provided in these expressions, will be discussed in Subsection 4.6, Transformation Methods.

A range of possible well configurations can also be considered, including wells with a finite diameter, wells with storage capacity, flowing wells, and partially penetrating wells. During the early periods of pumping, drawdowns for wells of finite diameter and/or with storage capacity will deviate from those predicted using an expression for a well of infinitesimal diameter and no storage. Thus, if piezometric

TABLE 4.1  INVENTORY OF SELECTED WELL HYDRAULICS SOLUTIONS (adapted from Walton, 1984a)

| Type | Aquifer Characteristics | | | Well Configuration | | | | Time Frame | Output | References |
|---|---|---|---|---|---|---|---|---|---|---|
| | Properties | Extent | Special Cases | Penetration | Storage | Diameter | | | | |
| C | H,I | IN | P | FP | NS | ID | | TV | D | Theis (1935) |
| C | H,I | IN | P | FP | NS | FD | | TV | D | Hantush (1964) |
| C | H,I | IN | P | FP | S | FD | | TV | D | Papadopulos (1967) |
| C | H,I | IN | P,VE | FP | NS | ID | | TV | D | Brutsaert and Corapcioglu (1976) |
| C | H,A | IN | P | PP | NS | ID | | TV | D | Hantush (1964) |
| C | H,I | IN | P | FP | NS | ID | | TV | D | Jacob and Lohmann (1952) |
| C | H,A | IN | F | FP | NS | ID | | TV | D | Boulton and Streltsova (1977b) |
| C | H,A | IN | F | PC | NS | ID | | TV | D | Boulton and Streltsova (1977a) |
| C | H,I | IN | P | FP | NS | ID | | TV | D | Moench and Prickett (1972) |
| C | H,A | IN | P | FP | NS | ID | | TV | D | Fapadopulos (1965) |
| L | H,I | IN | P | FP | NS | ID | | TV | D | Hantush and Jacob (1955) Hantush (1959) |
| L | H,I | IN | P | FP | S | FD | | TV | D | Lai and Chen Wu Su (1979) |
| L | H,I | IN | P,AS | PP | NS | ID | | TV | D | Witherspoon et al. (1967) |
| L | H,I | IN | P | FP | NS | ID | | TV | D | Corapcioglu (1976) |
| L | H,A | IN | P | PP | NS | ID | | TV | D | Hantush (1964) |
| L | H,I | IN | P | FP | NS | ID | | TV | D | Hantush (1967b) |

(continued)

TABLE 4.1 (continued)

| Type | Aquifer Characteristics | | | Well Configuration | | | Time Frame | Output | References |
|------|------------|--------|---------------|-------------|---------|----------|------|--------|------------|
| | Properties | Extent | Special Cases | Penetration | Storage | Diameter | | | |
| WT | H,I | IN | P | FP | NS | ID | TV | D | Neuman (1975) |
| WT | H,A | IN | P | PP | NS | ID | TV | D | Streltsova (1974) |
| WT | H,A | IN | P | PP | S | FD | TV | D | Boulton and Streltsova (1978) |
| WT | H,A | IN | F | PC | NS | ID | TV | D | Boulton and Streltsova (1978) |
| WT | H,A | IN | P,LB | PP | NS | ID | TV | D | Streltsova (1976) |

(1) Additional selected solutions not listed in this table include:

- Boulton (1954a) solution for non steady-state water table drawdown.

- Cooper, H. H., Jr., and C. E. Jacob (1946) give a straight line graphical solution to the Theis equation.

- Thiem (1906) gives a steady state solution for flow to a well in a confined aquifer.

LEGEND:
C  - Confined
WT - Water Table
L  - Leaky
H  - Homogeneous
I  - Isotropic
A  - Anisotropic
IN - Infinite
AS - Aquitard Storage
LB - Leaky Base
F  - Fractured Media
P  - Uniformly Porous
VE - Visco-Elastic Properties

FP - Fully Penetrating
PP - Partially Penetrating
PC - Partially Cased
NS - No Storage
S  - Storage
ID - Infinitesimal Diameter
FD - Finite Diameter

TV - Time Varying
D  - Drawdown

heads just after initiation of pumping are of concern, one of the appropriate solutions in Table 4.1 should be used.

Flowing wells are wells where the head in the well is held constant and flow rates are allowed to vary with time. If flowing wells are being considered, appropriate relationships must be used since the ones discussed earlier are for wells with constant flow rates and varying heads.

Partially penetrating wells are those that are screened only over a portion of the aquifer. Partial penetration creates vertical flow components that may preclude the use of expressions based on the assumption of complete penetration. As Bear (1979) states, the drawdown produced by a partially penetrating well is greater than that for a fully penetrating well. This difference is only significant for a distance 1.5 to 2.0 times the saturated thickness away from the well. As Table 4.1 shows, a number of analytical solutions have been developed for partially penetrating wells. Corrections to the drawdowns predicted with solutions for fully penetrating wells are also available in Bear (1979) for several different well configurations.

All of the analytical solutions shown in Table 4.1 are included in a handbook by Walton (1984a). It contains the actual expressions and supporting tables and graphs useful in estimating values for different well functions. It also discusses a number of other useful analytical solutions and some of the available hand-calculator programs. The handbook is a useful source for anyone planning to use analytical solutions for the evaluation of remedial action performance. It can be obtained from the International Ground Water Modeling Center (IGWMC), Holcomb Research Institute, Butler University in Indianapolis, Indiana (317-283-9458).

## 4.3 DRAIN HYDRAULICS

Drains are collection systems of finite length that can be used, like wells, to control plume movement, divert ground-water flow and depress water table levels. They can have a number of configurations ranging from fully penetrating, vertical trenches to partially penetrating ditches, to perforated pipes. Unlike wells, drains are almost always installed in water table aquifers. Rarely are they used in confined or leaky aquifers. For this reason, most of the analytical solutions that have been derived for drains are for water table conditions, although several solutions have been derived for confined and leaky systems. Due to the limited usefulness of solutions for confined or leaky systems for remedial action evaluation, the remainder of this subsection will focus on the theory, assumptions and limitations for solutions applicable to water table conditions.

The complete mathematical description of time-varying flow to drains in water table aquifers is nonlinear and intractable largely because of the effect of the moving water table boundary. As a result, several simplifications must be made before analytical expressions can be derived. The simplifications upon which most of the available expressions are based are the Dupuit-Forchheimer assumptions and linearization.

The Dupuit-Forchheimer assumptions are based on the observation that the slope of the water table in most aquifers is very small. In addition, under steady-flow conditions without accretion (i.e., recharge), the water table is a streamline. These observations lead to the following assumptions:

1) for small slopes on the water table, flow lines are horizontal and equipotentials are vertical and

2) the hydraulic gradient is equal to the slope of the water table and is invariant with depth

In effect, the Dupuit-Forchheimer assumptions make it possible to neglect vertical flow components. As a result, the mathematical description for steady flow in a horizontal, homogeneous, isotropic, water table aquifer without sources or sinks simplifies to

$$\frac{\partial}{\partial x}\left(h\frac{\partial h}{\partial x}\right) + \frac{\partial}{\partial y}\left(h\frac{\partial h}{\partial y}\right) = 0 \qquad (4.16)$$

Boussinesq (1904) extended the Dupuit-Forchheimer assumptions to include sources and sinks and time-varying conditions. The Boussinesq equation is

$$\frac{\partial}{\partial x}\left(h\frac{\partial h}{\partial x}\right) + \frac{\partial}{\partial y}\left(h\frac{\partial h}{\partial y}\right) + \frac{N}{K} = \frac{S}{K}\frac{\partial h}{\partial t} \qquad (4.17)$$

where N is the rate of accretion(L/T).

However, the Boussinesq equation is nonlinear. To simplify it further, the concept of linearization must be invoked. The most common linearization is to use a constant saturated thickness when the change in water table elevation is small compared to the saturated thickness. This is reasonable in many cases. Cohen and Miller (1983) present a relationship for estimating h, the constant saturated thickness

$$\overline{h} = d + \frac{D}{2} \tag{4.18}$$

where d and D are defined in Figure 4.5. This thickness can be used to estimate an average transmissivity, $\overline{T}$,

$$\overline{T} = K\overline{h} \tag{4.19}$$

which can be substituted into the Boussinesq equation to obtain

$$\frac{\partial^2 h}{\partial x^2} + \frac{\partial^2 h}{\partial y^2} + \frac{N}{\overline{T}} = \frac{S}{\overline{T}} \frac{\partial h}{\partial t} \tag{4.20}$$

This equation is linear in h, and can be solved for different boundary conditions to obtain a number of useful analytical solutions.

Before presenting these solutions, it is important to first identify those limitations affecting their use. The first set of limitations relate to the geometry of the water table aquifer. The Dupuit-Forchheimer assumptions are only valid in situations where D<<d and d<<L (see Figure 4.5). In addition, Bear (1979) notes that errors in predicted heads will be small when the square of the water table slope is much less than 1.0 (i.e., $(\Delta h/\Delta x)^2 << 1.0$). It is important to note that if accuracy in the rate of discharge is more important than water table elevations, the above geometry limitations can be neglected in many cases (Bear, 1979).

The second set of limitations relate to specific conditions where vertical flow components are significant. The conditions include: 1) the seepage face near a drain, 2) a ground-water divide, 3) near an impermeable barrier, 4) regions of significant accretion and 5) partially penetrating drains. Bear (1979) states that at distances greater than two times the saturated thickness away from these conditions, the assumptions are valid.

Few of the expressions that will be presented later incorporate corrections for the effect of partial penetration. Huisman (1972) presents a series of formulas for partially penetrating drains of different geometries. Cohen and Miller (1983) note that Hooghoudt (1940) provided a means of

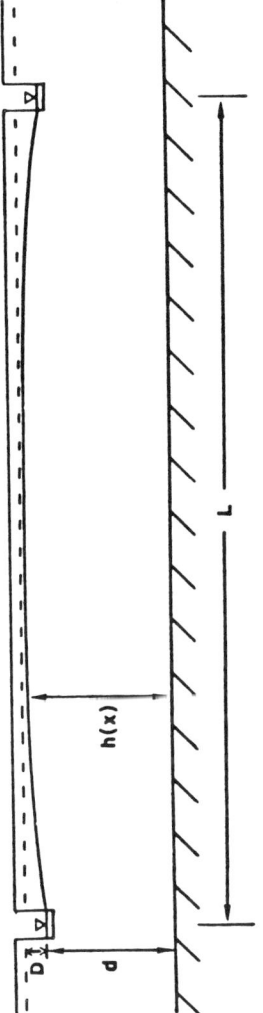

Figure 4.5  Ideal conditions for applying a drain hydraulics method based on Dupuit - Forchheimer assumptions - $D \ll d$ and $h \ll L$ (taken from Cohen and Miller, 1983).

correcting predicted heads through the use of an equivalent depth, $d_e$. Hooghoudt produced a table of equivalent depths that can be substituted for d in many of the available expressions. According to work by Wesseling (1964), the equivalent depth approach is accurate to within 5% of results obtained using a more rigorous mathematical approach. Figure 4.6 shows the graphical relationship between $d_e$ and d for different drain spacings, L. Moody (1966) provides a direct relationship:

$$\frac{d}{d_e} = 1 + \frac{d}{L}\left(\frac{8}{\pi}\ln\frac{d}{r} - a\right), \qquad 0 < \frac{d}{L} \leq 0.3 \qquad (4.21)$$

where

$$a = 3.55 - 1.6\frac{d}{L} + 2\left(\frac{d}{L}\right)^2$$

For

$$\frac{d}{L} \geq 0.3$$

$$\frac{L}{d_e} = \frac{8\,[\,\ln(L/r) - 1.15\,]}{\pi} \qquad (4.22)$$

When only a single drain is being considered, L goes to infinity and $d_e = d$ according to Cohen and Miller (1983). Huismann and Olsthoorn (1983) note that the additional drawdown due to partial penetration is negative when

$$\Omega > d \qquad (4.23)$$

where $\Omega$ is the wetted circumference of the drain.

One other limitation noted by Cohen and Miller (1983) is that the solutions for drain hydraulics do not consider flow in the unsaturated zone that may be induced by drains.

Finally, most of the available solutions for drains are one-dimensional. They are applicable only to those portions of drains where flow is <u>horizontal</u> and <u>perpendicular</u> to the axis of the drain. Near the ends of long drains and over most the length of short drains there are flow components that are perpendicular and parallel to the axis of the drain. This is depicted in Figure 4.7. As a result, there is a variation in flow and drawdown along the length of most drains. Relationships that can be used to evaluate how both change along the length of a drain are presented in Huisman (1972) and Huisman and Olsthoorn (1983).

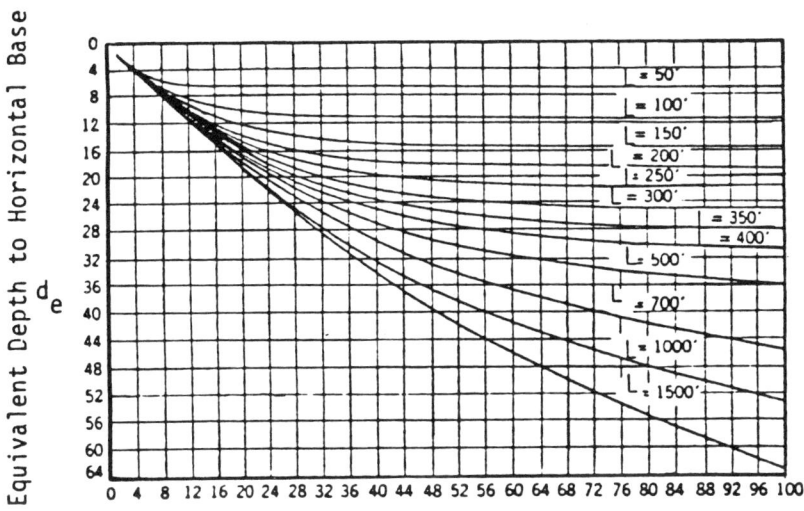

Figure 4.6 Relationship between equivalent depth and total depth for different drain separations (taken from Cohen and Miller, 1983).

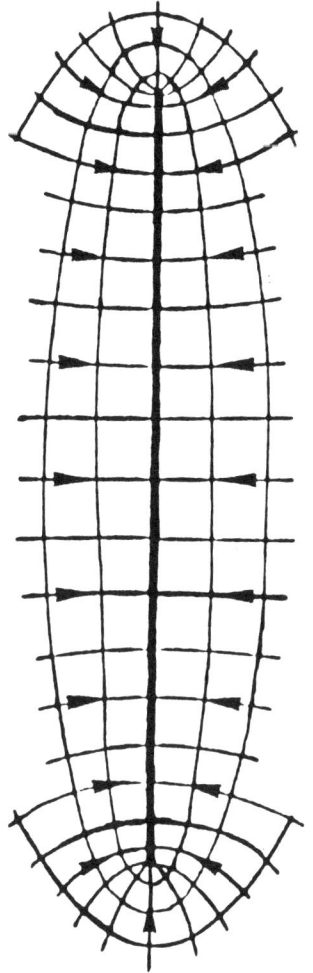

Figure 4.7 Plan view of flow to a drain of finite length (taken from Cohen and Miller, 1983).

Cohen and Miller (1983) recently compiled a large number of available analytical expressions for flow to drains. Most of them are for water table aquifers. Some were derived specifically for confined and leaky conditions. Tables 4.2 and 4.3 include a large number of steady state and transient drain hydraulic solutions, respectively. Each solution is again characterized according to: 1) the aquifer characteristics and drain configurations that can be analyzed, 2) whether the solution is for time-varying or steady-state conditions, and 3) the type of output that can be obtained.

Another useful compilation of analytical solutions for flow to drains is contained in Moore (1983). This technical resource document provides procedures for evaluating the effectiveness of sand or gravel drainage layers and drain pipes, as well as compacted clay liners.

## 4.4 GROUND-WATER MOUNDING ESTIMATION METHODS

Large quantities of leachate can produce a mound in the water table below certain types of waste disposal facilities and remedial action technologies. The ponding of waste in lagoons or impoundments can create large quantities of leachate, particularly if no liner system is used or if the liner fails. Certain landfill designs can also produce sufficient quantities of leachate to cause mounding. Remedial action technologies like seepage basins are generally used to dispose of water from pumping wells or drains following treatment. Since the objective of these technologies is to recharge water, mounding can also occur under these facilities.

Mounding of a water table aquifer can have a major impact on local ground-water flow patterns and the resultant movement of contaminants. This impact often needs to be considered when evaluating the effectiveness of different remedial action alternatives.

As was discussed in Subsection 4.2 and 4.3, well hydraulics and drain hydraulics methods can be used to represent point sources of recharge (e.g., injection wells) and line sources of recharge (e.g., seepage ditches or recharge ditches), respectively. They cannot be used directly, however, to represent areal sources of recharge (e.g, ponds, seepage basins and landfills). Multiple point or line sources have to be used to represent an areal source.

Several analytical methods have been developed for use in evaluating changes in water table elevations as a result of recharge from an areal source (Baumann, 1952; Glover, 1960;

TABLE 4.2  INVENTORY OF SELECTED STEADY-STATE DRAIN HYDRAULICS SOLUTIONS (adapted from Cohen and Miller, 1983)

| Type | Aquifer Characteristics | | | Special Cases | Drain Configuration | | | Time Frame | Output | References |
|---|---|---|---|---|---|---|---|---|---|---|
| | Properties | Extent | Dimensionality | | Number | Penetration | Length | | | |
| C | H,I | B | x | VT | 2 | FP | IH | SS | D,IF | Huisman (1972) |
| WT | H,I | B | x | RC | 2 | FP | IN | SS | D | Jacob (1943); Ferris et al. (1962); Bear (1979) |
| WT | H,I | B | x | RC | 2 | PP | IH | SS | D,IF | |
| WT | H,I | B | R | RC | 2 | PP | IN | SS | D | Kirkham (1958); Luthin (1973); Harr (1962); Moore (1983); Moore (1983) |
| WT | H,I | B | R | RC | 2 | EP | IN | SS | D | |
| WT | H,I | B | R | RC,SB | 2 | FP | IN | SS | D | |
| WT | H,I | B | R | RC,SB | 2 | FP | IN | SS | D | McBear et al. (1982) |
| WT | H,I | B | x | RC | 2 | FP | IN | SS | D | Kirkham (1967); Kirkham et al. (1974); Hooghoudt (1940); Luthin (1973); van Schilfgaarde (1970); Luthin (1973) |
| WT | H,I | B | x | RC | 2 | PP | IH | SS | D | |
| WT | IH,I | B | x | RC | 2 | PP | IN | SS | D | |
| WT | H,I | B | x | RC | 2 | PP | IN | SS | D | Youngs (1964, 1966a, 1966b); Kirkham et al. (1974); Bouwer (1974) |
| WT | H,I | SI | R | RC | 1 | PP | IH | SS | D | |
| L | H,I | IN | x | | 1 | FP | IN | SS | D,IF | Huisman (1972) |
| L | H,I | IN | x | LB,RC | 1 | FP | IN | SS | D,IF | Huisman (1972) |
| WT | H,I | IN | x | LB,RC | 2 | PP | IN | SS | D,IF | Huisman (1972) |
| WT | H,I | B | x | LB,RC | 2 | PP | IN | SS | D | Bear (1979) |
| WT,L | H,I | B | x | RC | 2 | FP | IN | SS | D | Huisman (1972) |

(continued)

TABLE 4.2 (continued)

LEGEND:
- C — Confined
- WT — Water Table
- L — Leaky
- H — Homogeneous
- IH — Inhomogeneous
- I — Isotropic
- B — Bounded
- SI — Semi-Infinite
- IN — Infinite
- R — Radial
- X — One-Dimensional
- VT — Varying Aquifer Thickness
- SB — Sloping Base
- LB — Leaky Base

- RC — Recharge
- FP — Fully Penetrating
- PP — Partially Penetrating
- IN — Infinite
- SS — Steady State
- TV — Time Varying
- IF — Inflow
- D — Drawdown

TABLE 4.3  INVENTORY OF SELECTED TRANSIENT DRAIN HYDRAULICS SOLUTIONS (adapted from Cohen and Miller, 1983)

| Type | Aquifer Characteristics | | | Special Cases | Drain Configuration | | | Time Frame | Output | References |
|------|------------|--------|----------------|---------------|--------------|-------------|--------|------------|--------|------------|
|      | Properties | Extent | Dimensionality |               | Drain Number | Penetration | Length |            |        |            |
| WT | H,I | SI | X |       | 1 | FP | IN | TV | D     | Venetis (1968) |
| WT | H,I | SI | X |       | 1 | FP | IN | TV | D,IF  | Moody and Ribbens (1965) U.S.D.I. (1981) |
| WT | H,I | B  | X |       | 1 | FP | IN | TV | D     | Cooper and Rorabaugh (1963) Ferris et al. (1962) |
| C  | H,I | SI | X |       | 1 | FP | IN | TV | D,IF  | Ferris et al. (1962) |
| C  | H,I | SI | X |       | 1 | FP | IN | TV | D     | Hufsman (1972) |
| WT | H,I | SI | X | LB    | 1 | FP | IN | TV | D,IF  | Glover (1974) Luthin (1973) van Schilfgaarde (1974) |
| WT | H,I | B  | X |       | 2 | PP | IN | TV | D,IF  | Brocks (1961) Glover (1966, 1974) |
| WT | H,I | B  | X |       | 2 | PP | IN | TV | D     | Glover (1966, 1974) |
| WT | H,I | B  | X |       | 2 | PP | IN | TV | D     | Glover (1974) |
| WT | H,I | B  | X | RC    | 2 | PP | IN | TV | D,IF  | van Schilfgaarde (1974) |
| WT | H,I | B  | R | RC    | 2 | PP | IN | TV | D     | Terzidis (1968) |
| WT | H,I | B  | R |       | 2 | PP | IN | TV | D     | van Schilfgaarde (1974) |
| WT | H,I | B  | R |       | 2 | PP | IN | TV | D     | van Schilfgaarde (1974) |
| WT | H,I | B  | R | SB    | 2 | FP | IN | TV | D     | van Schilfgaarde (1974) |
| WT | H,I | B  | X | SB    | 2 | FP | IN | TV | D     | Chauhan et al. (1968) |
| WT | H,I | IN | X | SB,RC | 2 | FP | IN | TV | D     | Singh and Jacob (1977) |

(continued)

TABLE 4.3 (continued)

LEGEND:
C – Confined
WT – Water Table
L – Leaky
H – Homogeneous
IH – Inhomogeneous
I – Isotropic
B – Bounded
SI – Semi-Infinite
IN – Infinite
R – Radial
X – One-Dimensional
VT – Varying Aquifer Thickness
SB – Sloping Base
LB – Leaky Base

RC – Recharge
FP – Fully Penetrating
PP – Partially Penetrating
IN – Infinite
SS – Steady State
TV – Time Varying
IF – Inflow
D – Drawdown

Hantush, 1967a; Hunt, 1971 and Rao and Sarma, 1981). Most of these methods were derived for an areal source, rectangular or circular in configuration. They can be used to estimate changes in water table elevations at different radial distances away from the center of the source area. They can be applied to sources with different areal configurations by first converting the actual source area into an equivalent rectangular or circular area.

In deriving solutions for mounding estimation, it is commonly assumed that the aquifer is homogeneous, isotropic, infinite in areal extent, and resting on a horizontal impermeable base. Further, it is assumed that the seepage rate is uniform and the water table remains below the base of the facility. Estimates of mounding using the method by Hantush (1967a) have been found to be reasonable if the rise of the water table is not more that 50 percent of the original saturated depth.

In addition to typical aquifer properties like hydraulic conductivity, specific yield and saturated thickness, the seepage rate for the areal source must be known. This rate is difficult to quantify without the use of relatively sophisticated models or field methods; McWhorter and Nelson (1980), for instance, present one approach to modeling seepage from lagoons. At best, only estimates can be obtained with relatively simple methods. Available methods are discussed below.

## 4.5 SEEPAGE/INFILTRATION ESTIMATION METHODS

Many of the methods for estimating seepage rates for ponded facilities are based on Darcy's Law. One example is a simple graphical method presented by Knight et al. (1980). It only requires an estimate of the permeability of the liner or sludge materials in the bottom of the pond. This method assumes a unit gradient (i.e., a gradient of one). It is applicable to situations where the soil beneath the ponded source is much more permeable than the liner or sludge materials and where the depth of the liquid is small compared to thickness of the liner or sludge.

In cases where this method is not applicable, a method developed by Witherspoon and Narasimhan (1973) can be used. It is a graphical technique based on results obtained from a finite element computer model. It requires estimates of the depth to the water table, pond depth, depth to an impervious layer, length of the flow domain and drop in hydraulic head. The first three parameters are generally easy to obtain. The last two can be estimated using the graphical method by Knight et al., (1980). Moore (1983) discusses the use of Darcy's Law for the purpose of estimating seepage rates and some of its limitations. Sandberg et al., (1981) note that Darcy's Law produces rates 2-5 times those calculated with numerical models by McWhorter and Nelson (1980).

Bicknell (1984) recently developed a computer code that can be used to estimate chemical emissions from ponded facilities. Both volatile emissions and leachate quality can be calculated. Seepage rates are estimated with Darcy's Law or can be input if they have been measured or calculated with another model.

Seepage rates for landfills or other areal sources without ponded surfaces can be estimated with several methods. Fenn et al., (1975) discuss the "water balance method." Given monthly values for precipitation and potential evapotranspiration, estimates of monthly evapo- transpiration, runoff and infiltration can be obtained for different types of soils. Seepage rates through multi-layered soil columns can be estimated through successive applications of the method. Thus, the water balance method is applicable to a landfill with or without a cap. Dass et al. (1977) reported on a similar method.

A somewhat more sophisticated method is incorporated in the Hydrologic Evaluation of Landfill Performance (HELP) model (Schroeder et al. 1984a and 1984b). HELP is a quasi-two-dimensional model that computes a daily water budget for a landfill represented as a series of horizontal layers. Each layer corresponds to a given element of a landfill design (e.g., cap, waste cell, leachate collection system, and liner). HELP considers a broad range of hydrologic processes including surface storage, runoff, infiltration, percolation, evapotranspiration, lateral drainage and soil moisture storage. The HELP model requires climatologic data, soil characteristics, and design specifications as inputs. Climatologic data consist of daily precipitation, mean monthly temperatures, mean monthly solar radiation, leaf area indices, root zone or evaporative zone depths, and winter cover factors. Soil characteristics include porosity, field capacity, wilting point, hydraulic conductivity, water transmissivity, evaporation coefficient and Soil Conversation Service (SCS) runoff curve numbers. Design specifications consist of the number of layers and their type, thickness, slope, and maximum lateral distance to a drain, if applicable, and whether synthetic membranes are to be used in the cover and/or liner.

While the water balance method described above can be solved by hand, the large number of calculations performed by HELP are most efficiently done on a computer. The program is operational on EPA's National Computer Center in Research

Triangle Park, North Carolina.

Bicknell (1984) recently modified HELP to include techniques for estimating chemical emissions. Volatile emissions and leachate quality can now be computed with HELP.

## 4.6 SUPERPOSITION

Many of the available analytical solutions were derived for single wells or drains with constant flow rates or heads in aquifers of infinite extent. Since few aquifers satisfy these conditions, it is often necessary to consider the hydraulics associated with and interactions between multiple wells and drains and nearby boundaries. This is particularly true for the evaluation of remedial action performance where wells, drains and impermeable barriers are often used conjunctively. It is the principle of superposition that makes it possible to combine the solutions for single wells and/or drains to obtain solutions for multiple well and drain systems with variable flow rates and head conditions. One special type of superposition, the method of images, makes it possible to add the effects of boundaries, like streams, ground-water divides and impermeable zones, to solutions for aquifers of infinite extent.

Superposition, as defined by McWhorter and Sunada (1977), is the method in which linear combinations of elementary solutions are formed to provide additional solutions. The method is valid for linear, homogeneous, partial differential equations. Since many of the solutions for wells and drains are linear, superposition can be used in most instances. The major exception is flow in water table aquifers. As was noted in Subsection 4.2, the governing equation is non-linear. However, if simplification through linearization is reasonable, superposition can even be used in water table aquifers. Bear (1979) provides a theoretical description of the principle of superposition and a procedure for determining when it can be used.

The most common use of superposition is in the analysis of multiple well systems. A multiple well problem can be decomposed into a series of individual well problems. The resultant draw-down at any point in the aquifer can be obtained by summing the drawdown produced at that point by each well. Figure 4.8 shows the drawdowns induced by pumping each well individually and the resultant drawdown for both wells together. The same procedure can be used to examine the effect of varying drawdown along the length of a drain. This problem can be decomposed onto a series of drains of different length with different drawdowns.

Theory Underlying Available Simplified Methods 109

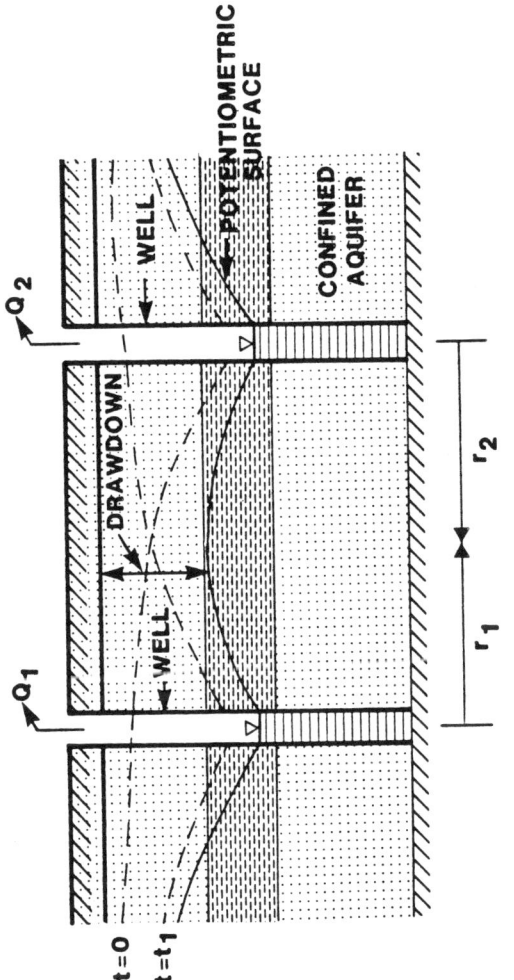

Figure 4.8 Superposition of drawdowns for two pumping wells in a confined aquifer.

Clearly, in some cases, a large number of tedious calculations could be required to evaluate a remedial action involving numerous wells and/or drains. Many of the hand-held calculator and micro-computer programs discussed in Section 5 do this automatically, thus reducing the amount of work required to evaluate a remedial action.

Superposition can also be used to evaluate time variable pumping rates. Again, if the equations are linear, the time variable solution for one pumping rate at a well can be added to that for another. Figure 4.9 shows this use of superposition for a well pumping at a rate of $Q_1$ from $t = 0$ to $t = t_2$ and then at a rate of $Q_2$ after $t = t_2$.

Another use of superposition is to include the effects of regional ground-water flow on the drawdowns induced by a well or drain system. As Huisman (1972) notes, this problem can be decomposed into two parts: 1) flow of ground water prior to pumping and 2) flow of ground water due to pumping. The use of superposition in this manner produces several analytical expressions useful for remedial action evaluation. The first of these is for steady flow to a single pumping well in a uniform steady regional flow. The flow system for this case is shown in Figure 4.10. The main features of importance in this flow system are a stagnation point and ground-water divide. Water within the envelope created by the ground-water divide will eventually be captured by the well. Water on the outside of the envelope will be affected only by the regional ground-water flow. The following relationships can be used to locate the stagnation point and to estimate the maximum width of the envelope (Bear 1979):

$$x_s = -\frac{Q}{2\pi q_o b} \qquad (4.24)$$

$$W = \frac{Q}{q_o b} \pm \tan\left(\frac{2\pi q_o b y}{Q}\right) \qquad (4.25)$$

where
$x_s$ = distance to stagnation point, L
$Q$ = well pumping rate, $L^3/T$
$q_o$ = specific discharge rate for aquifer, $L^2/T$
$b$ = aquifer thickness, L
$W$ = maximum width of envelope created by the ground-water divide, L

Both $x_s$ and $W$ are useful in determining what pumping rate would be required to capture a ground-water plume. Equations

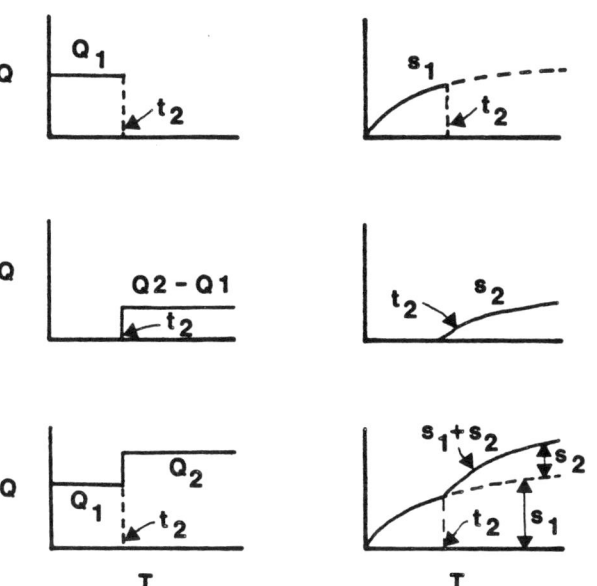

Figure 4.9 Superposition of drawdowns to obtain drawdown after a step change in discharge (taken from McWhorter and Sunada, 1977).

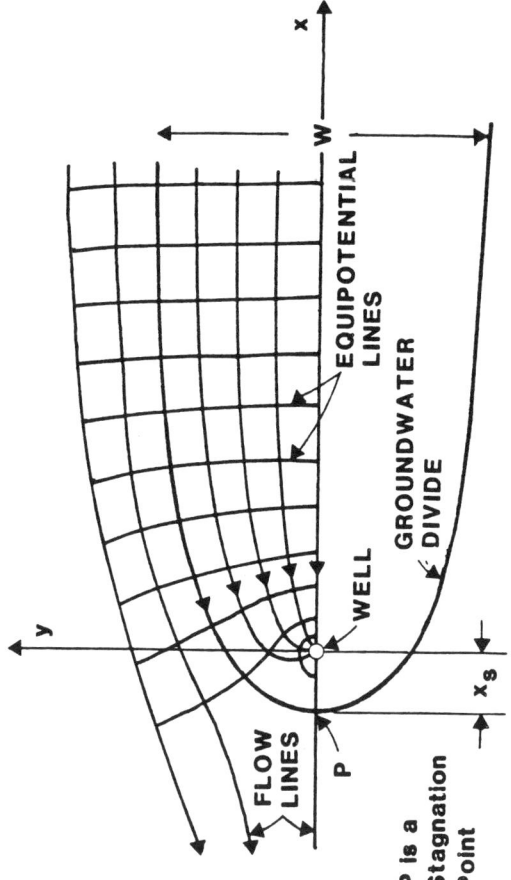

Figure 4.10 Flow pattern around a pumping well in a uniform regional flow (after Powers et al., 1981).

4.24 and 4.25 can also be used to examine the envelope that would eventually be occupied by water or chemicals discharged from an injection well in a uniform flow. In this case, $q_o$ has to be replaced by $-q_o$ and Q by $-Q$. This type of analysis would be useful in examining the portion of an aquifer that would be affected by a bioreclamation or chemical injection scheme.

A related set of expressions that can be obtained from the principal superposition is for a pair (doublet) of pumping and injection wells in uniform flow. Depending upon the orientation of the wells and their pumping rates relative to the regional flow, recirculation of water can be avoided or maximized. Figure 4.11 shows the envelope created by a doublet oriented along a line parallel to the direction of regional ground-water flow. Recirculation is maximized as a result of placing the injection well directly upgradient from the pumping well. This type of configuration may be desirable for chemical injection or bioreclamation actions where the intent is to perform in-situ treatment and, possibly, recapture the injected fluid. Powers et al. (1981) provide a relationship relating the maximum width of the envelope and the distance between the two wells to the pumping/injection rate and the properties of the aquifer:

$$\tan^{-1} \frac{c}{a} = \frac{\pi}{2} - \frac{\pi V c b}{Q} \qquad (4.26)$$

where
$c$ = half-width of the envelope, L
$a$ = half the distance between the wells, L
$V$ = pore water velocity of the regional flow component, L/T
$Q$ = pumping/injection rate, $L^3/T$

Figure 4.12 is a dimensionless plot of Equation 4.26.

In using this type of doublet configuration as a remedial action the time required for partial or complete recovery of the injected fluid is often of concern. Grove et al. (1970) provide a solution for estimating this time. The solution is provided graphically in Figure 4.13.

Recirculation can be minimized by reversing the position of the wells. This may be important for a ground-water pumping remedial action where the pumping well is used to capture the plume and the injection well is used to dispose of treated water. Bear (1979) shows that for wells with equal pumping rates recirculation can be avoided when

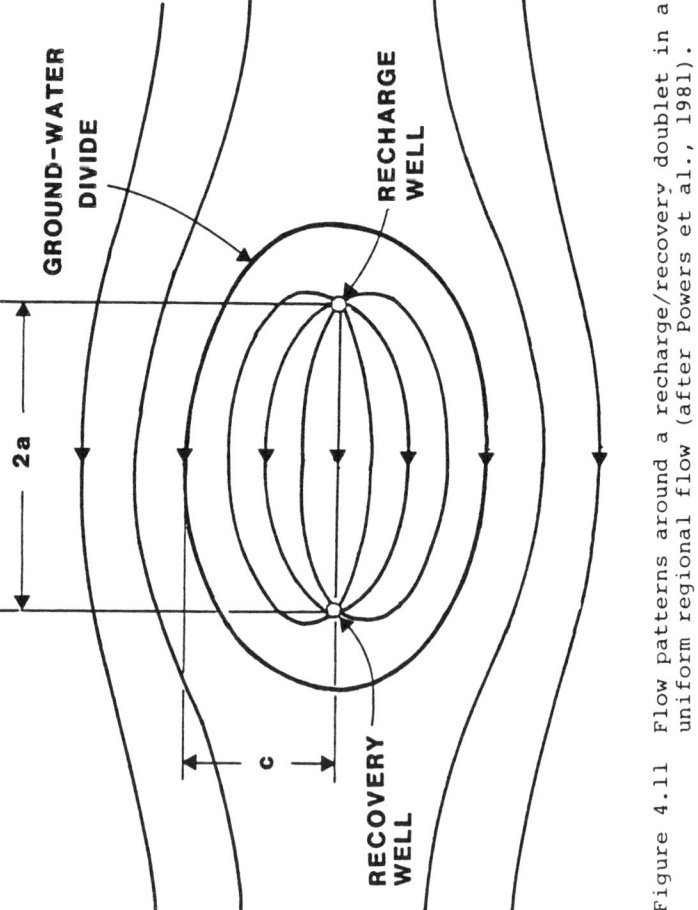

Figure 4.11  Flow patterns around a recharge/recovery doublet in a uniform regional flow (after Powers et al., 1981).

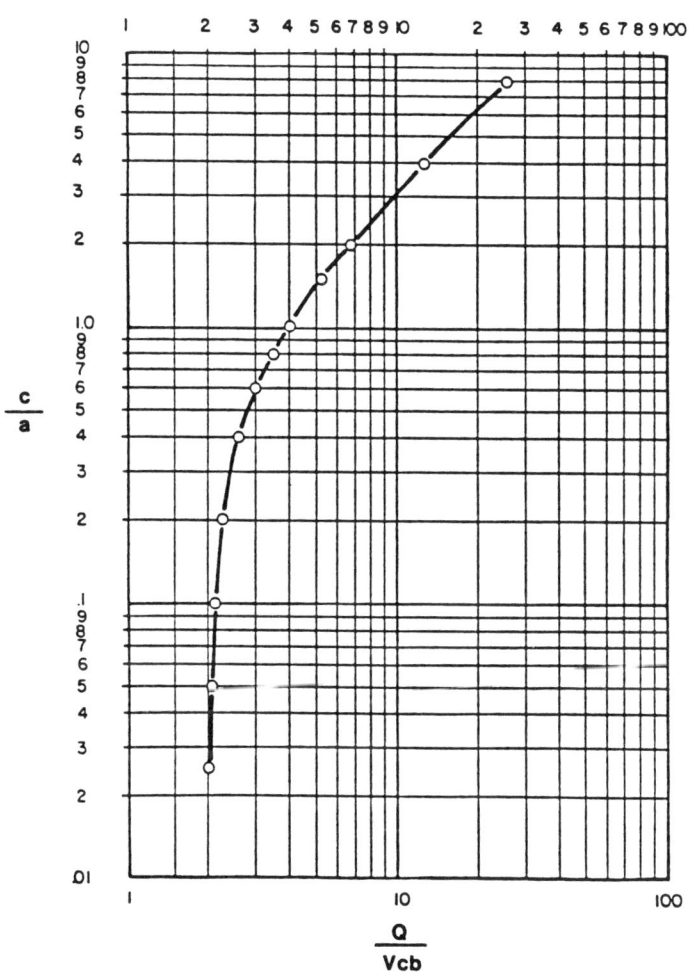

Figure 4.12  Dimensionless plot of doublet width (taken from Powers et al., 1981).

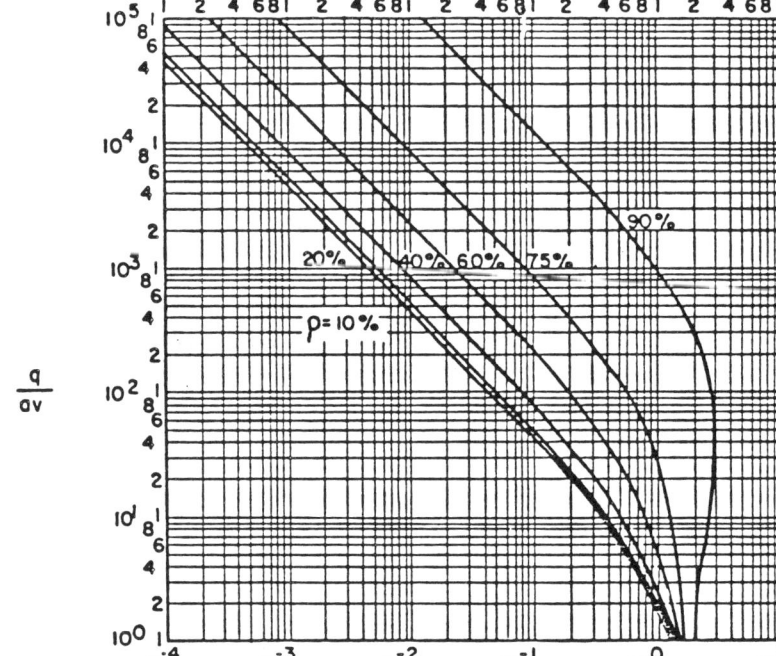

- q = DISCHARGE/RECHARGE RATE PER UNIT PENETRATION.
- a = HALF THE DISTANCE BETWEEN THE WELLS.
- v = DARCY VELOCITY = TRANSPORT VELOCITY TIMES POROSITY.
- n = POROSITY.
- t = TIME REQUIRED TO OBTAIN PERCENTAGE P.
- p = PERCENTAGE OF PUMPED FLOW DERIVED FROM RECHARGE.

Figure 4.13 Percent recharge being discharged in a doublet (after Grove et al., 1970).

$$Q_w \leq \pi d\, b\, q_o \qquad (4.27)$$

where d is one-half the distance between the wells. If, for some reason, this condition cannot be met, the proportion of recirculation, $Q_{wr}$, can be estimated by

$$\frac{Q_{wr}}{Q_w} = \frac{2}{\pi}\left\{ \tan^{-1}\sqrt{\frac{Q}{dq_o b}-1} - \frac{\pi d q_o b}{Q_w}\sqrt{\frac{Q}{dq_o b}-1}\right\} \qquad (4.28)$$

Wilson (1984) presents analytical solutions for an extension to the pumping/injection doublet. This extension involves the use of two pumping wells and two injection wells oriented so as to create inner and outer recirculation cells that effectively capture a plume. The pumping and injection wells are oriented as shown in Figure 4.14. As Wilson notes, the outer cell reduces the time required to capture the plume and the amount of water that must be treated. The outer cell also provides a back-up should the chemical escape the inner cell. Analytical solutions and type curves are provided by Wilson to determine cell discussions and plume flushing times.

As was mentioned earlier, most of the available analytical expressions are based on the assumption of infinite areal extent. Real world aquifer systems are normally bounded, however, by streams, lakes and geologic formations. Aquifers also contain natural ground-water divides. Finally, the use of impermeable barriers and drains as remedial actions can also act as boundaries. A special type of superposition, known as the method of images, can be used to examine the effects of different boundaries.

The method of images is discussed in some detail by Ferris et al., (1962), and is generally discussed in most ground-water text books. The method involves the use of "imaginary" wells placed in strategic locations to duplicate hydraulically the effects of physical boundaries. To hydraulically duplicate the effect of a no-flow boundary, an imaginary pumping well is used. As Figure 4.8 shows, at the intersection of the cones of depressions for two pumping wells, a ground-water divide equivalent to a no-flow boundary is created. A stream can be hydraulically duplicated by using an injection well as the image well. In this case, the intersection of the cones of depression act like a source of recharge. Image wells normally have the same pumping rates as the real wells, and are situated on a common line perpendicular to the boundary. Under these conditions the boundary is located at a distance

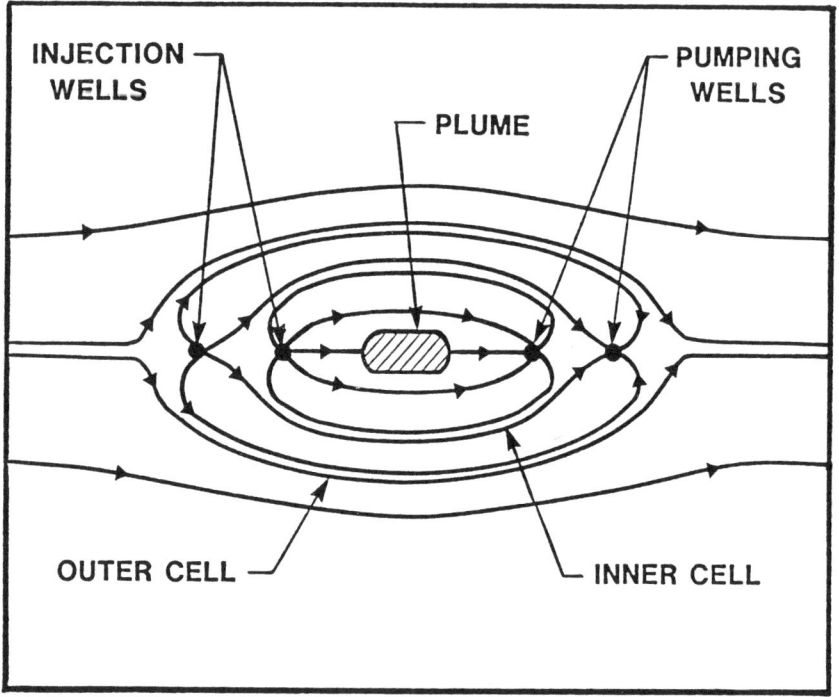

Figure 4.14 Inner and outer recirculation cells created by pairs of pumping and injection wells (after Wilson, 1984). Copyrighted by National Water Well Association.

halfway between the image and real wells.

In using the method of images, it is important to recognize several assumptions and limitations. First, when the method is used to represent a stream, it is assumed that the real pumping well does not lower the head in the stream. Second, when it used to represent a no-flow boundary, the barrier that is created is assumed to be vertical and fully penetrating. Thus, situations like shallow streams, partially penetrating drains or hanging slurry walls cannot be considered. Finally, the barrier that is created is assumed to be infinite in length. As a result, the types of aquifer geometries that can be considered are limited to infinite-strips, semi-infinite strips, wedge-shapes and rectangles. Figures 4.15 and 4.16 show the placement of image wells for these aquifer geometries.

As Figure 4.15 shows, in addition to image wells on common lines perpendicular to each boundary, an additional image well is needed to balance the effect of the other two image wells when evaluating "wedge-shaped" geometries. The former are known as primary image wells, the latter is a secondary image well (Ferris et al., 1962).

As Figure 4.16 shows, a long line of image wells is required to evaluate a infinite-strip or semi-infinite strip geometries. The wells on one side of the strip are required to balance the effects of the wells on the other. While the line of wells should be carried to infinity, in practice it is only necessary to add wells until the next pair has a negligible influence on the sum of the image well effects at a point (Walton, 1984a). The number of image wells required to evaluate a rectangular geometry can be large also (see Figure 4.16). Such a geometry can lead to a large number of tedious calculations. Again, many of the available hand-held calculator and micro-computer programs will automatically sum the drawdowns for a large number of image wells, thus reducing the work required to use the method of images.

## 4.7 TRANSFORMATION METHODS

Most of the available analytical solutions are based on the assumption that flow occurs in isotropic and homogeneous media. This assumption is often limiting because all real systems exhibit some degree of heterogeneity and anisotropy. Fortunately, there are practical ways to circumvent this limitation through the use of different "transformation" methods: 1) equivalent sections, 2) incremental methods, and 3) corrections for anisotropy.

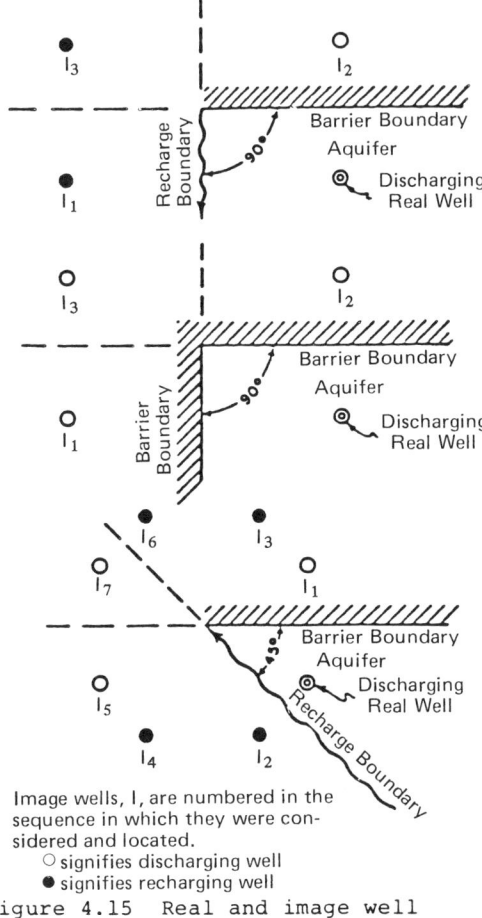

Image wells, I, are numbered in the sequence in which they were considered and located.
○ signifies discharging well
● signifies recharging well

Figure 4.15  Real and image well configurations for wedge-shaped aquifers (taken from Walton, 1984a).

Theory Underlying Available Simplified Methods 121

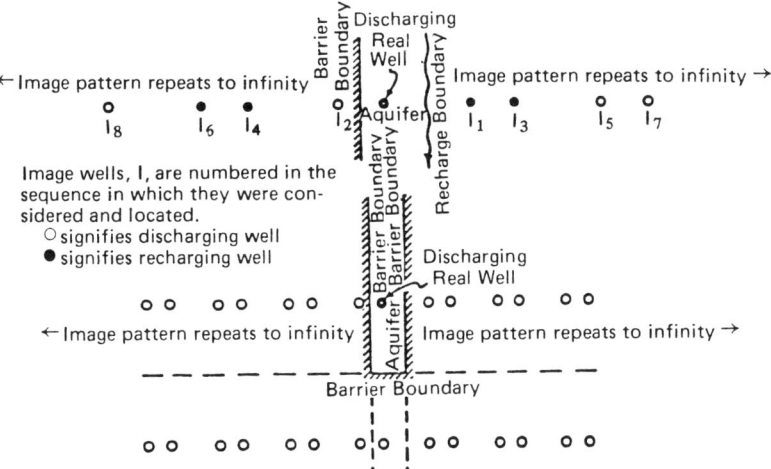

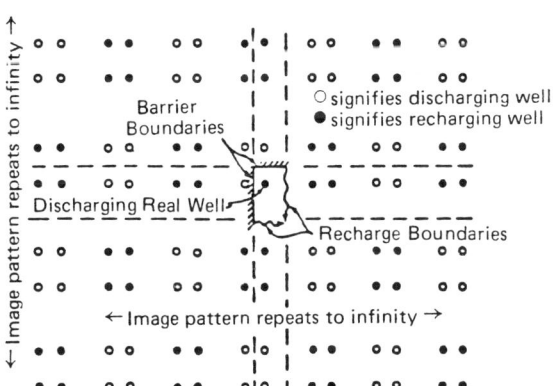

Figure 4.16  Real and image well configurations for strip and rectangular aquifers (taken from Walton, 1984a).

The use of equivalent sections basically involves converting the irregular geometry of a real world aquifer system into an equivalent system with a regular geometry. The geometries typically used are those that can be obtained from the use of the method of images (i.e., strips, rectangles and wedges). This conversion is required because most analytical solutions are derived for regular geometries.

In making the conversion to an equivalent system it is often necessary to account for layered heterogeneities. Layered heterogeneities are vertical changes in media properites. A hydraulically equivalent vertical conductivity for a layered system can be obtained by

$$K_z = \frac{d}{\sum_{i=1}^{n} \frac{d_i}{K_i}} \qquad (4.29)$$

where

$K_z$ = equivalent vertical hydraulic conductivity, L/T
$K_i$ = vertical hydraulic conductivity of each layer, L/T
$d_i$ = thickness of each layer, L
$d$ = total thickness, L

A hydraulically equivalent horizontal conductivity can be obtained by

$$K_x = \sum_{i=1}^{n} \frac{K_i d_i}{d} \qquad (4.30)$$

where $K_x$ = equivalent horizontal hydraulic conductivity, L/T

Horizontal changes in media properties, or <u>trending</u> heterogeneities, also have to be considered when converting to equivalent systems. Hydraulically equivalent horizontal and vertical hydraulic conductivities can be obtained by

$$K_x = \frac{d}{\sum_{i=1}^{n} \frac{d_i}{K_i}} \qquad (4.31)$$

$$K_z = \sum_{i=1}^{n} \frac{K_i d_i}{d} \qquad (4.32)$$

respectively, where d now is a horizontal distance rather than the total depth. Walton (1984a) and Ferris et al., (1962) recommend that when transient well or drain analyses are being conducted in systems with trending heterogeneities, the hydraulic conductivity be adjusted as the cone of depression moves outward. The initial value of the hydraulic conductivity would be equal to that for the media adjacent to the well or drain. When the cone of depression encounters another media, the hydraulic conductivity should be adjusted. This procedure continues until drawdowns stabilize. Walton (1984a) refers to this approach as the incremental method.

Another type of incremental method is to divide the aquifer into regions with relatively uniform properties and then apply the analytical solutions in a step-wise fashion to each region. Bear (1979) recommends this approach for water table aquifers with appreciable variations in head.

In many systems there may be distinct differences between horizontal and vertical hydraulic conductivities. In these cases, corrections need to be made before solutions based on isotropic conditions can be used. Huisman and Olsthoorn (1983) present a series of formulas for making corrections

$$K' = \sqrt[3]{K_x K_y K_z} \qquad (4.33)$$

$$x' = \sqrt{\frac{K'}{K_x}}\, x \qquad (4.34)$$

$$y' = \sqrt{\frac{K'}{K_y}}\, y \qquad (4.35)$$

$$z' = \sqrt{\frac{K'}{K_z}}\, z \qquad (4.36)$$

The resistance factor for leaky systems becomes:

$$c' = \frac{K'}{\sqrt{K_x K_y}} c \qquad (4.37)$$

where c is the ratio of the leaky layer thickness to its hydraulic conductivity.

The specific yield becomes:

$$S_y' = \sqrt{\frac{K'}{K_z}} S_y \quad \text{or} \quad S_y' = \sqrt{\frac{K_x K_y}{(K')^2}} S_y \qquad (4.38)$$

Huisman (1972) suggests the following correction for the radius, r:

$$r' = \sqrt{\left(\frac{K'}{K_x}\right) x^2 + \left(\frac{K'}{K_y}\right) y^2} \qquad (4.39)$$

He notes also that when the vertical hydraulic conductivity is similar to the horizontal hydraulic conductivity and the influence of the well is large compared to the saturated thickness, there is no need to consider the effect of differences in the vertical and horizontal hydraulic conductivities.

## 4.8 CONFORMAL MAPPING

Conformal mapping is a method for deriving analytical solutions by transforming a problem from one geometrical domain for which a solution is needed to one for which a solution can be obtained. This method has been used to derive expressions for selected two-dimensional ground-water flow problems involving relatively complicated geometries (e.g., seepage under cut off walls and through earthen dams). The theory behind the method is discussed by Harr (1962).

A major disadvantage of the method is that it is mathematically involved, and often produces fairly complex analytical solutions. The major advantage is that it provides solutions for one class of flow problems that cannot be considered with the methods discussed so far: flow under partially penetrating impermeable barriers or barriers that are keyed into leaky formations. As was noted in Subsection 4.6, the method of images requires that an impermeable barrier be fully penetrating and that no seepage occurs beneath the barrier. In addition to the solutions presented below, Knox (1984) recently developed an analytical technique for estimating the flow under a partially-penetrating barrier. A technique for developing breakthrough curves for contaminanted ground-water is also presented.

Figure 4.17 shows a barrier (i.e., grout curtain, slurry wall or sheet piling) that partially penetrates a horizontal water table aquifer. The quantity of seepage under this barrier can be estimated with the following relationship:

$$q = \frac{khK'}{2K} \qquad (4.40)$$

where
- $q$ = flow rate per unit width, $L^2/T$
- $k$ = hydraulic conductivity, $L/T$
- $h$ = head difference, $L$
- $K', K$ = values of complete elliptic integral of the first kind

Values of $K'/K$ have been tabulated for a range of values of $m^2$, the modulus (see Table 4.4). The modulus for this case is

$$m = \sin\left(\frac{\pi s}{2T}\right) \qquad (4.41)$$

where s and T are defined in Figure 4.17.

The above expression is for a single layered, homogeneous, isotropic aquifer. In many situations, aquifers with two layers of differing permeabilities are encountered. Harr (1962) provides a method for estimating the quantity of flow under a barrier for a two-layered system. It involves the calculation of a dimensionless parameter using the following relationship:

$$\tan \pi \epsilon = \sqrt{\frac{k_2}{k_1}} \qquad (4.42)$$

where $k_1$ = hydraulic conductivity of the upper

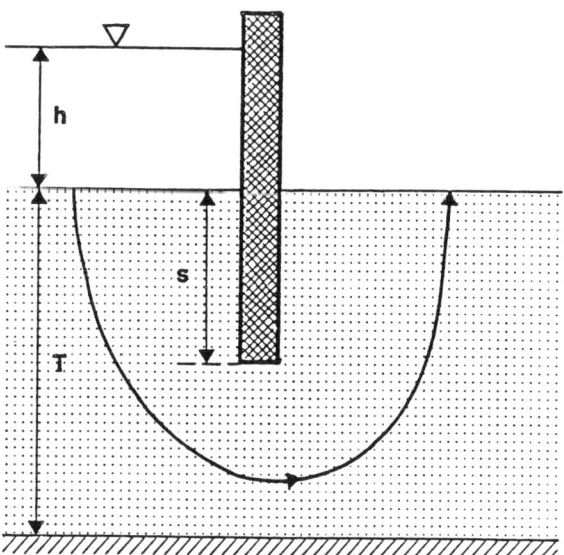

Figure 4.17  Configuration of an impermeable barrier that partially penetrates into a single-layered aquifer (taken from Harr, 1962). Copyrighted by McGraw-Hill.

TABLE 4.4 COMPLETE ELLIPTIC INTEGRALS OF THE FIRST KIND* (taken from Harr, 1962) Copyrighted by McGraw-Hill

| $m^2$ | $K$ | $K'$ | $\dfrac{K}{K'}$ | $\dfrac{K'}{K}$ | $m'^2$ | $m^2$ | $K$ | $K'$ | $\dfrac{K}{K'}$ | $\dfrac{K'}{K}$ | $m'^2$ |
|---|---|---|---|---|---|---|---|---|---|---|---|
| 0.000 | 1.571 | ∞ | 0.000 | ∞ | 1.000 | 0.21 | 1.665 | 2.235 | 0.745 | 1.34 | 0.79 |
| 0.001 | 1.571 | 4.841 | 0.325 | 3.08 | 0.999 | 0.22 | 1.670 | 2.214 | 0.754 | 1.33 | 0.78 |
| 0.002 | 1.572 | 4.495 | 0.349 | 2.86 | 0.998 | 0.23 | 1.675 | 2.194 | 0.763 | 1.31 | 0.77 |
| 0.003 | 1.572 | 4.293 | 0.366 | 2.73 | 0.997 | 0.24 | 1.680 | 2.175 | 0.773 | 1.29 | 0.76 |
| 0.004 | 1.572 | 4.150 | 0.379 | 2.64 | 0.996 | 0.25 | 1.686 | 2.157 | 0.782 | 1.28 | 0.75 |
| 0.005 | 1.573 | 4.039 | 0.389 | 2.57 | 0.995 | 0.26 | 1.691 | 2.139 | 0.791 | 1.26 | 0.74 |
| 0.006 | 1.573 | 3.949 | 0.398 | 2.51 | 0.994 | 0.27 | 1.697 | 2.122 | 0.800 | 1.25 | 0.73 |
| 0.007 | 1.574 | 3.872 | 0.406 | 2.46 | 0.993 | 0.28 | 1.702 | 2.106 | 0.808 | 1.24 | 0.72 |
| 0.008 | 1.574 | 3.806 | 0.413 | 2.42 | 0.992 | 0.29 | 1.708 | 2.090 | 0.817 | 1.22 | 0.71 |
| 0.009 | 1.574 | 3.748 | 0.420 | 2.38 | 0.991 | 0.30 | 1.714 | 2.075 | 0.826 | 1.21 | 0.70 |
| 0.01 | 1.575 | 3.696 | 0.426 | 2.35 | 0.99 | 0.31 | 1.720 | 2.061 | 0.834 | 1.20 | 0.69 |
| 0.02 | 1.579 | 3.354 | 0.471 | 2.12 | 0.98 | 0.32 | 1.726 | 2.047 | 0.843 | 1.19 | 0.68 |
| 0.03 | 1.583 | 3.156 | 0.502 | 1.99 | 0.97 | 0.33 | 1.732 | 2.033 | 0.852 | 1.17 | 0.67 |
| 0.04 | 1.587 | 3.016 | 0.526 | 1.90 | 0.96 | 0.34 | 1.738 | 2.020 | 0.860 | 1.16 | 0.66 |
| 0.05 | 1.591 | 2.908 | 0.547 | 1.83 | 0.95 | 0.35 | 1.744 | 2.008 | 0.869 | 1.15 | 0.65 |
| 0.06 | 1.595 | 2.821 | 0.565 | 1.77 | 0.94 | 0.36 | 1.751 | 1.995 | 0.877 | 1.14 | 0.64 |
| 0.07 | 1.599 | 2.747 | 0.582 | 1.72 | 0.93 | 0.37 | 1.757 | 1.983 | 0.886 | 1.13 | 0.63 |
| 0.08 | 1.604 | 2.684 | 0.598 | 1.67 | 0.92 | 0.38 | 1.764 | 1.972 | 0.895 | 1.12 | 0.62 |
| 0.09 | 1.608 | 2.628 | 0.612 | 1.63 | 0.91 | 0.39 | 1.771 | 1.961 | 0.903 | 1.11 | 0.61 |
| 0.10 | 1.612 | 2.578 | 0.625 | 1.60 | 0.90 | 0.40 | 1.778 | 1.950 | 0.911 | 1.10 | 0.60 |
| 0.11 | 1.617 | 2.533 | 0.638 | 1.57 | 0.89 | 0.41 | 1.785 | 1.939 | 0.920 | 1.09 | 0.59 |
| 0.12 | 1.621 | 2.493 | 0.650 | 1.54 | 0.88 | 0.42 | 1.792 | 1.929 | 0.929 | 1.08 | 0.58 |
| 0.13 | 1.626 | 2.455 | 0.662 | 1.51 | 0.87 | 0.43 | 1.799 | 1.918 | 0.938 | 1.07 | 0.57 |
| 0.14 | 1.631 | 2.421 | 0.674 | 1.48 | 0.86 | 0.44 | 1.806 | 1.909 | 0.946 | 1.06 | 0.56 |
| 0.15 | 1.635 | 2.389 | 0.684 | 1.46 | 0.85 | 0.45 | 1.814 | 1.899 | 0.955 | 1.05 | 0.55 |
| 0.16 | 1.640 | 2.359 | 0.695 | 1.44 | 0.84 | 0.46 | 1.822 | 1.890 | 0.964 | 1.04 | 0.54 |
| 0.17 | 1.645 | 2.331 | 0.706 | 1.42 | 0.83 | 0.47 | 1.829 | 1.880 | 0.973 | 1.03 | 0.53 |
| 0.18 | 1.650 | 2.305 | 0.716 | 1.40 | 0.82 | 0.48 | 1.837 | 1.871 | 0.982 | 1.02 | 0.52 |
| 0.19 | 1.655 | 2.281 | 0.726 | 1.38 | 0.81 | 0.49 | 1.846 | 1.863 | 0.991 | 1.01 | 0.51 |
| 0.20 | 1.660 | 2.257 | 0.735 | 1.36 | 0.80 | 0.50 | 1.854 | 1.854 | 1.000 | 1.00 | 0.50 |
| $m'^2$ | $K'$ | $K$ | $\dfrac{K'}{K}$ | $\dfrac{K}{K'}$ | $m^2$ | $m'^2$ | $K'$ | $K$ | $\dfrac{K'}{K}$ | $\dfrac{K}{K'}$ | $m^2$ |

*From V. I. Aravin, and S. Numerov, "Seepage Computations for Hydraulic Structures," Stpoitel'stvu i Arkhitekture, Moscow, 1955.

$k_2$ = hydraulic conductivity of the lower layer, L/T

The ratio of s/T is also calculated where s and T are defined in Figure 4.18. It is important to note that the thickness of each layer is assumed to be equal. Given values for $\epsilon$ and s/T, the seepage rate can be obtained for using Figures 4.19 and 4.20.

In cases where $k_2 >> k_1$, the seepage rate can be calculated directly by

$$q = \frac{k_1 h}{2} \sqrt{\frac{k_1}{k_2}} \qquad (4.43)$$

One key assumption behind both of the above methods is that the flow is occuring in two-dimensions only, the horizontal and vertical dimensions. This is equivalent to assuming the impermeable barrier is infinitely long and, therefore, no flow occurs around the ends. Since impermeable barriers used as remedial actions will always be of finite length, care must be exercized in using the above methods.

## 4.9 CONTAMINANT TRANSPORT

All of the analytical methods discussed in the previous subsections are useful for evaluating the changes only in ground-water flow patterns and hydraulic heads associated with wells, drains, mounds and impermeable barriers. Another area of interest is the effect of these actions on contaminant movement.

A number of analytical solutions for contaminant transport have been developed. Most of them are based on the classical convection-dispersion equation. In addition to these analytical solutions, several semi-analytical methods have also been developed. The theory behind the available analytical and semi-analytical methods is discussed below.

One form of the classical partial differential equation for contaminant transport in two-dimensions is

$$D_L \frac{\partial^2 C}{\partial x^2} + D_T \frac{\partial^2 C}{\partial y^2} - v \frac{\partial C}{\partial x} - \lambda RC = R \frac{\partial C}{\partial t} \qquad (4.44)$$

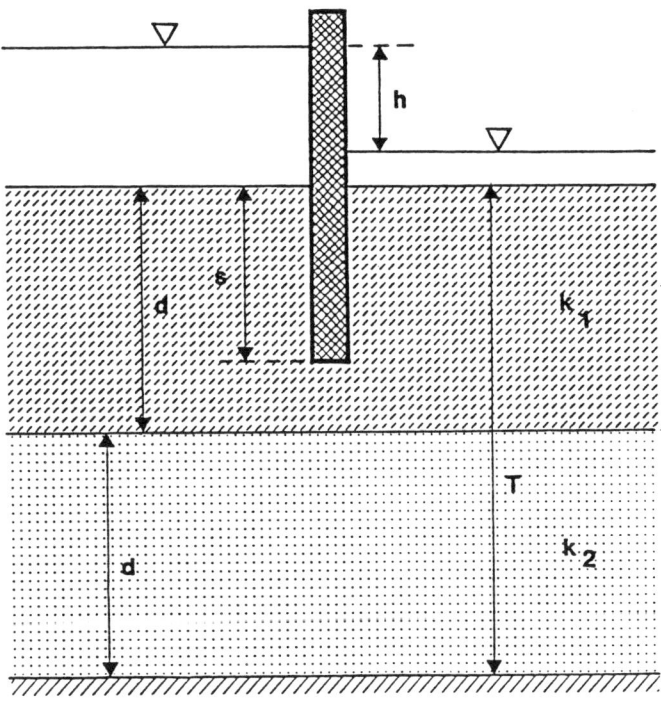

Figure 4.18  Configuration of an impermeable barrier that partially penetrates into a two-layered aquifer (taken from Harr, 1962). Copyrighted by McGraw-Hill.

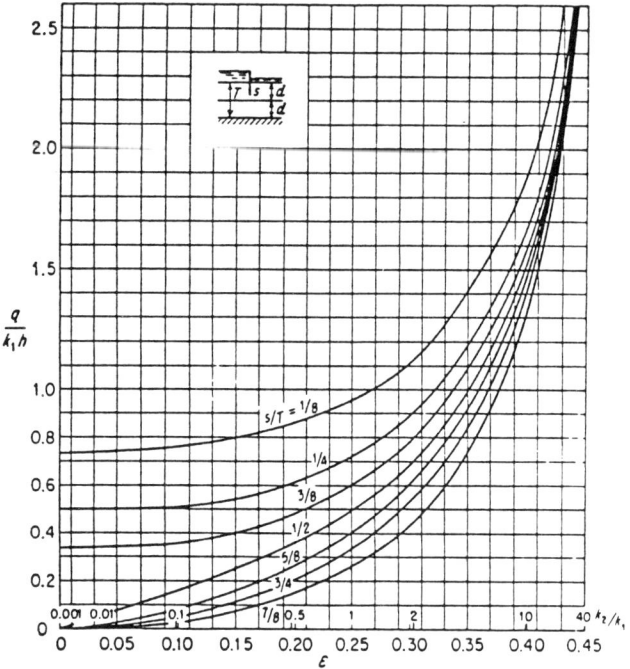

Figure 4.19 Relationship between ε and the flow under a partially penetrating barrier in a layered aquifer (taken from Harr, 1962; after Polubarinova-Kochina, 1952). Copyrighted by McGraw-Hill.

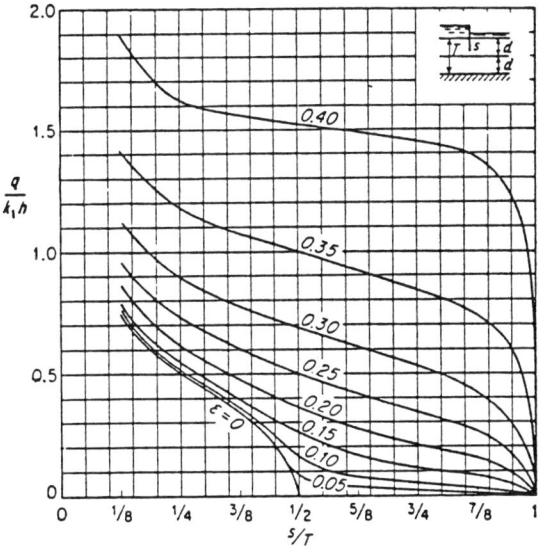

Figure 4.20  Relationship between depth of penetration and flow under a partially penetrating barrier in a layered aquifer (taken from Harr, 1962; after Polubarinova-Kochina, 1952). Copyrighted by McGraw-Hill.

where
- $D_L$ = longitudinal dispersion coefficient, $L^2/T$
- $D_T$ = transverse dispersion coefficient, $L^2/T$
- $C$ = contaminant concentration, $M/L^3$
- $V$ = average pore water velocity, $L/T$
- $\lambda$ = contaminant degradation rate, $/T$
- $R$ = $1+\rho K_d/n_e$ = retardation coefficient, dimensionless
- $\rho$ = bulk density, $M/L^3$
- $n_e$ = effective porosity, dimensionless
- $K_d$ = equilibrium partitioning coefficient, $L^3/M$

In deriving this equation it is assumed that flow is steady and uniform in the x direction, and that the aquifer is composed of homogeneous and isotropic media; the contaminant is assumed to have a constant density and viscosity. Almost all of the available analytical solutions based on the convection-dispersion equation are for steady and uniform flow conditions, except for a few that were derived for radial flow problems (i.e., flow to wells). Further, it is generally assumed that contaminant adsorption/desorption can be described as a linear and completely reversible process, and that contaminant degradation can be described as a first-order process. Finally, it is assumed that the dispersion and diffusion can be grouped together and described as a Fickian process (i.e., obeys Fick's first law). The coefficient for the dispersion component is assumed to be directly proportional to the pore water velocity and does not vary in time or space.

The available analytical solutions based on the convection-dispersion equation are consolidated in several key publications. Van Genuchten and Alves (1982) provide derivations for a relatively complete set of one-dimensional analytical solutions. Walton (1984a) presents several one-dimensional solutions, and a number of radial flow solutions involving single injection and withdrawal wells with and without regional flow. Other good sources of available solutions include Bear (1979) and Javandel et al. (1984), and Cleary and Ungs (1978).

Donigian et al., (1983) developed a methodology for evaluating the potential for ground-water contamination under emergency response conditions. The methodology uses a nomograph-based solution to the one-dimensional, convection-dispersion equation. Detailed guidance on parameter estimation is provided as part of the methodology.

While a large number of analytical solutions are available, their use in the evaluation of remedial action performance is limited to two types of analysis. First, the solutions can be used to estimate the rate and direction of plume migration away from an uncontrolled disposal site. This type of information is useful when determining how the extent of ground-water contamination may change with time and the type of remedial action that may be needed sometime in the future. If a remedial action needs to be implemented immediately, this type of information will be of limited value since monitoring and site characterization will have already determined the extent of contamination. The second type of analysis applies to those remedial action technologies that involve injection and recovery (e.g., bioreclamation and chemical injection). Analytical solutions for radial flow can be used to examine the portion of an aquifer that will be affected by an injected fluid and the time required for the injected fluid to reach a recovery well. Both are needed in the evaluation and design of in-situ treatment technologies. Some of the available expressions also include the effect of a regional flow component.

The first major type of semi-analytical method for contaminant transport is based on the complex velocity potential concept. Like superposition, this concept involves separating a complex flow field that itself is intractable into a series of simple flow fields for which tractable solutions are available. The velocity potentials and stream functions for each simple flow field are combined to obtain a complex velocity potential. Javandel et al., (1984) provide a procedure for constructing complex velocity potentials. Bear (1979) discusses the theory behind the concept; he refers to it as the sharp front approximation.

The complex velocity potential concept has several advantages and disadvantages. Its major advantage is that it is generally more powerful than analytical methods, largely because more complex flow systems can be considered. Its major limitation is that the concept generally applies only to the transport of water-coincident contaminants. That is, contaminants that move at the same velocity as the ground water. As a result, the effects of dispersion are not considered. In a few cases, retardation and decay can be considered (see Nelson and Schur, 1980). Another limitation is that some of the solutions developed using this concept are mathematically complex. While they can be solved by hand with the aid of tables or graphs of appropriate mathematical functions, they are generally programmed for use on computers or hand-held calculators to reduce the work involved in their application. Those expressions that have been programmed for use on hand-held calculators or micro-computers are discussed in Section 4.

Javandel et al. (1984) and Bear (1979) overview some of the available solutions that have been derived using the complex velocity potential concept. These expressions are applicable to homogeneous, isotropic, saturated aquifers with uniform flow. They can be used to predict contaminant transport in two-dimensions (lateral and longitudinal). A large number of injection wells, withdrawal wells and circular sources of finite radius (e.g., ponds and lagoons) can be evaluated. Thus, these expressions are applicable to a wide range of subsurface and waste control remedial action technologies.

The other type of semi-analytical methods for contaminant transport is based on a simple numerical technique discussed by Bear (1979). This technique involves tracking the movement of one or more particles of water with time. The rate and direction of particle movement at any location in the ground-water flow field is estimated by calculating the component pore water velocity towards each pumping well and away from each injection well. These component velocities are summed together along with the regional component of pore water velocity to obtain a resultant velocity vector. The particle of water is then moved in the direction of the resultant velocity. The distance it is moved is determined by the magnitude of the resultant velocity and the time interval selected by the analyst. The accuracy of the method improves as the length of the time interval decreases, particularly as the particle approaches a well.

The simple numerical method or particle mover method can be used to analyze contaminant transport in two ways. The first way involves tracking the movement of individual particles released from the perimeter of a waste site or plume to determine whether all particles will be recovered and how long it will take. The second way is to determine the location of a number of particles at the end of selected time increments. If a sufficient number of particles are tracked, the position of the perimeter of a plume can be estimated at the end of each time increment.

In using the simple numerical technique it is assumed that the contaminant is water-coincident. For contaminants that are not water-coincident, the travel times obtained from the technique can be adjusted by multiplying by the retardation factor for the contaminant of interest.

Due to the numerous repetitive calculations that must be performed when using this method, programs have been written for hand-held calculators to reduce the effort involved in their use. These programs allow the user to adjust the time interval and/or distance a particle is allowed to move. They also allow the user to consider a large number of injection and recovery wells. As a result, a range of remedial action technolgies can be considered.

# 5. Available Hand-Held Calculator and Micro-Computer Programs

## 5.1 OVERVIEW

The use of many of the analytical methods discussed in Section 4 to solve ground-water problems of practical interest will generally require numerous, repetitive calculations. An evaluation of the cone of depression for a single pumping well, for instance, will involve solving one of the well hydraulics equations for a number of radial distances away from the well at different points in time. If the cone of depression for a number of wells is of interest, the drawdown for each well will have to be calculated and then summed to obtain the total drawdown. This type of problem can lead to a large number of calculations.

With the recent development of relatively powerful, programmable hand-held calculators and micro-computers, the work involved in using many of the analytical methods is greatly reduced. Calculators and micro-computers can rapidly perform a large number of repetitive calculations in minutes that would otherwise require hours or days. As a result, the level of manpower required to solve a given problem is greatly reduced.

Calculators and microcomputers also have several other advantages. First, they can reduce the need for tables and graphs that are commonly required to solve analytical expressions. Values for the "well functions" in most well hydraulics equations generally have to be obtained from tables or graphs. While the required tables and graphs can be found in many ground-water textbooks and related publications, some of the "well functions" can be approximated by series expansions or mathematical functions that are easily solved on a calculator. Many of the functions contained in other types of analytical methods can also be approximated or solved directly with calculators. In addition, simple integration schemes that would require numerous, tedious calculations to solve by hand can also be used to quickly solve certain analytical equations.

A second advantage is that calculators and micro-computers offer peripherals that aid in the analysis of remedial actions. The programs required to solve different analytical expressions can be stored on magnetic cards, magnetic tape or disks. When an analysis is required the programs can be loaded rapidly. This reduces the level of effort required to key in a program or to make repetitive key strokes on a non-programmable calculator. Results of different analyses can be stored for use later or printed immediately. This reduces the level of effort involved in transcribing results.

The final advantage is that programmable calculators and micro-computers are readily available. Most site contractors and many state and Federal Superfund staff have access to them. In addition, software availability is increasing rapidly, particularly for micro-computers.

This section will identify what programs are currently available for solving the analytical expressions discussed in Section 4. It is important to note that there are a large number of programs currently available, particularly for hand-held calculators, and more are being written all the time. These programs have been written to meet a number of different needs, ranging from the solution of simple numerical ground-water flow problems, to the solution of well hydraulics equations, to the analysis of pump test data. This section will focus only on those written for analytical or semi-analytical methods that are of value in the evaluation of remedial action performance. Those readers interested in programs available for other types of analyses should consult the International Ground Water Modeling Center (IGWMC), Holcomb Research Institute, Butler University, Indianapolis, Indiana. IGWMC provides a clearinghouse of available programs.

It is also important to note that while an attempt was made to be comprehensive in the identification of available programs, resources were not available to consult every possible source. Therefore, the programs identified herein should be considered as representative of those that are available.

## 5.2 AVAILABLE PROGRAMMABLE, HAND-HELD CALCULATOR PROGRAMS

The large number of programs currently available for programmable, hand-held calculators is an indicator of their wide-spread use in solving practical ground-water problems. Despite the large number of programs that are available, they have only been written for a relatively small number of analytical methods. This is, in part, due to the fact that hydrologists have found that most problems can be solved with just a few methods. It is also due to the fact that programs are difficult, if not impossible, to write for certain methods. These methods are generally ones where reasonable approximations are not possible or graphical solutions are required.

The largest proportion of available programs are for well hydraulics, mainly because hydrologists are commonly faced with problems involving wells and because many of the well hydraulics equations are easily programmable. Table 5.1 provides a summary of a selected group of available programs for well hydraulics. It shows some of the basic assumptions and limitations for each program in terms of aquifer characteristics and well configurations. It also shows whether steady-state or time-varying analyses can be performed and the output provided by the program. Finally, the table lists the calculator for which the program is written and a reference or source for the program. Again, these programs were selected from the many that are currently available because of their usefulness for remedial action evaluation.

As Table 5.1 shows, most of the programs were developed for confined, homogeneous, isotropic aquifers of infinite extent. These programs can also be applied to water table aquifers as long as the assumptions discussed in Subsection 4.2.3 are not violated. Some of the programs were written explicitly for leaky aquifer systems. Corrections for heterogeneities and anisotropy have to be made using the methods discussed in Subsection 4.7 since only one of the programs considers other than homogenous and isotropic conditions. Aquifers that are bounded or remedial actions that include impermeable barriers cannot be analyzed explicitly with these programs unless the method of images is used. Since the method of images requires at least one real well and one image well, those programs that consider more than one well are particularly well suited to the analysis of bounded aquifers. These programs automatically sum the drawdowns attributable to each well.

Most of the well hydraulics programs are for fully penetrating wells. Corrections for partial penetration need to be made, if they are not explicitly considered. Walton (1984a) and others discuss methods for making the needed corrections.

Time-varying estimates of drawdown at different locations are typically provided by the available programs. Only a few provide steady-state results or inflows to a well.

Finally, well hydraulics programs are available for both Texas Instruments (mainly model 59) and Hewlett Packard (mainly model 41C or 41CV) calculators. Listings for some of these programs have been published in the open literature (e.g.,

TABLE 5.1 AVAILABLE HAND-HELD CALCULATOR PROGRAMS FOR WELL HYDRAULICS

| Program Title | Aquifer Characteristics | | | Well Configuration | | Time-frame | Program Output | Calculator Type | Reference |
|---|---|---|---|---|---|---|---|---|---|
| | Type | Properties | Extent | Number | Penetration | | | | |
| General Aquifer Analysis for Nonsteady Theis Conditions | C | H,I | IN | 24 | FP | TV | D | TI-59 | Sandberg et al. 1981; Prickett and Vorhees 1981 |
| Multiple Well, Variable Pumping Rate Problems | C | H,I | IN | 1 | FP | TV | D | HP-29C | Picking 1979 |
| Constant or Variable Pumping (Injection) Rate, Single or Multiple Fully Penetrating Wells | C | H,I | IN | - | FP | TV | D | TI-59 | Warner and Yow 1979 |
| Constant or Variable Pumping (Injection) Rate, Single or Multiple Fully Penetrating Wells | C | H,I | IN | - | FP | TV | D | HP-97 | Rayner 1981 |
| Dewatering Well Design | C | H,I | IN | 24 | FP | TV | D | TI-59 | Loch and Associates (1) |
| Theis Condition Well Field | C | H,I | IN | 57 | FP | TV | D | HP-41 | IGWMC (2) |
| Point Sink Aquifer Model | C | H,I | IN | 50 | FP | TV | D | HP-41 | Jirick (3) |
| Nonsteady State Nonleaky Artesian-Single Production Well | C | H,I | IN | 1 | FP | TV | D | TI-59 | Walton 1983 |
| AQMODL (4) | C | H,I | IN | 60 | FP | TV | D | HP-41 | Rayner 1983 |
| Nonsteady State Nonleaky Artesian-Partially Penetrating Wells | C | H,I | IN | 1 | PP | TV | D | TI-59 | Walton 1983 |

(continued)

TABLE 5.1 (continued)

| Program Title | Aquifer Characteristics | | Well Configuration | | Time-frame | Program Output | Calculator Type | Reference |
|---|---|---|---|---|---|---|---|---|
| | Type | Properties | Extent | Number | Penetration | | | |
| Constant Pumping (Injection) Rate, Fully Confined Aquifer, Partially Penetrating Well | C | H,I | IN | 1 | PP | TV | D | TI-59 | Warner and Yow 1980b |
| Radial Flow to a Constant Drawdown Hemisphere | C | H,I | IN | 1 | PP | TV | IF | TI-59 | Koch and Associates (1) |
| Analysis fo Source or Sink Flow Rates with Drawdown as a Given | C | H,I | IN | 7 | FP | TV | IF | TI-59 | Sandberg et al. 1981 Prickett and Vorhees 1981 |
| Nonsteady Discharge of a Flowing Well | C | H,I | IN | 1 | FP | TV | IF | TI-59 | Koch and Assocaites (1) |
| Anisotropic Confined Aquifers | C | H,A | IN | 1 | PP | TV | D | TI-59 HP-41 | Parr et al. 1983 |
| Jacob Leaky Artesian Steady-State | L | H,I | IN | 25 | FP | SS | D | TI-59 | T.A. Prickett and Associates (5) |
| Steady State Leaky Artesian - Single Production Well | L | H,I | IN | 1 | FP | SS | D | TI-59 | Walton 1983 |
| Nonsteady State Leaky Artesian - Single Production Well | L | H,I | IN | 1 | FP | TV | D | TI-59 | Walton 1983 |
| Leaky Aquifer Drawdown | L | H,I | IN | 1 | FP | TV | D | HP-41 | Ulrick (3) |
| Constant Pumping (Injection) Rate, Single Fully Penetrating Well, Semiconfined Aquifer | L | H,I | IN | 1 | FP | TV | D | TI-59 | Warner & Yow 1980a |

(continued)

TABLE 5.1 (continued)

| Program Title | Aquifer Characteristics | | | Well Configuration | | Time-frame | Program Output | Calculator Type | Reference |
|---|---|---|---|---|---|---|---|---|---|
| | Type | Properties | Extent | Number | Penetration | | | | |
| Hantush "Well Function" | L | H,I | IN | 1 | FP | TV | D | HP-41 | IGWMC(2) |
| Nonsteady State Two Mutually Leaky Artesian Aquifers - Single Production Well | L | H,I | IN | 1 | FP | TV | D | TI-59 | Ma ton 1983 |
| Steady Radial Ground-Water Flow in a Finite Leaky Aquifer | L | H,I | B | 1 | FP | SS | D | HP-41 | IGWMC(2) |
| Successive Steady States - Constant Head Points - Unconfined Aquifer | WT | H,I | IN | 7 | FP | TV | IF,D | TI-59 | Koch and Associates (1) |

(1) Programs available as of October 1983 from Koch and Associates, 1660 S. Fillmore Street, Denver, Colorado, 80210

(2) Programs available as of May 1984 from the International Ground Water Modeling Center, Holcomb Research Institute, Butler University, 4600 Sunset Avenue, Indianapolis, Indiana, 46208

(3) Programs available as of August 1983 from James S. Ulrick and Associates, 2100 Los Angeles Avenue, Berkeley, California, 94707

(4) Programs can also consider regional water level changes with time and the effects of a regional gradient

(5) Programs available as of July 1983 from Thomas A. Prickett and Associates, Inc., 8 Montclair Road, Urbana, Illinois, 61801

LEGEND:  
C - Confined  
L - Leaky  
WT - Water Table  
H - Homogeneous  
I - Isotropic  
A - Anisotropic  
IN - Infinite  
B - Bounded  

FP - Fully Penetrating  
PP - Partially Penetrating  

TV - Time Varying  
SS - Steady State  
D - Drawdown  
IF - Inflow

Warner and Yow, 1979, 1980a, 1980b; and Rayner, 1981, 1983). The rest are available for purchase from different sources. Both documentation and pre-programmed magnetic cards are available when they are purchased.

The available programs for drain hydraulics are summarized in Table 5.2. It essentially follows the same format as Table 5.1, except this table shows the drain configurations that can be considered.

Again, most of the available programs are for confined, homogeneous, isotropic aquifers of infinite extent. While only a few were explicitly written for water table conditions, most of the others can be used as long as the assumptions discussed in Subsection 4.3 are valid. The dimensionality column in Table 5.2 refers to whether or not the drain is assumed to be finite or infinite in length. If it is assumed to be infinite, differences in drawdown and inflow along the length of the drain cannot be considered.

As with the well hydraulics programs, the method of images is required if a bounded aquifer is being analyzed. Programs which can consider the drawdowns for multiple drains will facilitate such analyses. They will also facilitate analyses of multiple drains or drains with irregular boundaries. Sandberg et al., (1981) and Prickett and Voorhees (1981) provide a number of useful examples of how programs for multiple drains can be used to represent a number of ground-water conditions of practical interest (e.g., meandering river, lake shoreline or canal system).

All of the drain programs assume full penetration. The equivalent depth correction discussed in Subsection 4.3 or a different analytical expression (see Tables 4.4 and 4.5) will need to be used for partially penetrating drains.

The available programs can be used to estimate both steady-state and time-varying drawdown and inflow.

Table 5.3 lists the available programs for evaluating ground-water mounding. They are all based on the theory by Hantush (1967a). As was mentioned in Subsection 4.4, a seepage rate estimate is needed to evaluate the potential for mounding. For landfills this rate can be obtained from a computer code like the HELP model (Schroeder et al., 1984 a and 1984b) or the other simple techniques discussed in Subsection 4.5. There are also several simple techniques for ponds and impoundments (see Subsection 4.5). None of these simple techniques have been programmed for hand-held calculators. The only exception is a Hewlett-Packard 41 program written by James S. Ulrick and Associates in Berkeley, California. It performs monthly water balance calculations

TABLE 5.2  INVENTORY OF SELECTED HAND-HELD CALCULATOR PROGRAMS FOR DRAIN HYDRAULICS

| Program Title | Aquifer Characteristics | | | | Drain Configuration | | | | Time-frame | Program Output | Calculator Type | Reference |
|---|---|---|---|---|---|---|---|---|---|---|---|---|
| | Type | Properties | Extent | Dimensionality | Number | Penetration | Length | | | | | |
| Steady-State Drawdown Around Finite Line Sinks | C | H,I | IN | X-Y | 10 | FP | F | | SS | D | TI-59 | Sandberg et al. 1981; Prickett and Vorhees 1981 |
| Successive Steady States - Constant Head Finite Line Sinks - Compute Drawdowns | C | H,I | IN | X-Y | 10 | FP | F | | SS | D | TI-59 | Koch and Associates (1) |
| Finite Line Sinks for Nonsteady Conditions | C | H,I | IN | X-Y | 15 | FP | F | | TV | D | TI-59 | Sandberg et al. 1981; Prickett and Vorhees 1981 |
| Line Sink Aquifer Model | C | H,I | IN | X-Y | 15 | FP | F | | TV | D | HP-41 | Ulrick (2) |
| Study of Steady-State Flow to Finite Line Sources or Sinks with Drawdown as the Given | C | H,I | IN | X-Y | 6 | FP | F | | SS | IF | TI-59 | Sandberg et al. 1981; Prickett and Vorhees 1981 |
| Successive Steady States - One Dimensional Inflow to a Line | C | H,I | IN | X | 1 | FP | IN | | SS | IN | TI-59 | Koch and Associates (1) |
| Successive Steady States - Constant Head Finite Line Sinks - Compute Inflows | C | H,I | IN | X-Y | 6 | FP | F | | SS | IF | TI-59 | Koch and Associates (1) |

(continued)

TABLE 5.2 (continued)

| Program Title | Aquifer Characteristics | | | | Drain Configuration | | | Time-Frame | Program Output | Calculator Type | Reference |
|---|---|---|---|---|---|---|---|---|---|---|---|
| | Type | Properties | Extent | Dimensionality | Number | Penetration | Length | | | | |
| One Dimensional, Nonsteady Flow to a Constant Drawdown, Infinite Line Sink or Source | C | H,I | IN | x | 1 | FP | IN | TV | IF,D | TI-59 | Koch and Associates (1) |
| One Dimensional, Nonsteady Flow to an Increasing Drawdown, Infinite Line Sink or Source | C | H,I | IN | x | 1 | FP | IN | TV | IF,D | TI-59 | Koch and Associates (1) |
| Boussinesq Solution | WT | H,I | B | x | 1 | FP | IN | TV | IF,D | TI-59 | Koch and Associates (1) |
| One Dimensional, Nonsteady Flow to a Constant Drawdown, Infinite Line Sink or Source with Recharge | WT | H,I | IN | x | 1 | FP | IN | TV | IF,D | TI-59 | Koch and Associates (1) |
| One Dimensional Non-Steady Ground Water Flow (3) | WT | H,I | IN | x | 1 | FP | IN | TV | IN,D | HP-41 | Olsthoorn (4) |

(1) Programs available as of October 1983 from Koch and Associates, 1660 S. Fillmore Street, Denver, Colorado, 80210

(2) Programs available as of August 1983 from James S. Ulrick and Associates, 2100 Los Angeles Avenue, Berkeley, California, 94707

(3) Program can consider four boundary conditions for drain: constant head, constant flux, linearly varying head and linearly varying flux(Edelman cases).

(4) Programs available as of May 1984 from the International Ground Water Modeling Center, Holcomb Research Institute, Butler University, 4600 Sunset Avenue, Indianapolis, Indiana, 46208

(continued)

TABLE 5.2 (continued)

LEGEND:
C  - Confined
WT - Water Table
H  - Homogeneous
I  - Isotropic
IN - Infinite
B  - Bounded
X  - Longitudinal
Y  - Lateral

FP - Fully Penetrating
F  - Finite
IN - Infinite

TV - Time Varying
SS - Steady State
D  - Drawdown
IF - Inflow

TABLE 5.3 INVENTORY OF SELECTED HAND-HELD CALCULATOR PROGRAMS FOR GROUND-WATER MOUNDING ESTIMATION

| Program Title | Aquifer Characteristics | | | | Pond Configuration | Time-frame | Program Output | Calculator Type | Reference |
|---|---|---|---|---|---|---|---|---|---|
| | Type | Properties | Extent | Dimensionality | | | | | |
| Analysis of Ground Water Mounding Beneath Tailings Ponds | WT | H,I | IN | R | CI | TV | HH | TI-59 | Sandberg et al. 1981 Prickett and Vorhees 1981 |
| Circular Recharge Area | WT | H,I | IN | R | CI | TV | HH | TI-59 | Walton 1983 |
| Circular Basin Recharge Mound | WT | H,I | IN | R | CI | TV | HH | HP-41 | Ulrick (1) |

(1) Programs available as of August 1983 from James S. Ulrick and Associates, 2100 Los Angeles Avenue, Berkeley, California, 94707

LEGEND:  WT - Water Table         CI - Circular
         H  - Homogeneous         TV - Time Varying
         I  - Isotropic           HH - Hydraulic Head
         IN - Infinite
         R  - Radial

using the Thornthwaite method, but does not separate infiltration and runoff. Another method would have to be used to obtain an estimate of infiltration for landfills.

Available programs for contaminant transport are listed in Table 5.4. The basic assumptions and limitations regarding aquifer characteristics are similar to those for the other types of programs. The available programs fall into two groups. The first group includes those programs based on the simple numerical technique discussed in Subsection 4.9. This technique involves the tracking of particle movement over time in response to injection/pumping wells and regional ground-water flow. The only transport process considered in these programs is advection. The output of programs in this group is particle location with time. The second group of programs includes those based on analytical solutions. These programs consider advection and dispersion, and in some cases, retardation and degradation. The output of these programs is contaminant concentration at selected locations and points in time. In situations where a programmable calculator is not available, a nomograph-based solution to the convection-dispersion equation can be used (see Donigian et al., 1983).

All of the available programs are for point sources and sinks (i.e., wells). None of them consider line sources and sinks (i.e., drains) or area sources (i.e., ponds and landfills) explicitly. A line of wells is often used to aproximate a drain. A cluster of wells if often used to represent an area source. Some of the programs consider regional ground-water flow.

## 5.3 AVAILABLE PROGRAMS FOR MICRO-COMPUTERS

Access to micro-computers is increasing rapidly within consulting firms and governmental agencies involved in the evaluation of remedial action performance. These tools provide capabilities that go far beyond those available on programmable, hand-held calculators and which previously were available only on mini-computers or large mainframes. Many micro-computers are capable of running reduced versions of some of the more sophisticated numerical models used to study ground-water flow. For instance, there are a number of versions of the Prickett-Lonnquist model (Prickett and Lonnquist, 1971) that can be run on micro-computers.

Recognizing the benefits of micro-computers, many of the programs written for hand-held calculators have been expanded and programmed for use on a number of micro-computers. Many new programs that take advantage of the computational

TABLE 5.4  INVENTORY OF SELECTED HAND-HELD CALCULATOR PROGRAMS FOR CONTAMINANT TRANSPORT

| Program Title | Type | Aquifer Characteristics ||| Transport Processes | Timeframe | Program Output | Calculator Type | Reference |
|---|---|---|---|---|---|---|---|---|---|
| | | Properties | Extent | Dimensionality | | | | | |
| Advective Mass Transport Theis Particle Mover | C | H,I | IN | X-Y | AD | TV | PL | TI-59 | Sandberg et al. 1981 Prickett and Vorhees 1981 |
| Streamlines and Travel Times for Regional Ground-Water Flow affected by Sources and Sinks | C | H,I | IN | X-Y | AD | TV | PL | HP-41 | Olsthoorn (4) |
| Advective Transport Model | C | H,I | IN | X-Y | AD | TV | PL | HP-41 | Ulrick (5) |
| Advection and Dispersion Regional Flow | C | H,I | IN | X-Y | AD, DS, RD, DG | TV | CN | TI-59 | Walton 1983 |
| Ground Water Dispersion | C | H,I | IN | X-Y | AD, DS, RD, DG | TV | CN | TI-58/59 | Kelly 1982 |
| Plume Management Model | C | H,I | IN | X-Y | AD, DS, RD, DG | TV | CN | TI-59 | Sandberg et al. 1981 Prickett and Vorhees 1981 |
| Calculator Code for Evaluating Landfill Leachate Plumes | C | H,I | IN | X-Y | AD, DS, RD, DG | TV | CN | TI-59 | Pettyjohn et al. 1982 |
| Dissipation of a Concentrated Slug of Contaminant | C | H,I | IN | X-Y | AD, DS, RD, DG | TV | CN | TI-59 | T.A. Prickett and Associates (6) |
| Advection and Dispersion from a Stream | C | H,I | IN | X | AD, DS | TV | CN | TI-59 | Walton 1983 |
| Advection and Dispersion from a Single Pumping Well | C | H,I | IN | R | AD, DS | TV | CN | TI-59 | Walton 1983 |

(continued)

TABLE 5.4 (continued)

| Program Title | Aquifer Characteristics | | | | Transport Processes | Timeframe | Program Output | Calculator Type | Reference |
|---|---|---|---|---|---|---|---|---|---|
| | Type | Properties | Extent | Dimensionality | | | | | |
| Advection and Dispersion from a Single Solute Injection Well | C | H,I | IN | R | AD, DS | TV | CN | HP-41 | Van der Heijde (4) |
| S-Paths | C | H,I | IN | X-Y | AD, RD, DG | TV | ML | HP-41 | Oberlander and Nelson 1984 |

(1) Considers 23 injection or pumping wells

(2) Considers 63 injection or pumping wells

(3) Considers 45 injection or pumping wells

(4) Programs available as of May 1984 from the International Ground Water Modeling Center, Holcomb Research Institute, Butler University, 4600 Sunset Avenue, Indianapolis, Indiana, 46208

(5) Programs available as of August 1983 from James S. Ulrick and Associates, 2100 Los Angeles Avenue, Berkeley, California, 94707

(6) Programs available as of July 1983 from Thomas A. Prickett and Associates, Inc., 8 Montclair Road, Urbana, Illinois, 61801

LEGEND:
C - Confined
H - Homogeneous
I - Isotropic
IN - Infinite
X - Longitudinal
Y - Lateral
R - Radial

AD - Advection
DS - Dispersion
RD - Retardation
DG - Degradation

TV - Time Varying
PL - Particle Location with Time
CN - Concentration
ML - Mass Loading

TABLE 5.5 (continued)

| Program Title | Aquifer Characteristics | | | Well Configuration | | Timeframe | Program Output | Computer Type | Reference |
|---|---|---|---|---|---|---|---|---|---|
| | Type | Properties | Extent | Dimensionality | Number | Penetration | | | |
| Nonsteady State Two Mutually Leaky Artesian Aquifers - Single Production Well | L | H,I | IN | R | 1 | FP | TV | D | TRS-80 (5) | Walton 1983 |

(1) Programs available as of October 1983 from Koch and Associates, 1660 S. Fillmore Street, Denver, Colorado, 80210

(2) All programs from IGWMC available for Osborne, Kaypro, Superbrain and IBM

(3) GWFLOW is a series of eight flow solutions, including one for mounding estimation

(4) Programs available as of May 1984 from the International Ground Water Modeling Center, Holcomb Research Institute, Butler University, 4600 Sunset Avenue, Indianapolis, Indiana, 46208

(5) Osborne, Kaypro, Superbrain, IBM, Radio Shack PC-1 and PC-2, and Sharp PC 1250 and 1500 programs available from IGWMC; many of the programs can handle multiple wells

LEGEND:
C - Confined
L - Leaky
H - Homogeneous
I - Isotropic
IN - Infinite
R - Longitudinal

FP - Fully Penetrating
PP - Partially Penetrating

TV - Time Varying
SS - Steady State
D - Drawdown

capabilities of micro-computers have been written as well.

In expanding existing programs or writing new ones, many developers have also taken advantage of the interactive features inherent in micro-computers. The programs have been designed to query the user for input data and to generate different types of graphical outputs. These added features greatly enhance the usefulness of the programs and further reduce the level of effort required to perform an analysis.

Representative programs for well hydraulics are summarized in Table 5.5. The basic characteristics of these programs are similar to those in Table 5.1. The main difference is in the number of wells that can be considered. As Table 5.5 shows, programs are available for a number of different types of micro-computers. Walton (1984 b and c) has recently developed a series of programs for several pocket and desk top micro-computers; many of these programs are the same as those programs listed in Table 5.5. Documentation on the programs can be obtained from the International Groundwater Modeling Center.

Only one program was found to be available for drain hydraulics. It is called Mine Hydrology (FINITE). It is for a confined, homogeneous, isotropic aquifer of infinite extent. Up to 20 fully penetrating, finite length line sinks can be considered. The program predicts time-varying drawdown and inflow. It is available from Koch and Associates in Denver, Colorado for the TRS-80, Apple, IBM-PC and Osborne computers.

Table 5.6 shows the available programs for mounding. These programs are equivalent to those in Table 5.3. They are based on the theory of either Hantush (1967a) or Glover (1960). Similar programs for other micro-computers can be found in Walton (1984 b and 1984 c).

The available programs for transport are summarized in Table 5.7. Of particular interest are the programs titled "Plume Cross Section" and "Random Walk." The former provides a method for evaluating the vertical mixing of a contaminant plume. The latter is a micro-computer version of the transport model developed by Prickett et al. (1981). Both programs offer capabilities that are not available in existing hand-held calculator programs. Again, similar programs can be found in Walton, (1984 b and 1984 c).

All of the programs listed in Table 5.7 are for micro-computers. It should be noted, however, there are also a number of available programs based on analytical or semi-analytical transport methods that can be run on mini-computers and large mainframes. Since they are analytical or semi-analytical, they generally require limited

TABLE 5.5 INVENTORY OF SELECTED MICRO-COMPUTER PROGRAMS FOR WELL HYDRAULICS

| Program Title | Aquifer Characteristics | | | | Well Configuration | | Timeframe | Program Output | Computer Type | Reference |
|---|---|---|---|---|---|---|---|---|---|---|
| | Type | Properties | Extent | Dimensionality | Number | Penetration | | | | |
| General Aquifer Analysis (THEIS) | C | H,I | IN | X-Y | 100 | FP | TV | D | TRS-80 Apple IBM-PC Osborne | Koch and Associates (1) |
| THWELLS | C | H,I | IN | X-Y | - | FP | TV | D | (2) | IGWMC (4) |
| GWFLOW (3) | C,L | H,I | IN | X-Y | 1 | FP,PP | TV | D | (2) | IGWMC (4) |
| Nonsteady State Nonleaky Artesian - Single Production Well | C | H,I | IN | R | 1 | FP | TV | D | TRS-80 (5) | Walton 1983 |
| Nonsteady State Nonleaky Artesian - Partially Penetrating Wells | C | H,I | IN | R | 1 | PP | TV | D | TRS-80 (5) | Walton 1983 |
| | | H,I | IN | X-Y | 100 | FP | TV | D | TRS-80 | Koch and Associates (1) |

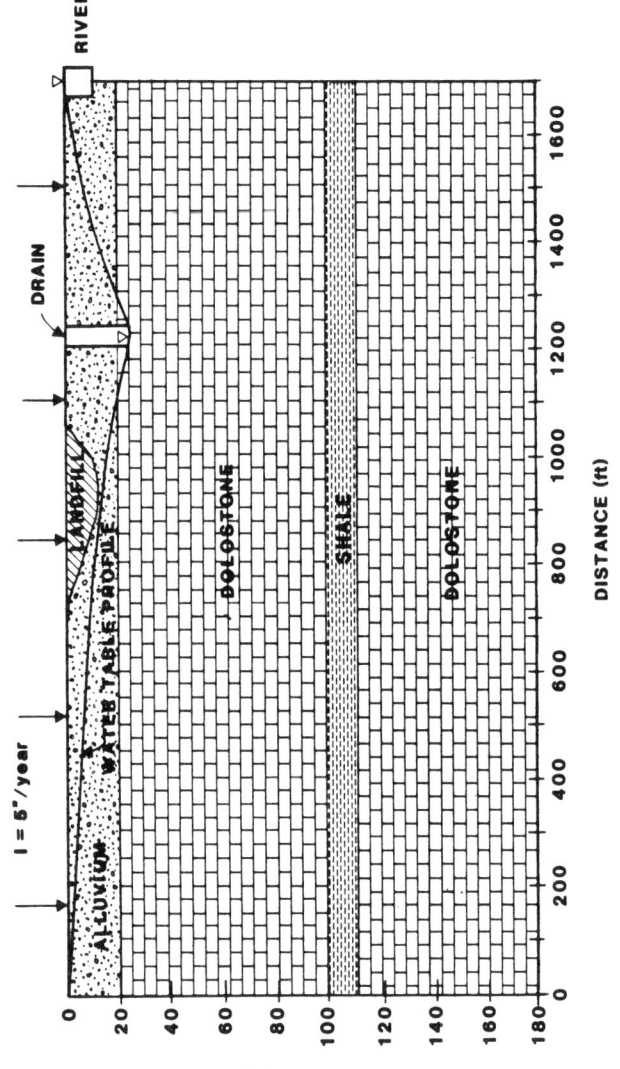

Figure 6.3 Steady-state water table profile for a partially penetrating drain located 200 ft downgradient from the landfill.

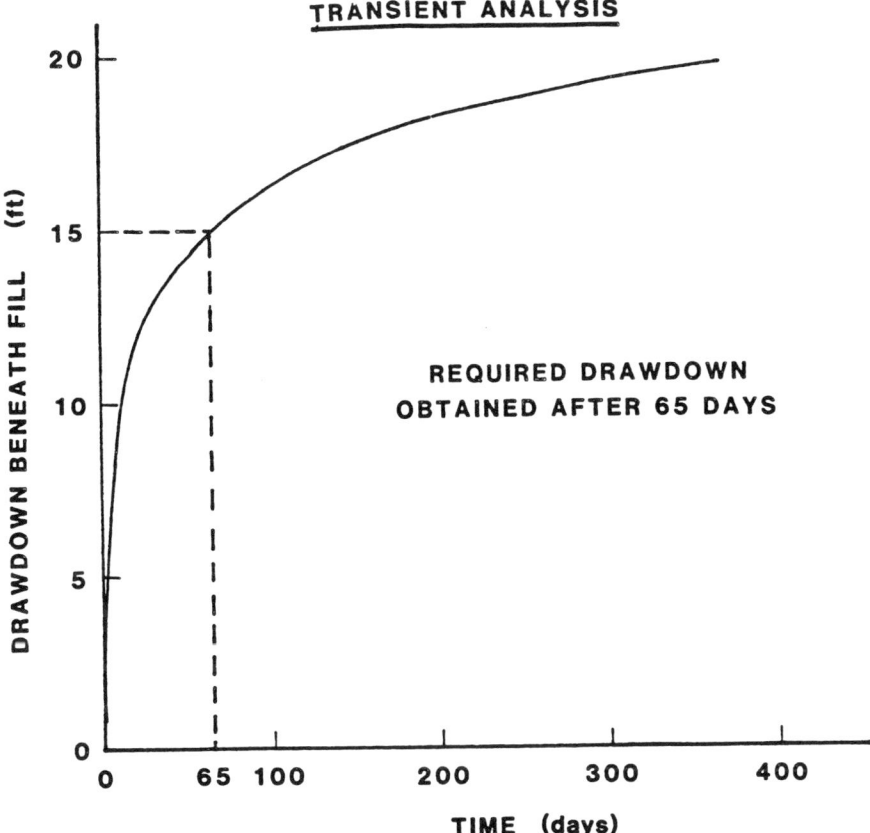

Figure 6.4 Change in water table elevation below the landfill following installation of the drain.

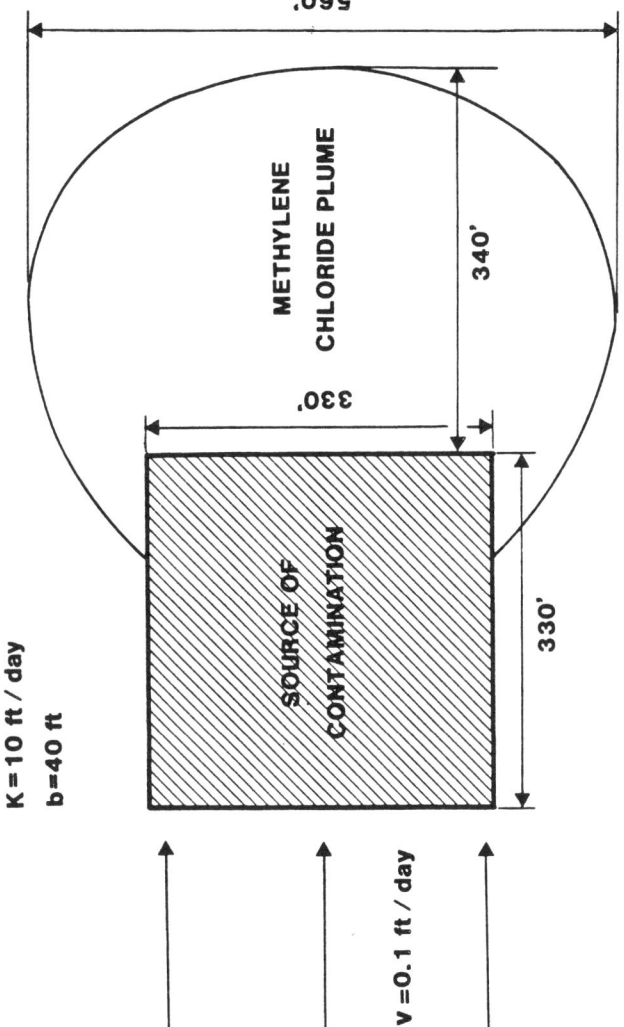

Figure 6.5 Aquifer characteristics and current extent of methylene chloride plume.

In evaluating available remedial action technologies, a pumping/injection doublet was identified as a potentially feasible alternative. The intent of the doublet would be to create a ground-water divide that would completely encompass the plume. This can be accomplished by locating the pumping well downgradient from the lagoon and the injection well just upgradient of the lagoon. If a line connecting the two wells is parallel to the major direction of flow an envelope similar to that shown in Figure 4.11 will be created. All water and contamination within the envelope will be directed towards the pumping well. The exact configuration of the envelope depends upon the distance between the wells, the regional ground-water velocity, and the pumping/injection rates.

The injection/withdrawal rate and location of the doublet wells can be estimated using Equation 4.26 or the graphical solution in Figure 4.12; both were discussed in Subsection 4.6. As Figure 6.5 shows, the maximum width of the plume is currently 560 ft. To ensure complete capture, an envelope of approximately 660 ft in width is assumed to be required. This translates to a value of 660 $\div$ 2 or 330 ft for c in Equation 4.26 or in Figure 4.12.

The other dimension that is required is the overall length of the envelope. This length is determined by the distance between the injection and pumping wells and the pumping/injection rate. The most appropriate location for the injection well is just upgradient from the lagoon. The pumping well can be located in the plume, but has to be near enough to the leading edge of the plume to ensure its capture. The exact location involves an iterative procedure wherein the distance between the wells is selected and a pumping/injection rate is calculated using Equation 4.26 or Figure 4.12; a is the parameter corresponding to one-half the distance between the wells. This rate is then used to estimate the distance, $x_s$, from the pumping well to the stagnation point or edge of the ground-water divide using the relationship for a single pumping well in uniform flow (Equation 4.24). If this distance is not long enough to ensure plume capture, the pumping well needs to be moved closer to the edge of the plume. Using this procedure a distance of 330 ft between wells and a pumping/injection rate of about 27 gpm was found to be adequate. Figure 6.6 shows the dimensions of the ground-water divide using this distance and pumping rate.

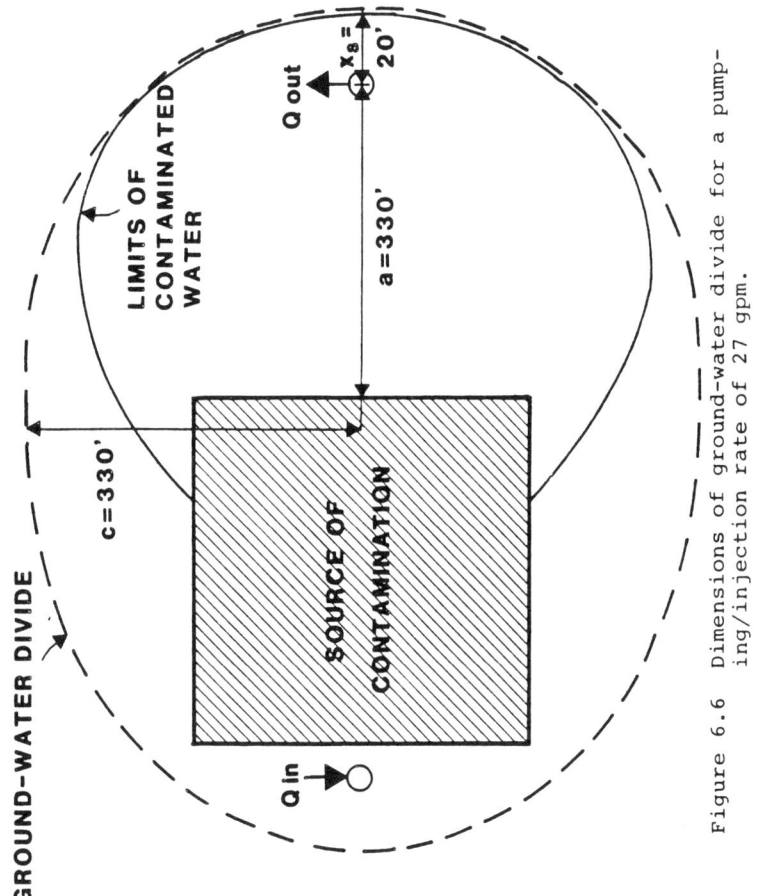

Figure 6.6 Dimensions of ground-water divide for a pumping/injection rate of 27 gpm.

## 6.4 EXAMPLE 3: GROUND-WATER PUMPING WITH AND WITHOUT AN IMPERMEABLE BARRIER

In this example a number of abandoned underground storage tanks were found to have lost their contents over a period of years. A detailed field sampling program found that the ground-water system was extensively contaminated. Figure 6.7 shows the location of the plume and the characteristics of the ground-water system.

The screening of remedial actions during the Engineering Feasibility Study suggested that the plume could be captured with a line of pumping wells located near the leading edge of the plume. It also suggested that an impermeable barrier completely surrounding the plume might act to expedite the clean-up action of the pumping wells. Thus, it was decided to analyze the time required for plume extraction with and without an impermeable barrier.

The technique selected for conducting the analysis was the simple numerical technique (i.e., particle mover method) discussed in Subsection 4.9. This technique involves tracking the movement of a particle of water with time. The rate and direction of particle movement are controlled by the pumping/injection action of wells and the regional ground-water flow.

The initial well configuration selected for analysis was a line of three pumping wells located 100 ft upgradient from the leading edge of the plume. Each well was assumed to be pumped at a rate of 20 gpm.

Particles were released from a number of locations along the perimeter of the plume. Their movement was tracked over time until each particle arrived at one of the wells. The location of each particle at the end of selected time increments was noted. These locations were then used to estimate the approximate location of the perimeter of the plume at the end of each time increment.

Figure 6.8 shows the position of the plume 0, 10, 20, 40, 80 and 120 days after the initiation of pumping. The results show that it takes approximately the same length of time for contaminants to travel from the storage tanks to the wells as it does for contaminants to travel from the leading edge of the plume to the wells. This is because the net ground-water velocity upgradient of the wells is the sum of the regional velocity and the velocity induced by the pumping action of the wells. Downgradient the net velocity is smaller because it is

168  *Modeling Remedial Actions at Uncontrolled Hazardous Waste Sites*

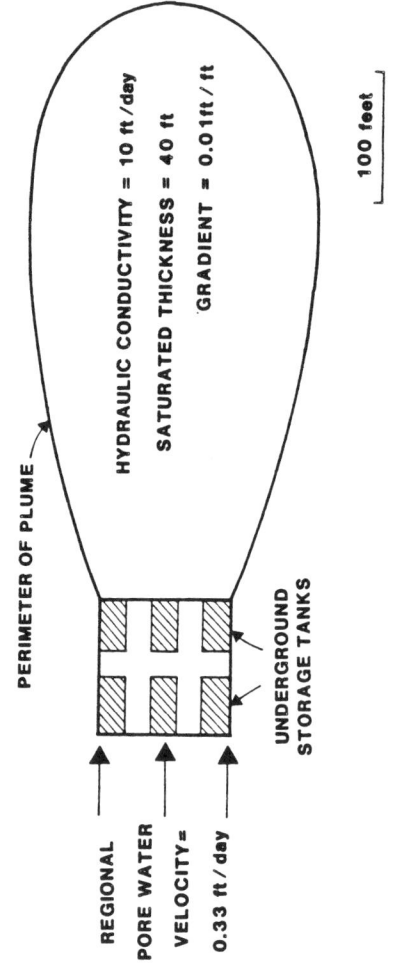

Figure 6.7  Plume location and aquifer characteristics for Example 3.

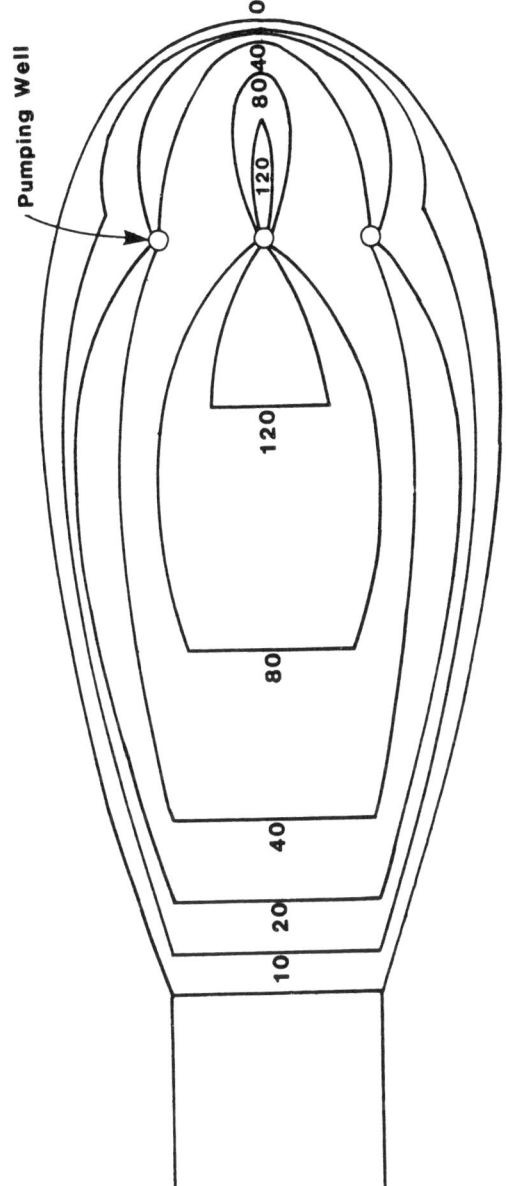

Figure 6.8  Plume position 0, 10, 20, 40, 80 and 120 days after initiation of pumping.

the difference between the two velocities.

The results also show that most of the plume can be captured in about 120 days. It is important to recognize, however, that this timeframe is based on the assumption that the contaminants are not retarded in their movement relative to the movement of the ground water. The timeframes in this example would have to be extended for contaminants that are retarded by using an appropriate retardation factor. It is also important to recognize that this analysis neglects the effects of dispersion. The perimeter of the plume is assumed to behave like a "sharp front."

The impact of installing an impermeable barrier around the plume was examined with the same technique. Figure 6.9 shows the resulting configuration of the remedial action alternative.

The method of images was used to simulate the impact of installing the barrier. The analysis was simplified somewhat by only considering the effects of the upgradient and downgradient portions of the barrier. The sides were neglected. Figure 6.10 shows the image well configuration used to create these barriers. Since they are assumed to be infinite in extent, the "real aquifer" (i.e., the portion of the aquifer inside the barrier) has the configuration of a semi-infinite strip. A complete representation could be obtained by using a more complex image well configuration similar to that shown in the lower portion of Figure 4.15.

Particles were again released from the perimeter of the plume and their movement towards the recovery wells was tracked with time. Since the barrier eliminates the regional ground-water flow component, it was assumed that the pumping rate of the wells could be reduced. The wells were left in the same location. Figure 6.11 shows the estimated position of the plume after 0, 10, 20, 40, 80, 120, 160, 320, 480 and 640 days. These results show that the barrier wall reduces the time required to capture contaminants downgradient from the wells, but increases the time to capture contaminants between the wells and the storage tanks. In part, this is due to the use of a reduced pumping rate. However, it is also due to the fact that there is no regional component of velocity within the impermeable barrier. The velocity due to the pumping wells is all that is affecting contaminant movement. This velocity is very small near the facility.

The efficiency of several other well and impermeable barrier configurations were also evaluated using this approach. Figure 6.12 summarizes the results for four different alternatives. Alternative 1 is the initial configuration (i.e., no barrier and 3 wells pumping 20 gpm). Alternative 2

Example Applications 171

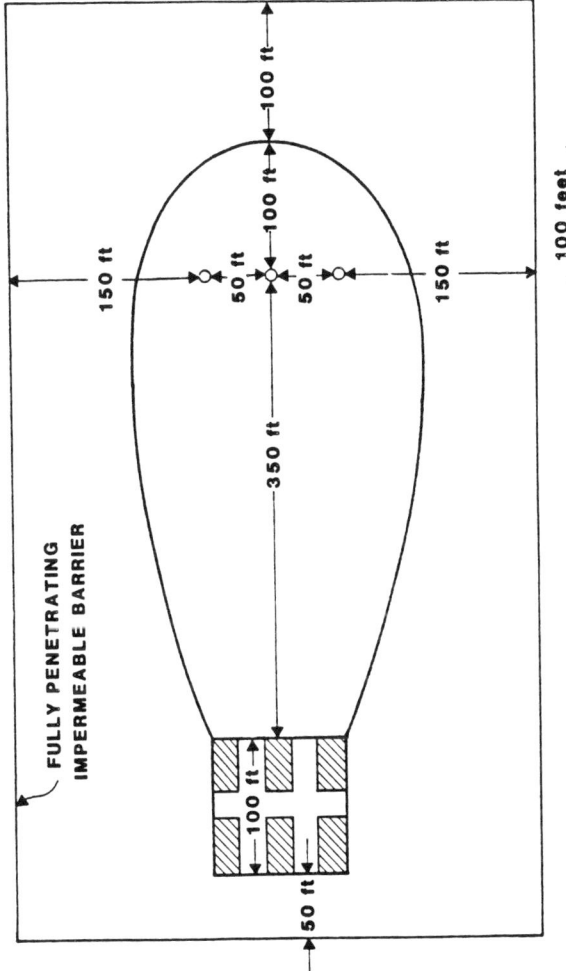

Figure 6.9  Impermeable barrier configuration.

172 *Modeling Remedial Actions at Uncontrolled Hazardous Waste Sites*

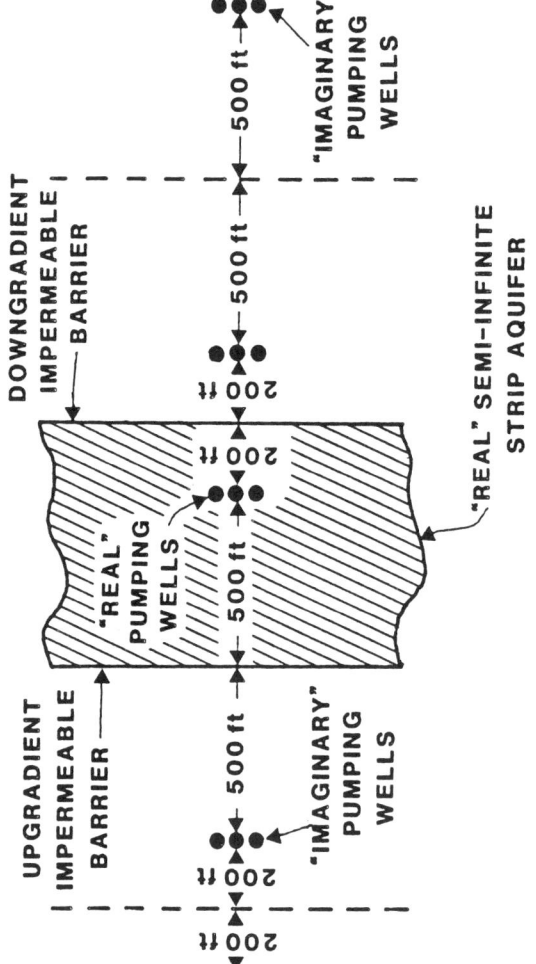

Figure 6.10 Image well configuration for impermeable barrier.

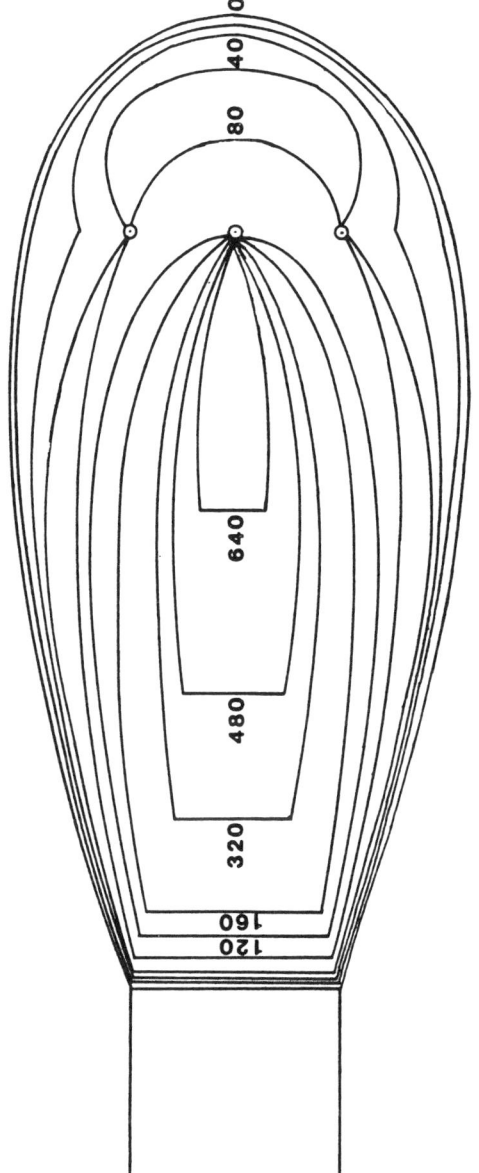

Figure 6.11  Plume position 0, 10, 20, 40, 80, 120, 160, 320 and 640 days after initiation of pumping.

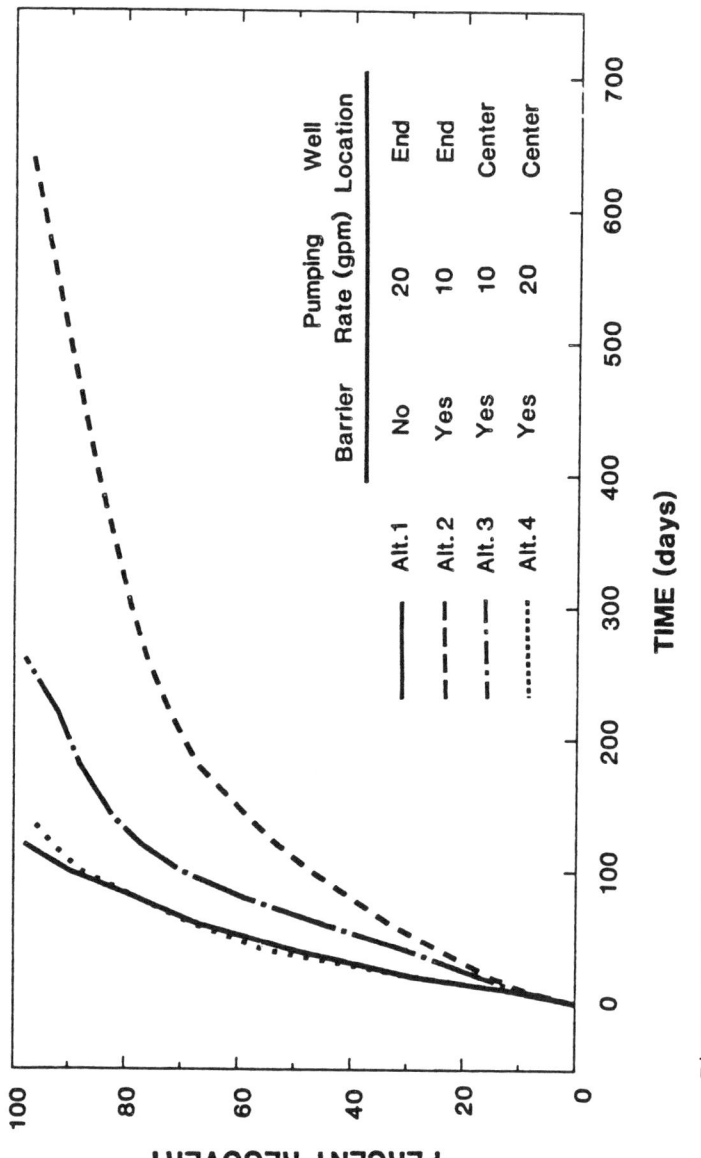

Figure 6.12 Percent recovery as a function of time for alternative well and barrier configurations.

is for the configuration shown in Figure 6.9. In Alternatives 3 and 4, the pumping wells were moved to the center of the plume along a line parallel to the main axis of the plume. Figure 6.12 shows the effectiveness of each alternative in terms of percent removal. Here, percent removel refers to the reduction in areal extent of contamination (i.e., plume size) relative to the initial areal extent of the contamination. Centering the wells reduces the recovery time by a factor of 3 for Alternative 3 and a factor of 6 for Alternative 4. In addition, Alternative 1 and Alternative 4 have approximately the same recovery time.

A large number of calculations were required to generate the results in Figures 6.8, 6.11 and 6.12. The Advective Particle Mover program by Ulrick (see Table 5.4) was used to reduce the level of effort involved in this example application.

## 6.5 EXAMPLE 4: RECIRCULATION SYSTEM FOR GROUND-WATER CLEAN-UP

This example is for a cooling water pond that fails and releases several thousand gallons of ethylene glycol solution. The affected ground-water system is a shallow, water table aquifer composed of silty-sand type materials.

During the site characterization effort, a biological feasibility study conducted on water samples indicated that the ethylene glycol could easily be degraded. Based on an analysis of the local hydrology and geology, a remedial action alternative was proposed. The alternative consists of a number of well points that would withdraw contaminated ground water. This water would then be aerated, inoculated with bacteria and then discharged back into the pond. As the treated water seeps downward to the saturated zone it would flush the remaining ethylene glycol towards the well points. The bacteria would act to degrade the residual ethylene glycol in place. The overall configuration of the remedial action is shown in Figure 6.13 along with the characteristics of the aquifer.

The analysis of the remedial action involved the use of several analytical methods. Given the proposed pumping rate for the well points of 2 gpm, the first step was to determine the amount of drawdown produced at each well point and halfway between well points. The drawdown at each well point needs to be estimated to determine whether the water table aquifer can be treated as a confined aquifer for purposes of the analysis. As was noted in Subsection 4.2, a method developed for confined conditions can only be used for water table conditions if drawdowns are small relative to the total saturated thickness. The drawdowns halfway between well points is of importance because without sufficient drawdown, the plume may migrate past the line of well points, particularly if there is a significant regional ground-water flow component.

Drawdown around a well in a water table aquifer can be estimated using a method by Neuman (1975). This method involves the evaluation of the unconfined well function. Values have been tabulated in a number of ground-water textbooks. For the conditions listed in Figure 6.13 and for a 2 gpm pumping rate, the drawdown after 30 days is 2.4 ft at the well point. This drawdown is relatively small compared to the total saturated thickness. In addition, for these conditions the unconfined well function is equivalent to the confined well function. Therefore, it is reasonable to use methods for confined aquifers in this analysis.

Using the same method the total drawdown halfway between well points is 1.2 ft considering the effects of superposition. This drawdown should be sufficient to direct contamination towards one of the well points.

The next step in the analysis is to determine whether the treated water discharged back into the pond will create a mound and, if so, whether it would impact the effectiveness of the well point system. The approximate height and extent of the mound can be estimated using the method developed by Hantush (1967a), as described in Subsection 4.6. Since the withdrawal system is composed of four well points, the total flow to the pond will be 8 gpm. Using Darcy's law it can be shown that the seepage rate through the base of the pond is approximately 2.4 gpd/ft$^2$. This rate produces a fairly extensive mound that is 4.3 ft in height just below the pond.

Mounding with this seepage rate is so extensive that no drawdown occurs at the well points. As a result, the well points would be totally ineffective. Additional calculations with different pumping rates and pond seepage rates showed that the only way to ensure that a cone of depression would occur around the well points was to discharge only a fraction of the total flow back in to the pond. Using the original pumping rate of 2 gpm, it was found that if 25 percent of the total was discharged to the pond a cone of depression would occur. At this discharge rate the maximum height of the mound was found to be only 1.1 ft.

Residual ethylene glycol flushed into the saturated zone by the treated water will tend to move radially away from the pond as a result of this mound. Thus, it is important to determine whether the well point system can effectively capture all of the ethylene glycol.

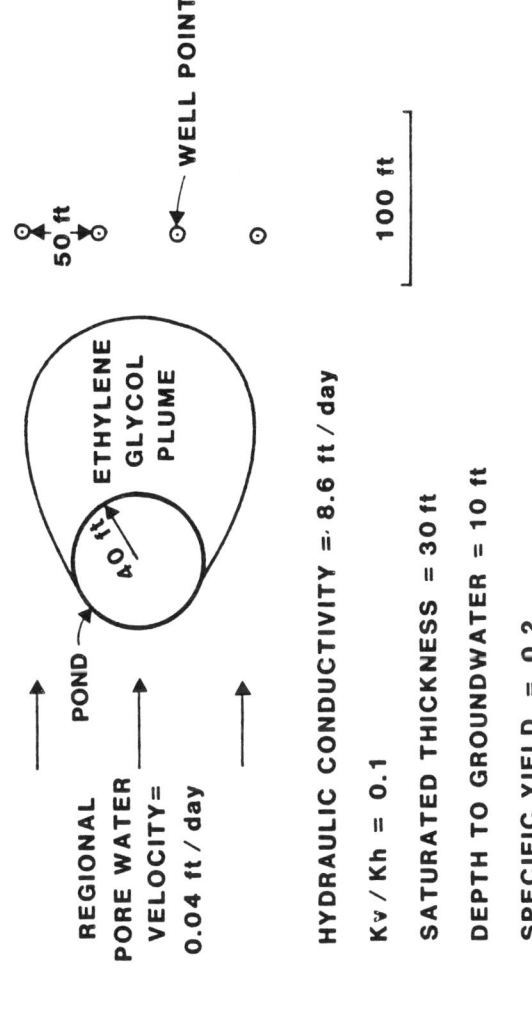

Figure 6.13 Aquifer characteristics and remedial action configuration for well point recirculation system.

The effectiveness of the well points was evaluated using the simple numerical technique discussed in Subsection 2.8. Particles were released from different locations along the perimeter of the mound to determine whether they would be captured by the well point system or whether they would be entrained in the regional flow field. Since this technique can only consider point sources or sinks (i.e., injection or withdrawal wells), an area source like the pond cannot be considered directly. Instead, the effect of the mound has to be simulated using one or more point sources (i.e., injection wells). Through an iterative procedure it was found that an injection well pumping at a rate of 3 gpm could produce a head distribution roughly equivalent to the level of mounding that would be expected, particularly beyond a distance of 40 ft, which is the radius of the pond and the location where the particles would be released. Figure 6.14 shows a comparison between the estimated water table elevations for the mound and for the injection well.

Using this injection rate, a well point pumping rate of 2 gpm and the aquifer characteristics given in Figure 6.13, particles were released from different positions along the perimeter of the mound. Figure 6.15 shows the pathway followed by each particle and the time in days for each particle to arrive at one of the well points. These results show how the glycol that is flushed into the saturated zone will initially move radially away from the pond until it is entrained in the regional flow and then directed towards one of the well points. These results also show that even the ethylene glycol that is initially on the far upgradient side of the pond will be captured by one of the outside well points. This ethylene glycol will take about seven times longer than that which is initially on the downgradient side of the pond. In fact, the results show that most of the contamination will initially be captured by the inside wells. As a result, it may be possible to initially just treat the water from the two inside wells and then shift to the outer wells when the remaining ethylene glycol ultimately arrives.

Due to the large number of calculations involved in tracking the movement of different particles, this analysis was conducted using a programmable hand-held calculator. The Advective Transport program by Ulrick (see Table 5.4) was used.

## 6.6 EXAMPLE 5: DRAIN RECIRCULATION SYSTEM

This final example involves the release of a solvent from a large number of drums in a waste storage yard. The release

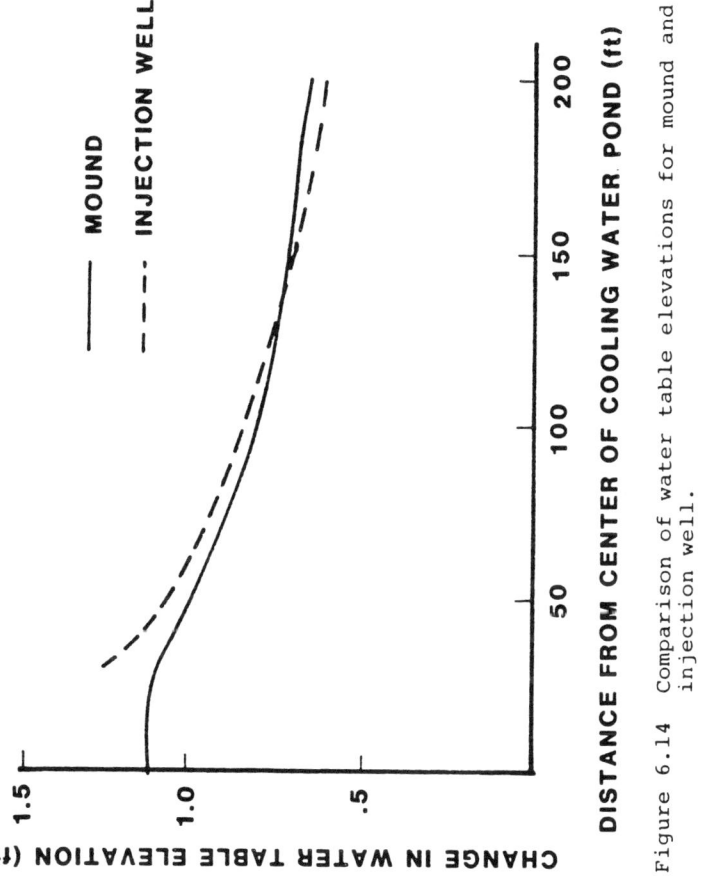

Figure 6.14  Comparison of water table elevations for mound and injection well.

180  *Modeling Remedial Actions at Uncontrolled Hazardous Waste Sites*

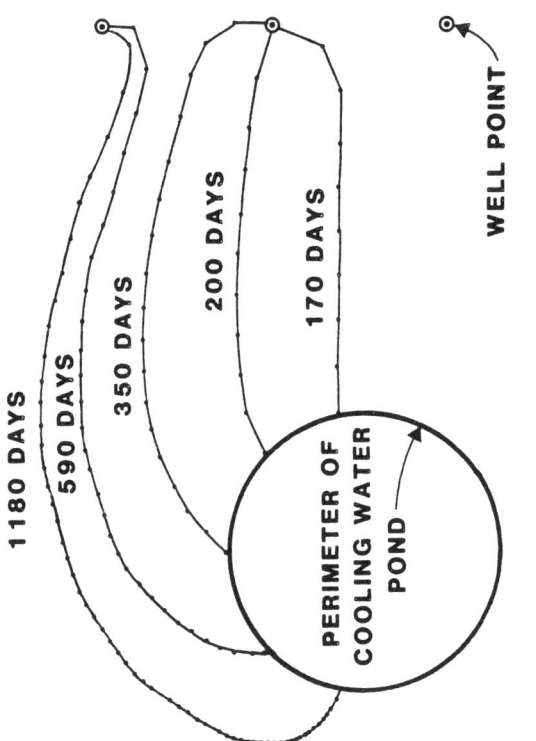

Figure 6.15  Particle movement from the perimeter of the cooling water pond to each well point.

occurred into an area where the ground water is near the land surface. The regional flow is negligible in the local area. Access to areas beyond the perimeter of the storage yard is limited on all sides by roads and buildings. Therefore, the selected remedial action has to be implemented within the perimeter of the storage yard.

In evaluating different remedial action technologies it was decided that a recirculating drain system could be used. This system would be composed of a fully penetrating interceptor trench installed along one side of the storage yard. Water withdrawn from the drain would be treated on-site and then reinjected through another drain located along the other side. This drain would create a mound that would direct the solvent towards the recovery drain.

The flow system created by the installation of such a drain system was evaluated using a hand-held calculator program that estimates the drawdown around line sources and sinks of finite length. The Line Sink program developed by Ulrick (see Table 5.2) was used.

The principle of superposition is used in this program to obtain the total drawdown due to multiple line sources and sinks. Using this program, the elevation of the water table can be evaluated rapidly at a large number of locations. These elevations can then be used to generate equipotential contours (i.e., contours of equal elevation). These contours can be used to generate flow lines (i.e., the direction ground water will move).

Figure 6.16 shows a plan view of the site. It also shows the location of the plume and the drains. Finally, equipotential contours and flow lines for a recovery/reinjection rate of 60 gpm are also shown.

These results indicate that most of the plume outside the perimeter of the yard is contained within the equipotential contour corresponding to 1 ft of drawdown. Given the negligible regional flow in the area, this drawdown should be sufficient to ensure that the entire plume is captured by the recovery drain. The remainder of the plume will be directed towards the recovery drain by the mounding action of the reinjection drain.

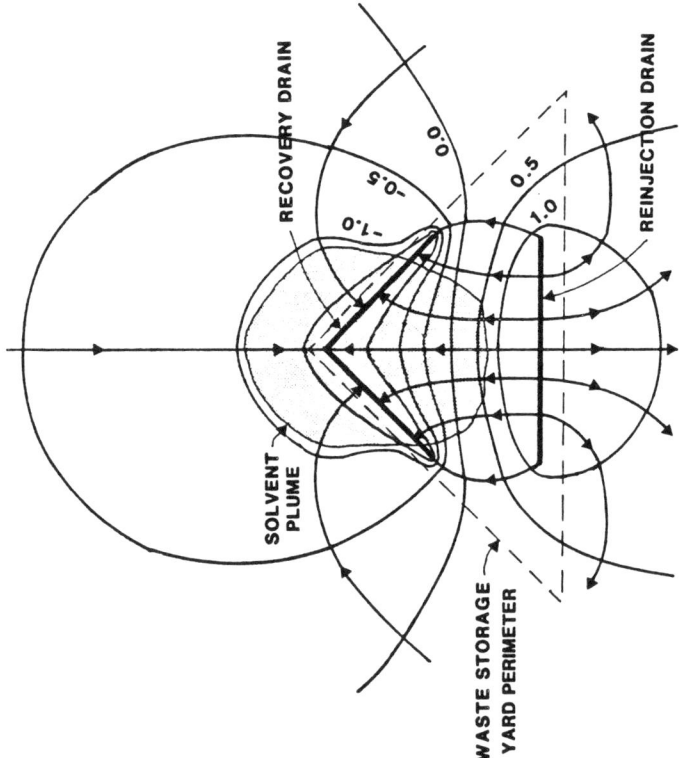

Figure 6.16  Equipotential contours (in feet) and flow lines produced by the drain recirculation system.

# References

Baumann, P. 1952. "Groundwater Movement Controlled through Spreading," Transactions, ASCE, Vol. 117.

Bear, J. 1979. Hydraulics of Groundwater, McGraw-Hill Inc., New York, NY.

Bentall, R. 1963. Shortcuts and Special Problems in Aquifer Tests, USGS Water-Supply Paper 1545-C.

Bicknell, B.R. 1984. Modeling Chemical Emissions from Lagoons and Landfills, Final Report, U.S. Environmental Protection Agency, Environmental Research Laboratory, Athens, GA.

Boulton, N. S. 1954. "The Drawdown of the Water Table under Non- Steady Conditions near a Pumped Well in an Unconfined Formation," Proc. Inst. Civil Engrs., 3.

Boulton, N. S. 1955. "Unsteady Radial Flow to a Pumped Well Allowing for Delayed Yield from Storage," Proc. Gen. Assembly Rome, Intern. Ass. Sci. Hydrol. Publ. 37.

Boulton, N. S. 1963. "Analysis of Data from Nonequilibrium Pumping Tests Allowing for Delayed Yield from Storage," Proc. Inst. Civil Engrs., 26.

Boulton, N. S., and T. D. Streltsova. 1977a. "Unsteady Flow to a Pumped Well in a Fissured Water-Bearing Formation," J. Hydrol., 35.

Boulton, N. S., and T. D. Streltsova. 1977b. "Unsteady Flow to a Pumped Well in a Two-Layered Water-Bearing Formation," J. Hydrol., 35.

Boulton, N. S., and T. D. Streltsova. 1978. "Unsteady Flow to a Pumped Well in a Fissured Aquifer with a Free Surface Level Maintained Constant, "Water Resources Research, 14(3).

Boussinesq, J. 1904. "Recherches Theoretiques sur l'Ecoulement des Nappes d'Eau Infiltrees dans le Sol et sur le debit des Sources," J. Math. Pure Appl., 10.

Bouwer, H., 1974. "Design and Operation of Land Treatment Systems for Minimum Contamination of Groundwater," Ground Water, 12(3).

Brooks, R. H. 1961. "Unsteady Flow of Groundwater into Drain Tile," ASCE. Proc. 87.

Brutsaert, W., and M. Y. Corapcioglu. 1976. "Pumping of Aquifer with Viscoelastic Properties," Proceedings American Society of Civil Engineering, Journal of the Hydraulics Division, 102 (HY11).

Chauhan, H. S., G. O. Schwab and M. Y. Hamdy. 1968. "Analytical and Computer Solutions of Transient Water Tables for Drainage of Sloping Land," Water Resources Research, 4(3).

Cleary, R. W. and M. J. Ungs. 1978. Groundwater Pollution and Hydrology, Mathematical Models and Computer Programs, Water Resources Program, Report No. 78-WR-15, Princeton University, Princeton, N.J.

Codell, R. B., K. T. Key and G. Whelan. 1982. A Collection of Mathematical Models for Dispersion in Surface Water and Groundwater, NUREG-0868, U.S. Nuclear Regulatory Commission, Washington, DC.

Cohen, R. M. and W. J. Miller, III. 1983. "Use of Analytical Models for Evaluating Corrective Actions at Hazardous Waste Sites," Proceedings of the 3rd National Symposium on Aquifer Restoration and Ground-Water Monitoring, National Water Well Association, May 25-27.

Cooper, H. H., Jr., and M. I. Rorabaugh, 1963. Ground Water Movements and Bank Storage Due to Flood Stages in Surface Streams, USGS Water Supply Paper 1536-J.

Corapcioglu, M. Y. 1976. "Mathematical Modeling of Leaky Aquifers with Rheological Properties," Proceedings of Anaheim Symposium on Subsidence, Int. Assoc. Hydrol. Sci. Publ. No. 121.

Dass, P. et al. 1977. "Leachate Production at Sanitary Landfills," Proceedings of the American Society of Civil Engineers, Journal of the Environmental Engineering Division, 103 (EE6).

Donigian, A. S., Jr., T. Y. R. Lo, and E. W. Shanahan. 1983. Rapid Assessment of Potential Ground-Water Contamination Under Emergency Response Conditions, EPA-600/8-83-030, U.S. Environmental Protection Agency, Office of Research and Development, Washington, D.C.

Fenn, D. G., K. J. Hanley and T. V. De Geare. 1975. "Use of the Water Balance Method for Predicting Leachate Generation from Solid Waste Disposal Sites," Report SW-168, U.S. Environmental Protection Agency, Cincinnati, OH.

Ferris, J. G., D. B. Knowles, R. H. Brown and R. W. Stallman. 1962. Theory of Aquifer Tests, USGS Water-Supply Paper 1536E.

Freeze, R. A. and J. A. Cherry. 1979. Groundwater, Prentice-Hall, Inc., Englewood Cliffs, N.J.

Glover, R. E. 1960. Mathematical Derivations as Pertain to Ground-water Recharge, Agricultural Research Service, USDA, Ft. Collins, CO.

Glover, R. E. 1966. Theory of Ground Water Movement, U.S. Bureau of Reclamation Engineering Monograph No. 13.

Glover, R. E. 1974. Transient Ground Water Hydraulics, Water Resources Publications, Fort Collins, CO.

Grove, D. B., W. A. Beetem, and F. B. Sower. 1970. "Fluid Travel Time between a Recharging and Discharging Well Pair in an Aquifer Having a Uniform Regional Flow Field," Water Resources Research, 6.

Hantush, M. S. 1956. "Analysis of Data from Pumping Tests in Leaky Aquifers," Transactions American Geophysical Union, 37(6).

Hantush, M. S. 1959. "Non-Steady Flow to Flowing Wells in Leaky Aquifers," Journal of Geophysical Research, 64(8).

Hantush, M. S. 1960. "Modification of the Theory of Leaky Aquifers," Journal Geophysical Research, 65(11).

Hantush, M. S. 1964. "Hydraulics of Wells," In Advances in Hydroscience, V. T. Chow (Editor). Academic Press, New York, N.Y.

Hantush, M. S. 1967a. "Growth and Decay of Ground water Mounds in Response to Uniform Percolation," Water Resources Research, 3(1).

Hantush, M. S. 1967b. "Flow to Wells in Aquifers Separated by a Semi-Pervious Layer," Journal of Geophysical Research, 72(6).

Hantush, M. S., and C. E. Jacob. 1955. "Non-Steady Radial Flow in an Infinite Leaky Aquifer and Nonsteady Green's Function for an Infinite Strip of Leaky Aquifer," Transactions of the American Geophysical Union, 36(1).

Harr, M. E., 1962. Ground Water and Seepage, McGraw-Hill Inc., New York, N.Y.

Hooghoudt, S. B. 1940. Bijdragen tot de Kennis van Eenige Natuurkundige Grootheden van den Grond, 7, Algemeene Beschouwing van het Probleem van de Detail ontwatering en de Infiltratie Door Middel van Parallel Loopende Drains, Greppels, Slooten, en Kanalen, Versl. Lanbouwk, Ond., 46.

Huisman, L. 1972. Groundwater Recovery, Winchester Press, New York, N.Y.

Huisman, L. and T. N. Olsthoorn. 1983. Artificial Groundwater Recharge, Pitman Publishing Inc., Marshfield, MA.

Hunt, B. W. 1971. "Vertical Recharge of Unconfined Aquifers," Journal of Hydraulics Division, ASCE, 97(HY7).

Jacob, C. E. 1943. "Correlation of Ground Water Levels with Precipitation on Long Island, N.Y.," Transactions of the American Geophysical Union, pt. 2.

Jacob, C. E. and S. W. Lohman. 1952. "Nonsteady Flow to a Well of a Constant Drawdown in an Extensive Aquifer," Transactions of the American Geophysical Union, 33(40).

Javandel, I., C. Doughty and C. F. Tsang. 1984. Ground-Water Transport: Handbook of Mathematical Models, Water Resources Monograph 10, American Geophysical Union, Washington, DC.

JRB Associates. 1982. Handbook of Remedial Actions at Hazardous Waste Disposal Sites, EPA 625/6-82-006, U.S. Environmental Protection Agency, Municipal Environmental Research Laboratory, Cincinnati, OH.

Kelly, W. E. 1982. "Field Reports - Ground-Water Dispersion Calculations with a Programmable calculator," Ground Water, 20(6).

Kirkham, D. 1958. "Seepage of Steady Rainfall through Soil into Drains," Transactions of the American Geophysical Union, 39.

Kirkham, D. 1967. "Explanation of Paradoxes in the Dupuit-Forchheimer Seepage Theory," Water Resources Research, 3.

Kirkham, D., S. Toksoz, and R. R. van der Ploeg. 1974. "Steady Flow to Drains and Wells," In Drainage for Agriculture, van Schilfgaarde, ed., American Society of Agronomy, Inc., Madison, WI.

Knight, R. G., E. H. Rothfuss and K. D. Yard. 1980. FGD Sludge Disposal Manual, Second Edition, CS-1515, Electric Power Research Institute, Palo Alto, CA.

Knox, R.C. 1984. "Assessment of the Effectiveness of Barriers for the Retardation of Pollutant Migration," Ground Water, 22(3).

Lai, R. Y. S., and Cheh Wu Su. 1974. "Nonsteady Flow to a Large Well in a Leaky Aquifer," J. Hydrol., 22.

Luthin, J. N. 1973. Drainage Engineering, R. E. Krieger Publishing Co., Huntington, NY.

McBean, E. A., R. Poland, F. A. Rovers and A. J. Crutcher. 1982. "Leachate Collection Design for Containment Landfills," Proceeding of the ASCE, Journal of the Environmental Engineering Division, 108(1).

McWhorter, D. B. and D. K. Sunada. 1977. Ground-Water Hydrology and Hydraulics, Water Resource Publications, Fort Collins, CO.

McWhorter, D. B. and J. D. Nelson. 1980. "Seepage in a Partially Saturated Zone Beneath Tailings Impoundments," Mining Engineering, April.

Moench, A. F. and T. A. Prickett. 1972. "Radial Flow in an Infinite Aquifer Undergoing Conversion from Artesian to Water Table Conditions," Water Resources Research, 9(2).

Molden, D., D. K. Sunada and J. W. Warner. 1984. "Microcomputer Model of Artificial Recharge Using Glover's Solution, " Ground Water, 22(1).

Moody, W. T. 1966. "Nonlinear Differential Equation for Drain Spacing," Proceedings of the ASCE, 92.

Moody, W. T. and R. W. Ribbens. 1965. Ground Water-Tetrama-Colusa Canal Reach No. 3, Sacramento Canal Units, Central Valley Project, Memorandum to Chief, Canals Branch, Bureau of Reclamation Office of Chief Engineer, Denver, CO.

Moore, C. A. 1983. Landfill and Surface Impoundment Performance Evaluation, SW-869, U.S. Environmental Protection Agency, Office of Solid Waste and Emergency Response, Washington, DC.

Nelson, R. W. and J. A. Schur. 1980. PATHS Groundwater Hydrologic Model, PNL-3162, Prepared by Battelle, Pacific Northwest Laboratories, Richland, WA for the Office of Nuclear Waste Isolation, U.S. Department of Energy.

Neuman, S. P. 1972. "Theory of Flow in Unconfined Aquifers Considering Delayed Response of the Water Table,"Water Resources Research, 8.

Neuman, S. P. 1973. "Calibration of Distributed Parameter Groundwater Flow Models Viewed as a Multiple-Objective Decision Process under Uncertainty," Water Resources Research, 9.

Neuman, S. P. 1975. "Analysis of Pumping Test Data from Anisotropic Unconfined Aquifers Considering Delayed Gravity Response," Water Resources Research, 11(2).

Neuman, S. P. and P. A. Witherspoon. 1969a. "Theory of Flow in a Confined Two-Aquifer System," Water Resources Research, 5.

Neuman, S. P. and P. A. Witherspoon. 1969b. "Applicability of Current Theories of Flow in Leaky Aquifers," Water Resources Research, 5.

Neuman, S. P. and P. A. Witherspoon. 1972. "Field Determination of the Hydraulic Properties of Leaky Multiple-Aquifer Systems," Water Resources Research, 8.

Oberlander, P.L. and R.W. Nelson. 1984. "An Idealized Ground-Water Flow and Chemical Transport Model (S-PATHS)," Ground Water, 22(4).

Ogata, A. 1970. Theory of Dispersion in a Granular Medium, U.S.G.S. Professional Paper 411-I.

Papadopulos, I. S. 1965. "Nonsteady Flow to a Well in an Infinite Anisotropic Aquifer," Symp. Int. Assoc. Sci. Hydrol., Dubrovnik.

Papadopulos, I. S. 1967. "Drawdown Distribution around a Large-Diameter Well," Proceedings of a Symposium on Ground Water, American Water Resources Association.

Parr, A. D., J. G. Melville and F. J. Molz. 1983. "HP41C and TI59 Programs for Anisotropic Confined Aquifers," Ground Water, 21(2).

Pettyjohn, W. A., D. C. Kent, T. A. Prickett, H. E. Le Grand and F. E. Witz. 1982. Methods for the Prediction of Leachate Plume Migration and Mixing, Draft Report, U.S. Environmental Protection Agency, Municipal Environmental Research Laboratory, Cincinnati, OH.

Picking, L. W. 1979. "Field Reports - Programming a Pocket Calculator for Solving Multiple Well, Variable Pumping Rates Problems," Ground Water, 18(2).

Polubarinova-Kochina, P. Ya. 1952. "Theory of the Motion of Ground Water," Gostekhizdat, Moscow.

Powers, M. A., P. P. Virgadamo and W. E. Kelly. 1981. Groundwater Control at Hazardous Waste Sites, American Society of Civil Engineers Meeting, St. Louis, MO.

Prickett, T. A. and C. G. Lonnquist. 1971. Selected Digital Computer Techniques for Groundwater Resource Evaluation, Bulletin No. 55, Illinois State Water Survey, Urbana, IL.

Prickett, T. A., T. G. Naymik and C. G. Lonnquist. 1981. A "Random-Walk," Solute Transport Model for Selected Groundwater Quality Evaluations, Bulletin No. 65, Illinois State Water Survey, Urbana, IL.

Prickett, T. A. and M. L. Vorhees. 1981. Selected Hand-Held Calculator Codes for the Evaluation of Cumulative Strip-Mining Impacts on Groundwater Resources, Prepared for the Office of Surface Mining, Region V, Denver, CO.

Rao, N. H. and P. B. S. Sarma. 1981. "Ground-Water Recharge from Rectangular Areas," Ground Water, 19(3).

Rayner, F. A. 1981. "Discussion of Programmable Hand Calculator Programs for Pumping and Injection Wells: I - Constant or Variable Pumping (Injection) Rate, Single or Multiple Fully Penetrating Wells," Ground Water, 19(1).

Rayner, F. A. 1983. "Aquifer Modeling with a Handheld Calculator - AQMODL," Ground Water, 21(1).

Sandberg, R., R. B. Scheibach, D. Koch and T. A. Prickett. 1981. Selected Hand-Held Calculator Codes for the Evaluation of the Probable Cumulative Hydrologic Impacts of Mining, Report H-D3004/030-81-1029F, Prepared for the Office of Surface Mining, Region V, Denver, CO.

Schoeller, H. 1959. Arid Zone Hydrology, Recent Developments, UNESCO, Paris.

Schroeder, P.R., J.M. Morgan, T.M. Walski and A.C. Gibson. 1984a. Hydrologic Evaluation of Landfill Performance (HELP) Model, Volume I. User's Guide for Version 1, EPA/530-SW-84-009, U.S. Environmental Protection Agency, Municipal Environmental Research Laboratory, Cincinnati, OH.

Schroeder, P. R., A. C. Gibson and M. D. Smolen. 1984b. Hydrologic Evaluation of Landfill Performance (HELP) Model, Version II. Documentation for Version 1, EPA/530-SW-84-010, U.S. Environmental Protection Agency, Municipal Environmental Research Laboratory, Cincinnati, OH.

SCS Engineers. 1982. Costs of Remedial Response Actions at Uncontrolled Hazardous Waste Sites, EPA 600/2-82-035, Prepared for the U.S. Environmental Protection Agency, Municipal Environmental Research Laboratory, Cincinnati, OH.

Singh, S. R. and C. M. Jacob. 1977. "Transient Analysis of Phreatic Aquifers Lying between Two Open Channels," Water Resources Research, 13(2).

Stallman, R. W. 1965. "Effects of Water Table Conditions on Water Level Changes near Pumping Wells," Water Resources Research, 1(2).

Streltsova, T. D. 1974. "Drawdown in Compressible Unconfined Aquifers," Proceedings of the American Society of Civil Engineers, Journal of the Hydraulics Division, 100(HY11).

Streltsova, T. D. 1976. "Analysis of Aquifer-Aquitard Flow," Water Resources Research, 12(3).

Terzidis, G. 1968. "Discussion, Falling Water Table Between Tile Drains," ASCE Proceedings, 94.

Theis, C. V. 1935. "The Relation Between the Lowering of Piezometrc Surface and the Rate and Duration of Discharge of a Well Using Ground-Water Storage," Transactions American Geophysical Union, 16th Annual Meeting, Part 2.

U.S. Department of the Interior. 1981. Ground Water Manual, U.S. Department of the Interior.

van Genuchten, M. Th. and W. J. Alves. 1982. Analytical Solutions of the One-Dimensional Convective-Dispersive Solute Transport Equation, Technical Bulletin 1661, USDA.

van Schilfgaarde, J. 1970. "Theory of Flow to Drains," In Advances in Hydroscience, Vol. 6, Academic Press, New York, NY.

van Schilfgaarde, J. 1974. "Nonsteady Flow to Drains," In Drainage for Agriculture, van Schilfgaarde, ed., American Society of Agronomy, Inc., Madison, WI.

Venetis, C. 1968. "On the Impulse Response of an Aquifer," Bulletin of Int. Assn. of Scientific Hydrology, Vol. 13.

Walton, W. C. 1970. Groundwater Resource Evaluation. McGraw-Hill, Inc. New York, NY.

Walton, W. C. 1979. "Progress in Analytical Groundwater Modeling," In: W. Back and D. A. Stephenson (Guest-Editors), Contemporary Hydrogeology - The George Burke Maxey Memorial Volume, J. Hydrol; 43.

Walton, W. C. 1983. Handbook of Analytical Ground Water Model Codes for Radio Shack TRS-80 Pocket Computer and Texas Instruments TI-59 Hand-Held Programmable Calculator, GWMI 83-02/2, International Ground Water Modeling Center, Holcomb Research Institute, Butler University, Indianapolis, IN.

Walton, W. C. 1984a. Handbook of Analytical Ground Water Models, GWMI 84-06/1, International Ground Water Modeling Center, Holcomb Research Institute, Butler University, Indianpolis, IN.

Walton, W. C. 1984b. 26 Basic Ground Water Model Programs for Pocket Microcomputers, GWMI 84-06/3, International Ground Water Modeling Center, Holcomb Research Institute, Butler University, Indianapolis, IN.

Walton, W. C. 1984c. 35 Basic Ground Water Model Programs for Desk Top Microcomputers, GWMI 84-06/4, International Ground Water Modeling Center, Holcomb Research Institute, Butler University, Indianapolis, IN.

Warner, D. L. and M. G. Yow. 1979. "Programmable Hand Calculator Programs for Pumping and Injection Wells: I - Constant or Variably Pumping (Injection) Rate, Single or Multiple Fully Penetrating Wells," Ground Water, 17(6).

Warner, D. L. and M. G. Yow. 1980a. "Programmable Hand Calculator Programs for Pumping and Injection Wells: II - Constant Pumping (Injection) Rate, Single Fully Penetrating Well, Semiconfined Aquifer," Ground Water, 18(2).

Warner, D. L. and M. G. Yow. 1980b. "Programmable Hand Calculator Programs for Pumping and Injection Wells: III - Constant Pumping (Injection) Rate, Fully Confined Aquifer, Partially Penetrating Well," Ground Water, 18(5).

Wesseling, J. 1964. "A Comparison of the Steady State Drain Spacing Formulas of Hooghoudt and Kirkham in Connection with Design Practice," Tech. Bull., Vol. 34, Inst. for Land and Water Management Research, Wageningen.

Wilson, J.L. 1984. "Double Cell Hydraulic Containment of Pollutant Plumes," Proceedings of the Fourth National Symposium and Exposition on Aquifer Restoration and Ground Water Monitoring, National Water Well Association, Columbus, OH.

Witherspoon, P. A., I. Javandel, S. P. Neuman and R. A. Freeze 1967. Interpretation of Aquifer Gas Storage Conditions from Water Pumping Tests, American Gas Association, New York, N.Y.

Witherspoon, P. A. and T. N. Narasimhan. 1973. "Seepage from Recharge and Landfill Ditches," Specialty Conference on Agricultural and Urban Considerations in Irrigation and Drainage, Colorado Section of ASCE.

Yeh, G. T. 1981. AT123D: Analytical Transient One-, Two-, and Three-Dimensional Simulation of Waste Transport in the Aquifer System, ORNL-5602, Oak Ridge National Laboratory, Oak Ridge, TN.

Youngs, E. G. 1964. "Horizontal Seepage through Unconfined Aquifers with Hydraulic Conductivity Varying with Depth," J. Hydrology, 3.

Youngs, E. G. 1966a. "Horizontal Seepage through Unconfined Aquifers with Nonuniform Hydraulic Conductivity," J. Hydrology, 4.

Youngs, E. G. 1966b. "Exact Analysis of Certain Problems of Ground Water Flow with Free Surface Conditions,"J. Hydrology, 4.

# VOLUME 3

## Numerical Modeling of Surface, Subsurface and Waste Control Actions

# 1. Introduction

1.1 PURPOSE OF REPORT

Recent studies at several uncontrolled hazardous waste sites have demonstrated the benefits of using numerical models to evaluate remedial action performance. Models have been used in the detailed analysis of alternative actions to identify those that would be ineffective or would fail to meet site clean-up goals. The quantitative measures of performance derived from simulation results provide a useful basis for comparison with other factors like remedial action costs. Models have also been used to refine and, in some cases, optimize conceptual designs prior to their implementation; post-implementation modeling studies have also been conducted to improve remedial action operation. Another beneficial use has been in the prediction of future contamination levels for purposes of exposure and risk assessment. Finally, the increased level of understanding gained regarding important processes/pathways and levels of uncertainty associated with parameters that require additional characterization have been a benefit to many model users.

The purpose of this volume is to provide guidance on the use of surface, unsaturated and saturated zone models to evaluate the performance of remedial actions. The guidance provided herein focuses on: 1) important considerations related to the <u>application</u> or <u>use</u> of numerical models and 2) <u>modeling requirements</u> for specific remedial actions or groups of actions. The guidance applies only to those actions commonly implemented at hazardous waste sites, namely surface, subsurface and waste control actions.

This volume will be of most value to two major groups: 1) EPA and state Superfund staff and 2) certain site contractors. EPA and state staff should gain an improved understanding of how numerical models can be used to assess remedial actions. This information should be of particular benefit when reviewing proposed site contractor plans to use numerical models. This volume will aid site contractors that have limited experience in using numerical modeling techniques.

1.2 REPORT ORGANIZATION

Brief conclusions regarding the use of numerical models for remedial action evaluation are presented in Section 2.

Section 3 discusses the processes that act to transport, transform and transfer hazardous waste constituents in the local environment surrounding a hazardous waste site. Section 4 discusses specific surface, subsurface, and waste control remedial action technologies and how their implementation affects these processes. Both sections are meant to provide the reader with a brief overview. They also set the stage for Sections 5 and 6.

Section 5 discusses a number of important considerations associated with the application or use of numerical models. The section starts by overviewing the general capabilities of surface, unsaturated and saturated zone models; brief reviews of several representative models are also provided along with sources of information on a number of other models. Considerations related to the linkage of models for different zones are presented for those situations involving the analysis of relatively complicated site and remedial action conditions. The process of "applying" a numerical model is also presented, followed by a discussion of user expertise and resources commonly required when using numerical models. The section concludes by describing alternative ways of using models to evaluate actions; a number of examples from published modeling studies are included.

Section 6 provides modeling requirements for different remedial actions or groups of actions. The requirements include: 1) the type of model(s) (i.e., surface, unsaturated or saturted zone), 2) dimensionality and grid configuration (i.e., two-dimensional, x-y), and 3) parameter adjustments. Where possible, parameter estimation guidance specific to a given action is provided. Where this is not possible general guidance is provided. The parameter guidance is meant to be used only when site-specific data are not available.

To provide EPA and its site contractors with a modeling capability that can be used to assess a broad range of site and remedial action conditions, three models were selected from the large number of available surface, unsaturated and saturated zone models. The selected models include: 1) the Hydrologic Simulation Program-FORTRAN (HSPF) model for the surface zone; 2) FEMWATER/FEMWASTE models for the unsaturated zone; and 3) the Finite Element, Three-Dimensional Ground Water (FE3DGW) and Combined Fluid, Energy and Solute Transport (CFEST) models for the saturated zone. Each model is being made available for use on the EPA National Computer Center (NCC) in Research Triangle Park, N.C.

Appendix A to this report describes the rationale for selecting each model, their capabilities, linkage considerations, their implementation on NCC, available documentation and user support, and specific parameters that must be adjusted to represent selected actions.

## 2. Conclusions

Numerical models are finding increased use in the analysis of remedial action performance. To date, most model applications have been for the purpose of evaluating alternative remedial action designs and the impacts associated with uncertain estimates of key model parameters and assumptions. Both types of uses have generally led to an improved understanding of site conditions and an ability to quantitatively evaluate the feasibility of different remedial actions. Numerical models also have been used to a lesser extent to estimate reductions in exposure levels associated with the implementation of remedial actions. Such estimates have been used directly in exposure assessment or as input to more comprehensive risk assessments. Future uses of models include the analysis of remedial action design life, the impacts associated with remedial action failure and optimal remedial action design and operation.

Limited field and laboratory data exist on the performance of certain remedial actions. As a result, only limited guidance can be provided on the model parameter adjustments required to properly simulate the effects of implementing these actions. In particular, data are lacking on: 1) in-place hydraulic conductivities for different impermeable barrier materials; 2) changes in chemical mobility resulting from chemical injection and solution mining; 3) hydraulic properties and sorption characteristics of permeable treatment bed materials; and 4) changes in chemical susceptibility to degradation resulting from bioreclamation. As these technologies are implemented at different sites, laboratory and field experimental work should be conducted to obtain data useful for future modeling studies.

The modeling requirements for remedial actions are highly variable. If site contractors decide to use numerical modeling for remedial action evaluation, the modeling requirements for all potentially feasible actions should be considered as _early_ as possible in the Feasibility Study/Remedial Investigation. Early consideration will allow for the selection of a numerical model with the appropriate capabilities and level of sophistication. Early consideration will also lead to more efficiency in terms of data collection to support model application.

# 3. Migration and Fate Processes

3.1 OVERVIEW

The local environment surrounding an uncontrolled hazardous waste site can be subdivided into four zones, as defined below and in Figure 3.1.

1. Atmospheric Zone: Segment of the total atmosphere just above the land surface extending to areas adjacent to the disposal site.

2. Surface Zone: Parcel of soil from the land surface down to the root zone covering the waste site and the surrounding drainage area.

3. Unsaturated Zone: Parcel of soil with boundaries at the surface zone and the ground-water table; soil pores may contain varying amounts of air and water.

4. Saturated Zone: Soil and rock below the ground-water table, where all pores are filled with water and extending down to impermeable basement rock.

There are a number of processes that act to control the movement of contaminants within and between zones. These processes can be grouped as follows: 1) processes controlling movement within a zone (intra-zone), 2) processes controlling transfers between zones (inter-zone) and 3) processes controlling the transformation of chemicals. Table 3.1 lists the specific environmental processes that fall into each group and defines the affected zones and key parameters that influence each process. Figure 3.2 provides a schematic overview of a waste site and the role that selected intra- and inter-zone processes play in controlling water and waste migration.

The following subsections provide brief descriptions of each of the key processes listed in Table 3.1. These descriptions are not meant to be comprehensive. Rather, they are meant to

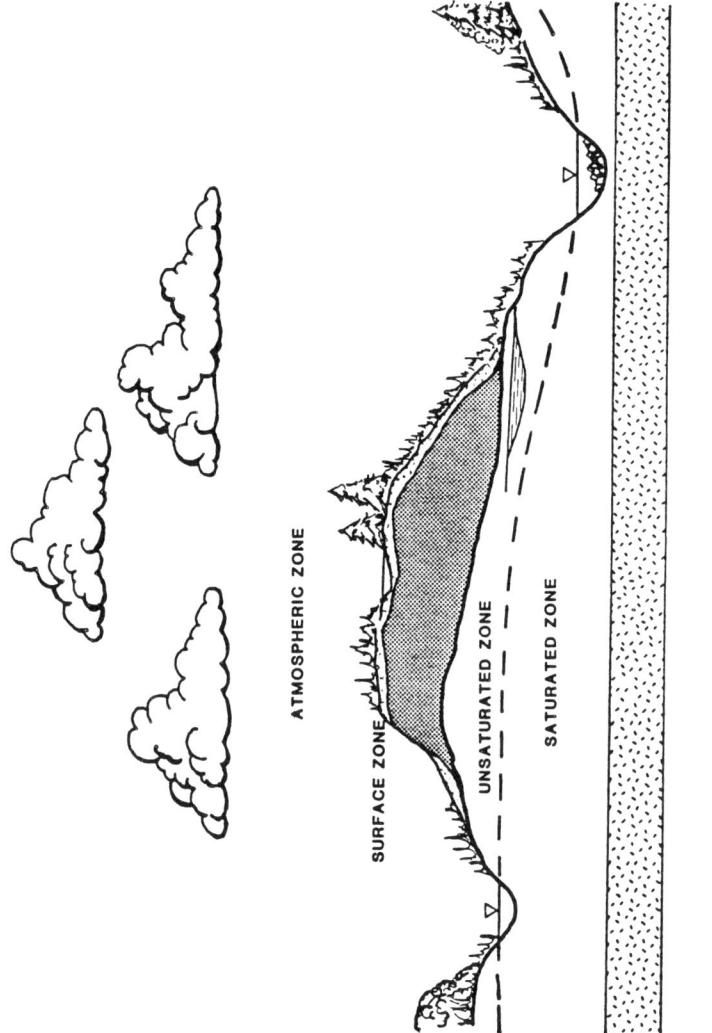

Figure 3.1 Local environment zones surrounding an uncontrolled hazardous waste site (adapted from JRB Associates, 1982).

TABLE 3.1 PROCESSES CONTROLLING THE MIGRATION AND FATE OF HAZARDOUS WASTE CONSTITUENTS

| Group | Processes | | Affected Zones | Key Parameters |
|---|---|---|---|---|
| Intra-Zone | Advection: | Runoff | Surface | Topography, vegetation, precipitation, soil moisture |
| | | Percolation | Unsaturated | Porosity, moisture content, infiltration rate |
| | | Ground-water flow | Saturated | Porosity, hydraulic conductivity, gradient |
| | Dispersion | | Unsaturated Saturated | Soil/rock heterogeneity, pore size distribution |
| | Erosion | | Surface | Topography, vegetation, soil type, precipitation |
| | Sorption/Retardation | | Surface | Organic content, sediment concentration |
| | | | Unsaturated Saturated | Organic content, porosity, chemical properties |
| Inter-Zone | Evapotranspiration | | Surface to air | Soil moisture, meteorologic conditions, vegetation |
| | Infiltration | | Surface to unsaturated | Soil moisture, precipitation, soil type, topography |
| | Drainage | | Unsaturated to saturated | Percolation rate, hydraulic conductivity, location of water table |
| | Volatilization | | Surface to air | Meteorologic conditions, chemical properties |
| Transformation | Photolysis | | Surface, air | Meteorologic conditions |
| | Hydrolysis, Oxidation, Chemical Reaction | | All | Chemical properties, soil properties |
| | Bio-degradation | | All except air | Chemical properties, bacterial activity |

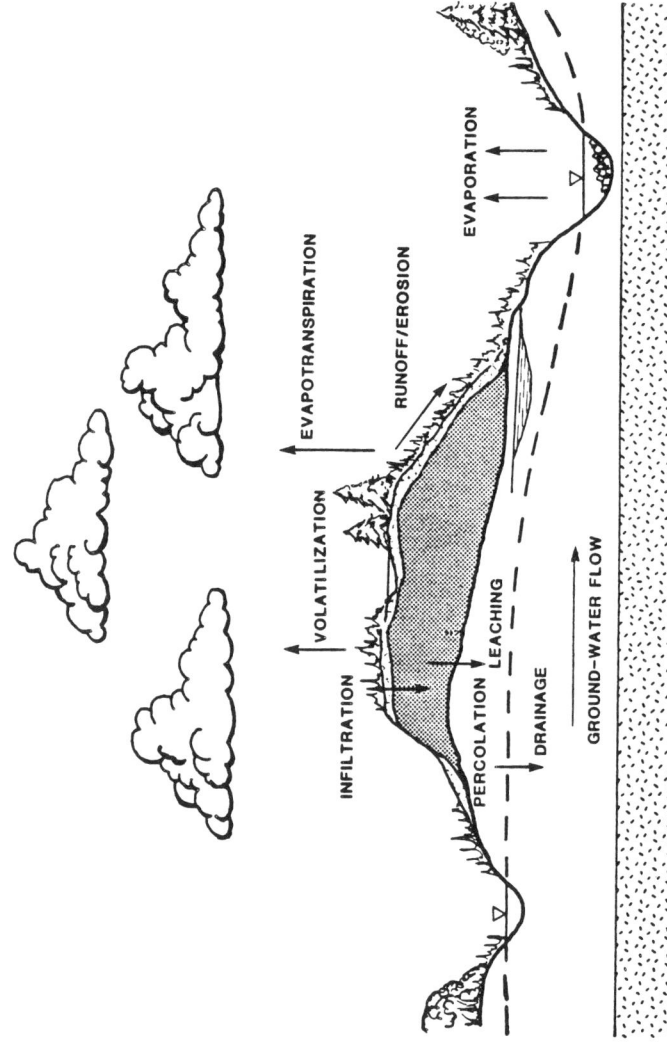

Figure 3.2  Schematic overview of a waste site and selected intra- and inter-zone processes affecting water and waste constituent migration (adapted from JRB Associates, 1982).

provide elementary definitions of processes to set the stage for subsequent discussions of remedial actions and the use of numerical models to evaluate them.

It is important to note that the overall focus of this report is on the use of models to analyze the effectiveness of surface, subsurface and waste control remedial action technologies, not surface water remedial action technologies. Volume 4 discusses the use of both simplified and numerical models to evaluate surface water technologies. Equivalent guidance on modeling of gas migration technologies does not currently exist.

3.2 PROCESSES CONTROLLING MOVEMENT WITHIN ZONES

3.2.1 Advection

Advection is the movement of a waste constituent as a result of bulk water movement. Water movement in the surface zone occurs in the form of overland flow or runoff, which can entrain chemicals and transport them to stream channels. Water movement in the unsaturated zone is primarily due to percolation or vertical movement through the soil profile. Passage of this water through waste materials can result in the leaching of waste constituents. Lateral movement occurs if the water reaches an impermeable strata or when the vertical flux exceeds the saturated permeability of a given strata. Water and associated contaminant movement in the saturated zone are largely in response to natural and man induced stresses (e.g., drainage from unsaturated zone and pumping).

3.2.2 Dispersion

Dispersion is a dilution process that occurs as a result of the spreading of a contaminant plume. In the surface zone, dispersion in overland flow is normally not considered due to the high velocities normally associated with runoff, and the relatively short distances runoff travels before entering some type of channel. In the unsaturated and saturated zones, dispersion can be of importance and occurs as a result of:

- o Molecular diffusion (in response to concentration gradients).

- o Mechanical dispersion: mechanical mixing on a microscopic scale due to tortuosity (erratic pattern of flow through pores), branching, and changes in pore size.

o  Heterogenous properties of the media: layering and differences in permeabilities and porosities on a megascopic scale.

Dispersion in the unsaturated and saturated zones is primarily a function of media properties and the scale at which the heterogeneities of an aquifer system are considered.

### 3.2.3 Erosion

Erosion is the detachment of soil particles by rain droplets and subsequent transport by overland flow originating upslope from or on a waste site. This process occurs only in the surface zone.

### 3.2.4 Sorption/Retardation

Sorption is the transfer of a portion of the soluble phase of a waste constituent to the surface of soil, rock or organic materials. In the surface zone, sorption is considered as a separate process which determines the amount of a waste constituent that will move with runoff, as opposed to with eroded soil materials. Thus, it is simply a partitioning process.

In the unsaturated and saturated zones, sorption is usually combined with a number of other processes to describe the delayed movement of certain waste constituents relative to that of water. The other processes include:

o  Filtration
o  Molecular diffusion into dead end pore spaces or fractures
o  Ion exchange
o  Reversible chemical reactions with other contaminants or the media
o  Precipitation/dissolution
o  Flocculation

Retardation is the general term used to describe the delay constituents will experience due to all of these processes.

## 3.3 PROCESSES CONTROLLING TRANSFERS BETWEEN ZONES

### 3.3.1 Evapotranspiration

Evapotranspiration collectively describes all processes which act to transfer water from the surface zone and unsaturated zone to the atmospheric zone. This includes evaporation from water, soil, snow, ice, and vegetation, as well as transpiration by plants.

### 3.3.2 Infiltration

Infiltration transfers water from the surface zone to the unsaturated zone through progressive wetting of underlying soils and movement due to hydrostatic pressure. The infiltration rate is usually high just after the onset of rainfall and decreases with time as soil pores become filled with water.

### 3.3.3 Drainage

Drainage is the transfer of water between the unsaturated and saturated zones. The hydraulics of drainage are complicated by the fact that the soil pores in the unsaturated zone contain both water and air. When the pores are almost completely filled with water (i.e., near saturation), water will tend to drain relatively freely in response to gravitational forces. As the water content decreases, capillary pressures increase and the amount of drainage that can occur decreases sharply. This inter-zone process also acts to transfer waste consitituents into the saturated zone.

### 3.3.4 Volatilization

The dominant mechanisms for vapor-phase transport of constituents from the surface zone to the atmospheric zone are gas phase molecular diffusion and convection by biogas venting and barometric pressure pumping. Emissions from ponded wastes are controlled primarily by volatilization at the air-water interface.

3.4 PROCESSES CONTROLLING TRANSFORMATION/DEGRADATION

Transformation refers to a number of chemical and biological processes that act to change or degrade a specific waste constituent. Quite often, the rate of transformation is controlled by one or two processes. Key transformation processes include photolysis, hydrolysis, oxidation, chemical reaction, and biological (microbial) degradation. Bio-degradation can also act to transfer contaminants into the atmospheric zone through respiration of the degrading organisms or changes from liquid to gas phase during chemical reactions.

# 4. Remedial Actions and Affected Processes

4.1 OVERVIEW

Remedial action technologies may be classified as surface control, subsurface control, and waste control. Surface control actions are directed at containing the waste. Subsurface control actions prevent contamination of the subsurface by directly containing the waste or by removal of contamination. Waste control actions are directed at reducing the source by direct removal or in-situ treatment. The remedial action technologies that are described herein were compiled from existing remedial action handbooks (JRB Associates, 1982 and SCS Engineers, 1982). Sample applications of many of these technologies to a hypothetical waste site, including costs, is provided by Tolman et al., (1978).

For the purpose of this report, the large number of available remedial action technologies have been condensed into fourteen "remedial measures" within the three control groups mentioned above. These measures are listed along the left axis of Table 4.1. This organization was based upon the similarity of design objectives of the individual technologies. For example, subsurface drains and ditches, as well as bottom liners, were grouped into one measure because they all are designed to control leachate migration (JRB Associates, 1982). Remedial actions designed to reduce airborne emissions, such as gas migration control and fugitive dust control, are not considered.

The purpose of this section is to: 1) briefly overview the design objectives of each of the measures listed in Table 4.1 and 2) identify which zones and processes are affected by these measures and how they are affected. This type of information is needed to support the guidance given in Sections 5 and 6.

Table 4.1 summarizes the discussion provided in this section. It lists each of the measures that will be discussed along one axis and the zones and processes discussed in the previous

210  Modeling Remedial Actions at Uncontrolled Hazardous Waste Sites

TABLE 4.1  PROCESSES AFFECTED BY DIFFERENT REMEDIAL MEASURES

Legend:
− = Mitigates Process in Relation to No Action
+ = Enhances Process in Relation to No Action

| REMEDIAL ACTIONS | SURFACE ZONE | | | | | INTER-ZONE TRANSFER | UNSATURATED ZONE | | | INTER-ZONE TRANSFER | SATURATED ZONE | | | |
|---|---|---|---|---|---|---|---|---|---|---|---|---|---|---|
| | Sorption | Degradation | Runoff | Evapotranspiration | Erosion | Infiltration / Percolation / Leaching | Dispersion | Retardation | Degradation | Drainage | Groundwater Movement | Dispersion | Retardation | Degradation |
| **SURFACE CONTROL** | | | | | | | | | | | | | | |
| Grading | | | ±  | | − | ± | | | | | | | | |
| Revegetation | | | − | + | − | − | | | | | | | | |
| Surface Water Diversion and Collection | | | − | | − | − | | | | | | | | |
| **SUBSURFACE CONTROL** | | | | | | | | | | | | | | |
| Capping and Top Liners | | | + | | | − | | | | | | | | |
| Seepage Basins and Ditches | | | − | | | + | − | − | − | | | | | |
| Subsurface Drains / Ditches / Bottom Liners | | | | | | | | | | + | − | | | |
| Impermeable Barriers | | | | | | | | | | + | − | | | |
| Ground Water Pumping | | | | | | | | | + | | | + | | |
| Interceptor Trenches | | | | | | | | | | | | + | + | |
| **WASTE CONTROL** | | | | | | | | | | | | | | |
| Permeable Treatment Beds | | | | | | + | + | | ± | | | | | +++ |
| Bioreclamation | | | | | | | | | | | | | | |
| Chemical Injection | | | | | | ± | + | + | ± | + | + | + | − | |
| Solution Mining Extraction | | | | | | | | | | | | | | |
| Excavation / Hydraulic Dredging | | | | | | | | | | | | | | |

section along the other axis. The extent of impact on the processes affected by each measure is denoted by either a (+) or (-). The former indicates that the measure tends to increase the effects of the process in terms of water and/or contaminant movement. The latter indicates that the measure tends to decrease the effects of the process. Figures 4.1 and 4.2 show, in plan view and cross-section, a hypothetical waste site prior to the installation of any remedial actions. Diagrams of each technology, including key process-related changes, are provided in Figures 4.3 - 4.11.

## 4.2 SURFACE CONTROL

Surface control measures such as grading, revegetation, and water diversion are designed to contain wastes reducing infiltration and limiting runoff from waste disposal sites. They can also reduce erosion, stabilize the surface of covered landfills, and protect receiving water quality. This is accomplished primarily by directing runoff away from a hazardous waste site or by containing contaminated runoff. Surface control measures mainly affect processes in the surface zone (i.e., runoff, evapotranspiration and erosion) and the transfer of water and waste constituents between the surface and unsaturated zones via infiltration. Figures 4.3 and 4.4 show how surface control actions affect different processes.

### 4.2.1 Grading

Grading is used to reshape the topography of landfills, affecting surface zone processes in one of two ways. Usually, the slope is increased and roughness is decreased to facilitate runoff and decrease infiltration. The higher velocities that result from these changes may cause increased erosion and entrainment of contaminated soil unless other measures are taken. A reduced slope and increased roughness may be desired in some arid environments where clay capping has been installed, to enhance infiltration and keep the cap pliable. Grading is often used in conjunction with surface sealing practices and revegetation.

### 4.2.2 Revegetation

This measure is used to stabilize the topsoil of a covered landfill. Revegetation decreases erosion by reducing the detachment of soil particles and reducing overland flow

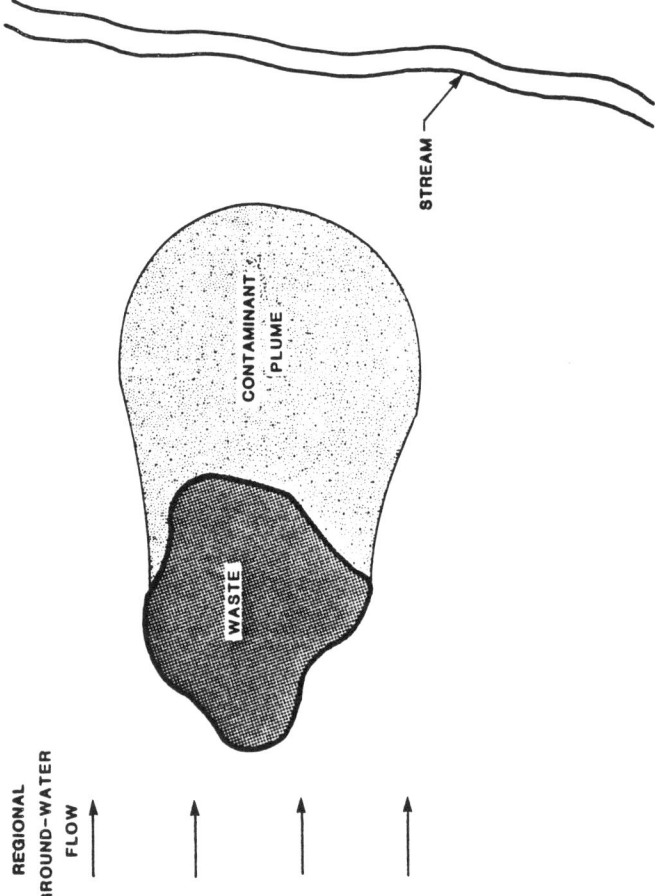

Figure 4.1 Hypothetical hazardous waste site (plan view).

Remedial Actions and Affected Processes    213

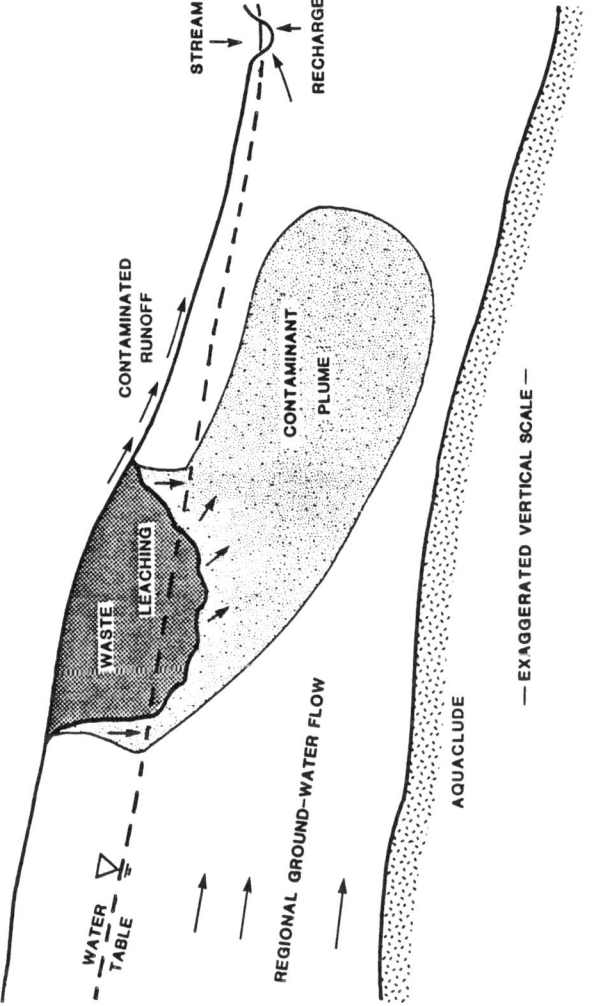

Figure 4.2  Hypothetical hazardous waste site (cross-section).

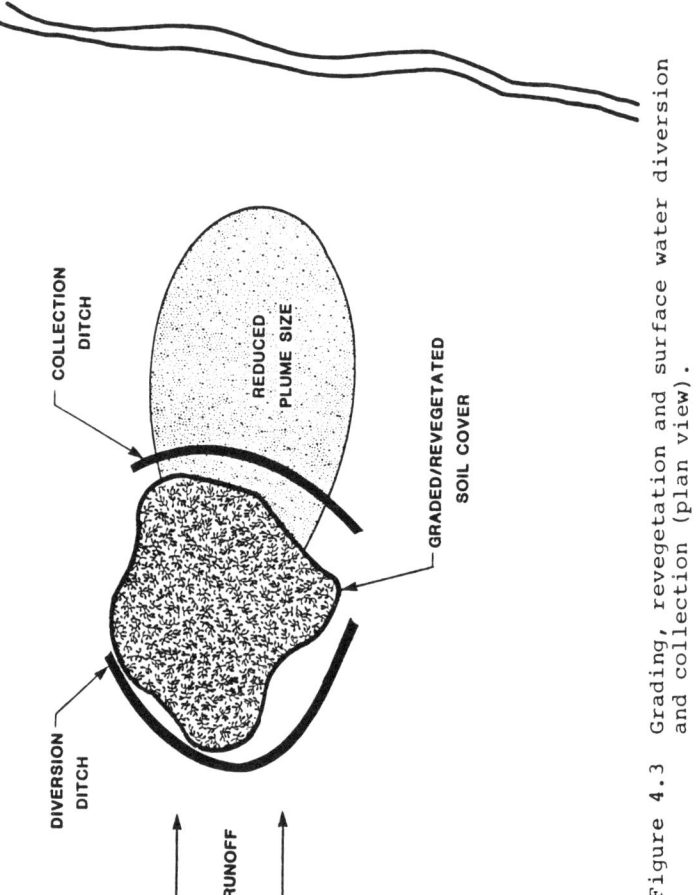

Figure 4.3 Grading, revegetation and surface water diversion and collection (plan view).

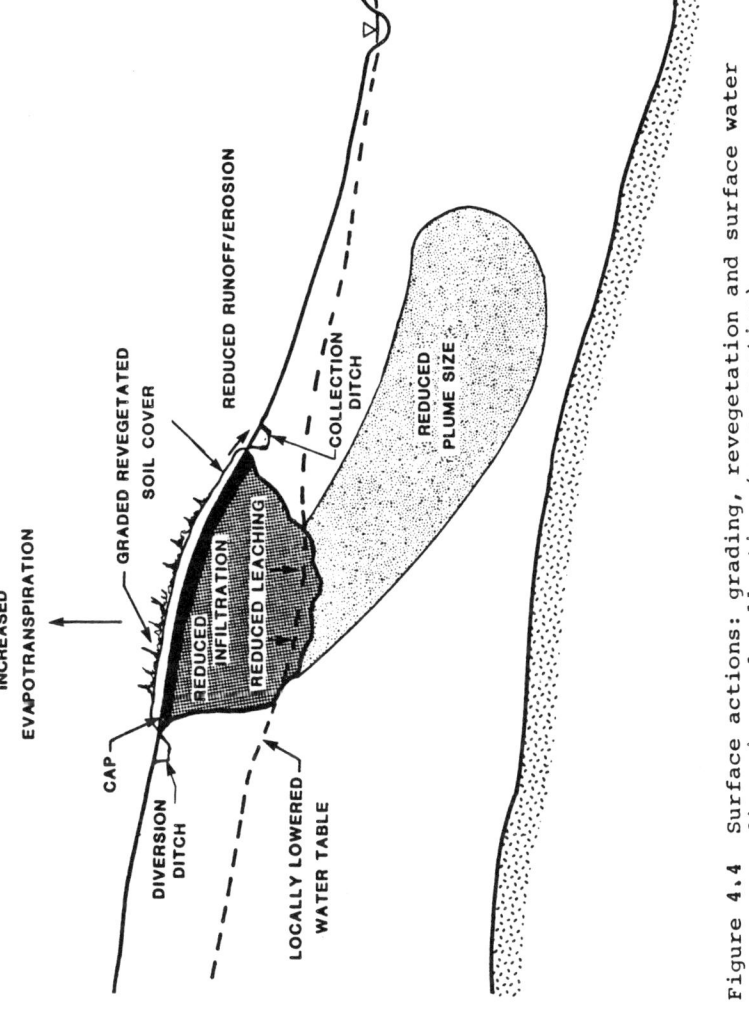

Figure 4.4 Surface actions: grading, revegetation and surface water diversion and collection (cross-section).

velocities. The introduction of vegetation also increases evapotranspiration and temporary water storage on the surface. The net effect on infiltration is site dependent, but will often be a decrease, particularly in less humid areas. Transpiration capacities, rooting depth characteristics, durability, preparation and planting characteristics affect the impacts of revegetation at a site.

### 4.2.3 Surface Water Diversion and Collection

This type of measure is designed primarily to route runoff away from a site. The techniques used to accomplish this include: dikes and berms, ditches, diversions, and waterways; terraces and benches; and chutes and downpipes. By removing surface water from the site, these measures reduce the depth of standing water on the surface, thereby limiting infiltration. Because overland flow is confined to collection channels, erosion can be controlled and transport of contaminated sediments can be eliminated.

### 4.3 SUBSURFACE CONTROL

Remedial measures that are included in this group primarily affect processes in the unsaturated and saturated zones, as well as processes acting to control the transfer of water and contaminants between the two zones. Two exceptions are the capping and top liner measure and the seepage basin and ditch measure, which also affect processes in the surface zone. The primary goals of subsurface control measures are to prevent leachate migration and ground-water contamination through diversion, containment or collection.

### 4.3.1 Capping and Top Liners

The placement of impermeable caps and top liners on waste disposal sites reduces infiltration, increases runoff, reduces erosion, and isolates the waste hydrologically. Cover soils, such as clay, that have low permeabilities and are erosion resistant are spread over the waste and then topsoil and vegetation are added to stabilize the cap as shown in Figures 4.3 and 4.4. Capping, because it reduces infiltration of water into the waste, minimizes the possibility that the waste might reach field capacity and subsequently begin to leach. Drainage to the saturated zone is also reduced. The water table beneath the site may be lowered and contaminant plume size may be reduced as a result of the decreased movement of leachate into the saturated zone.

### 4.3.2 Subsurface Drains and Bottom Liners

Subsurface drains are highly permeable trenches designed to collect leachate or infiltrating water in the unsaturated zone, thus reducing contamination of the aquifer. Drains may also be used to collect leachate trapped by bottom liners placed underneath the waste site. Bottom liners are low permeability barriers, usually composed of injected slurries or grout, that are installed underneath the waste site to retard the percolation of contaminants. Bottom liners may also be used to isolate the waste from a high ground-water table. As shown in Figure 4.5, leachate generation is minimized by these actions, resulting in a reduced plume area and lower concentrations.

### 4.3.3 Ground-Water Pumping

Figures 4.6 and 4.7 show several ways that ground-water pumping can be used alone or in combination with other measures. Pumping of ground water is designed to lower the ground-water table around the waste site or to contain a ground-water plume. The ground-water table may be lowered to: 1) prevent contaminated ground water from discharging to a receiving stream, 2) prevent direct contact between the waste and the aquifer (as shown in Figure 4.6), and 3) prevent leaky aquifers from contaminating other aquifers.

Ground-water pumping typically involves three steps: 1) pumping to remove contaminated water and/or depress the water table, 2) treatment of removed water to extract contaminants, and 3) recharge of treated water through either injection wells or seepage basins. A locally elevated ground-water table is often created as a result of recharging treated ground water. By depressing and elevating the ground-water table in the right locations, a plume of contaminated ground water can be isolated, as shown in Figure 4.7.

### 4.3.4 Interceptor Trenches

Interceptor trenches are used for the same purposes as ground-water pumping: to lower the water table around the site and to capture a plume by controlling the direction of ground-water flow (see Figures 4.6 and 4.7). They are characterized by high permeability material like gravel or a slotted drain pipe in a trench which intersects the saturated

218 *Modeling Remedial Actions at Uncontrolled Hazardous Waste Sites*

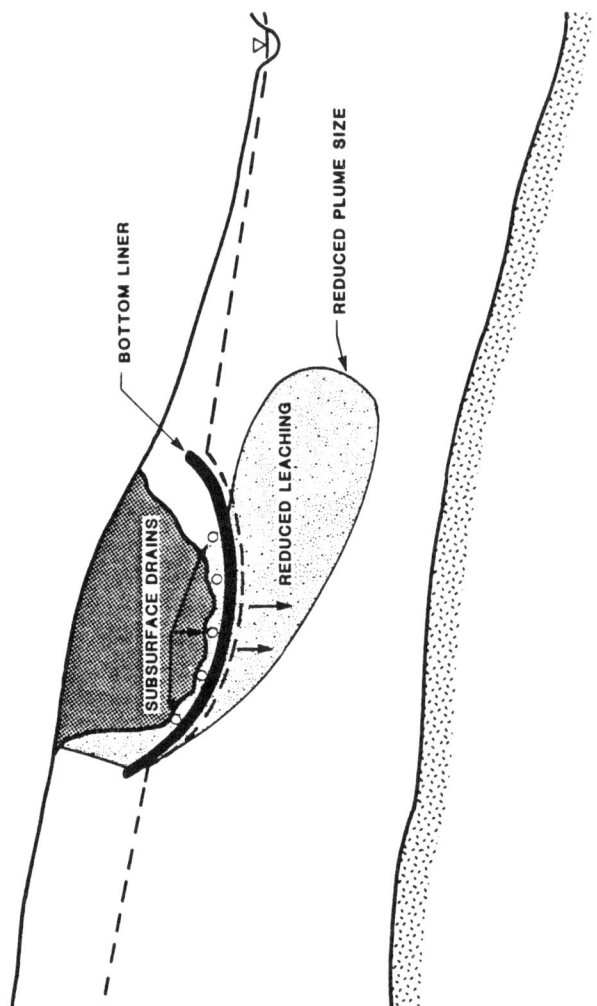

Figure 4.5 Subsurface drains and bottom liner.

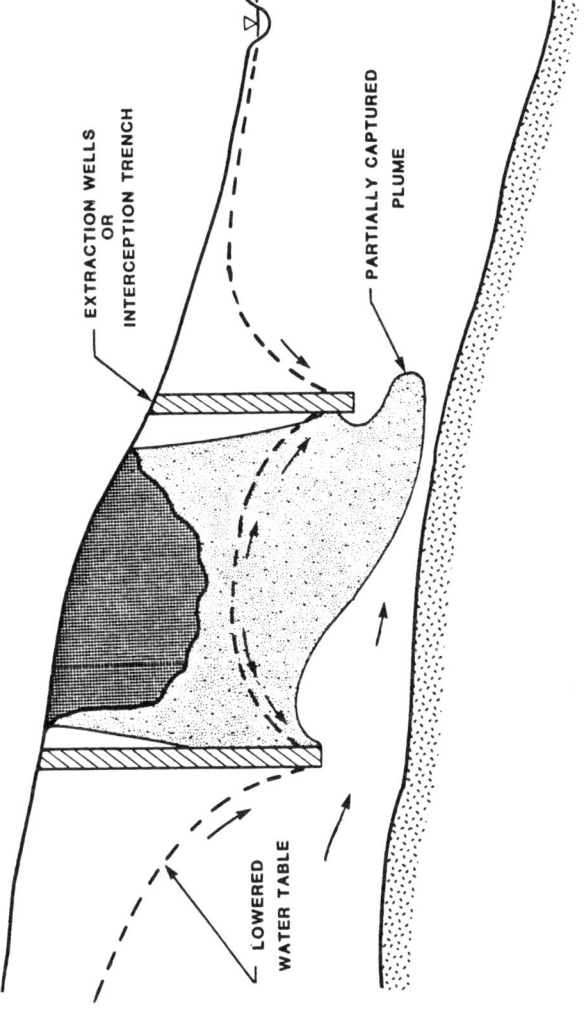

Figure 4.6 Extraction wells or interceptor trenches used to lower water table (cross-section).

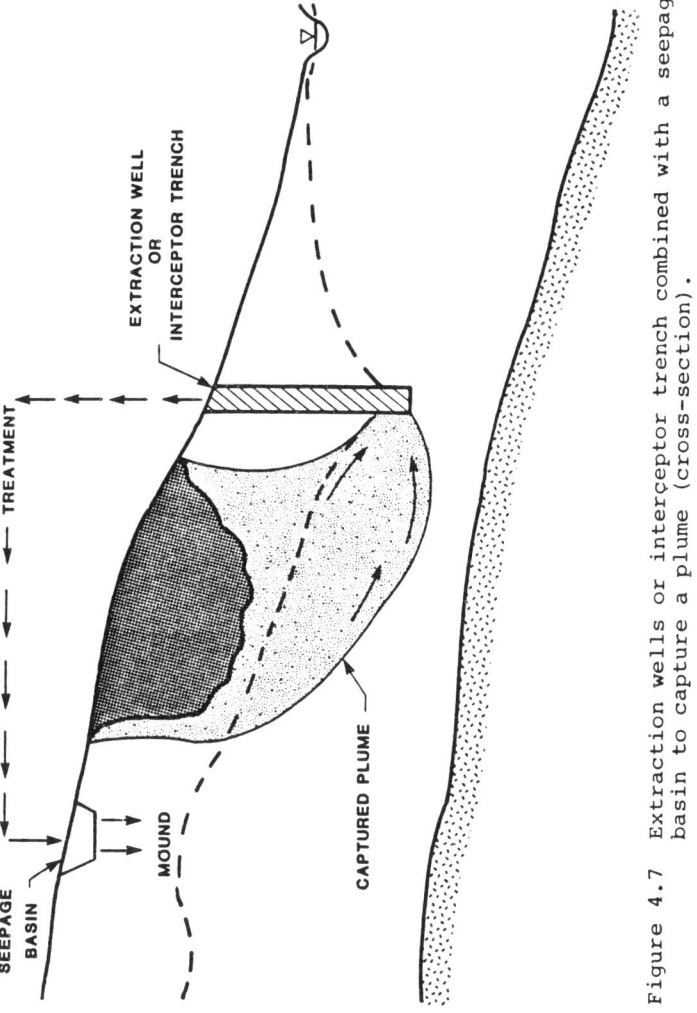

Figure 4.7 Extraction wells or interceptor trench combined with a seepage basin to capture a plume (cross-section).

zone. Water drains passively by gravitational forces into the
trench, thereby lowering the water table. They act in a
fashion similar to subsurface drains, but are used to capture
contaminated water in the saturated zone. Interceptor
trenches cause changes in processes similar to those caused by
extraction wells.

### 4.3.5 Seepage Basins and Ditches

Seepage basins and ditches are designed to recharge water from
surface collectors or extraction wells, drains and interceptor
trenches. They are sometimes used in conjunction with a
pumping system to change the ground-water table profile (see
Figure 4.7). The bottom of the basin itself is generally
constructed of highly pervious materials, allowing for
increased infiltration into the unsaturated zone. This
increase in infiltration leads to an increase in percolation
and drainage to the saturated zone. A localized rise in the
water table (i.e., mound) results. As a result, local changes
in ground-water flow directions can often be achieved.

### 4.3.6 Impermeable Barriers

Impermeable barriers are vertical walls of low permeability
material, such as bentonite slurry, cement, chemical grout, or
sheet piling, that are installed through the unsaturated zone
into the saturated zone. They are designed to either prevent
the migration of contaminated ground water away from a site or
to divert uncontaminated ground water away from a site.
Figures 4.8 and 4.9 show plan and cross-sectional views,
respectively, of a barrier completely surrounding the site.
Plume movement inside the barrier is reduced considerably.
However, the potential exists for the plume to escape if the
barrier is not keyed into an impermeable strata. Under
certain situations, partially penetrating or hanging barriers
can be used to reduce leachate generation by lowering the
water table.

### 4.4 WASTE CONTROL

Waste control measures are used to remove or treat wastes or
contaminated water and sediments. Removal may be accomplished
by excavation or hydraulic dredging. Treatment methods
include permeable treatment beds, bioreclamation, chemical
injection, and solution mining (extraction). These methods
are considered in-situ because treatment is accomplished

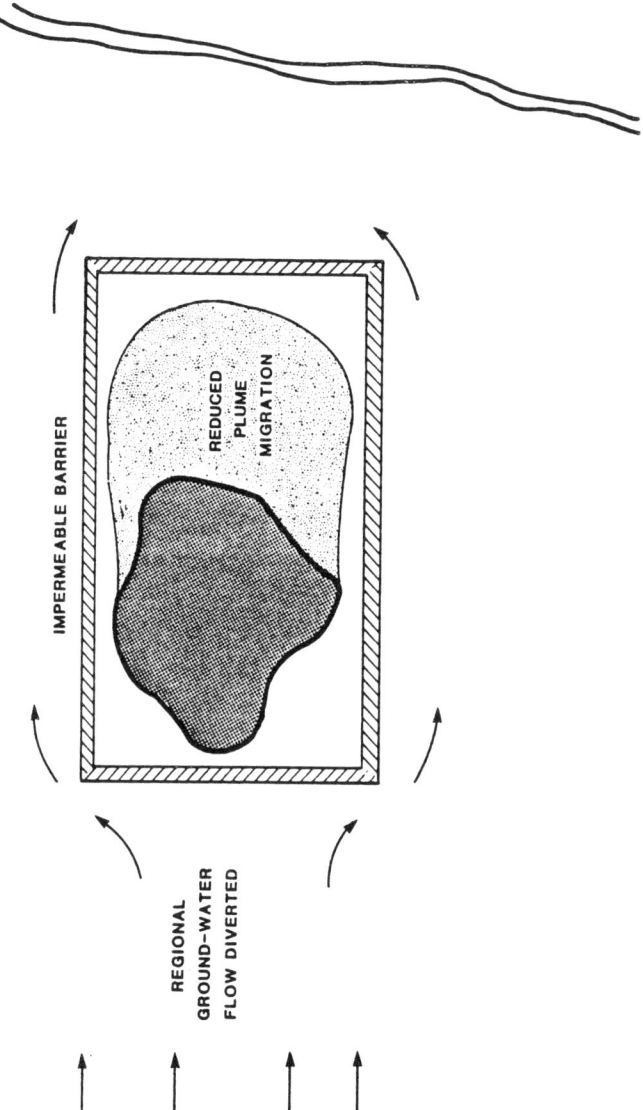

Figure 4.8 Circumferential impermeable barrier (plan view).

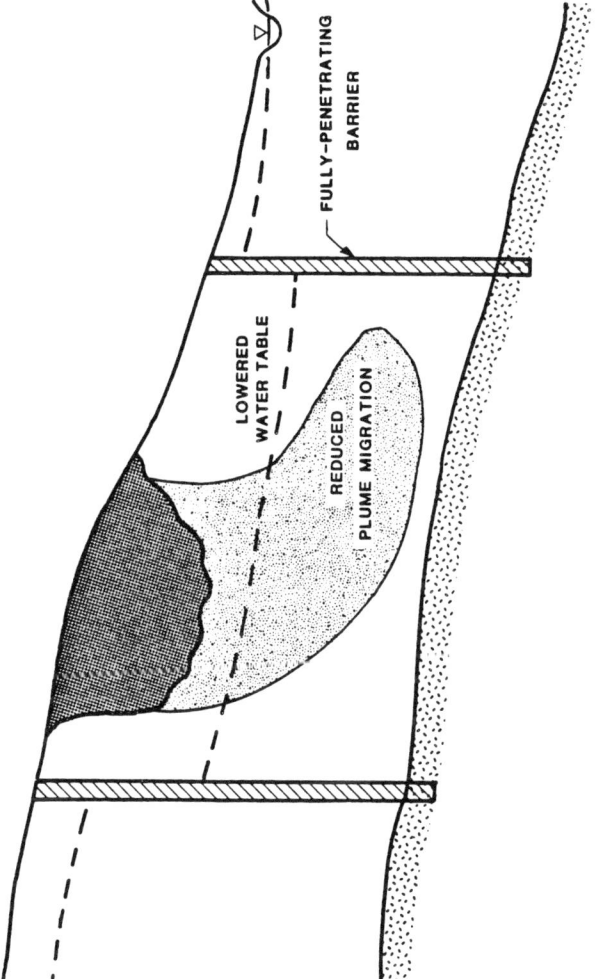

Figure 4.9 Circumferential impermeable barrier (cross-section).

within the landfill/lagoon or plume itself. On-site, as opposed to in-situ, treatment methods involve the extraction of contaminated water and above ground treatment, followed by disposal. Waste control measures have an effect on adsorption and degradation processes, as well as advection and dispersion processes.

### 4.4.1 Permeable Treatment Beds

Permeable treatment beds consist of limestone and/or activated carbon, and are placed vertically in the saturated zone, downgradient from a site, as shown in Figure 4.10. The objective is to remove contaminants from the ground water as it flows through the bed. Removal effectiveness may diminish with time, however, as the adsorptive capacity of the bed decreases or the bed becomes plugged. Permeable treatment beds mainly increase retardation and degradation processes in the bed itself.

### 4.4.2 Bioreclamation

In cases where the ground water has become contaminated with biodegradable pollutants such as hydrocarbons, bioreclamation may be considered as a remedial measure. It is an in-situ ground-water treatment method, involving the injection of microbial organisms, nutrients, and oxygen into a plume. The objective is to greatly accelerate the degradation of a pollutant. Bioreclamation acts to increase degradation processes. It can also locally affect ground-water movement and plume dispersion if injection and withdrawal rates are high enough to substantially modify the ground-water flows. Figure 4.11 shows the extent of in-situ treatment for a bio-reclamation system that includes an injection/withdrawal doublet.

### 4.4.3 Chemical Injection

Chemical injection is used to treat the waste in a landfill or lagoon, or in a contaminated saturated zone. It is usually applied to sites with wastes well defined in both location and chemical composition with shallow landfill or lagoon depths, and where the vertical and horizontal extent of the contamination is small (JRB Associates, 1982). The objective of the method is to immobilize or destroy a pollutant. The effect of this measure is to substantially increase local retardation and degradation processes. Figure 4.11 shows the

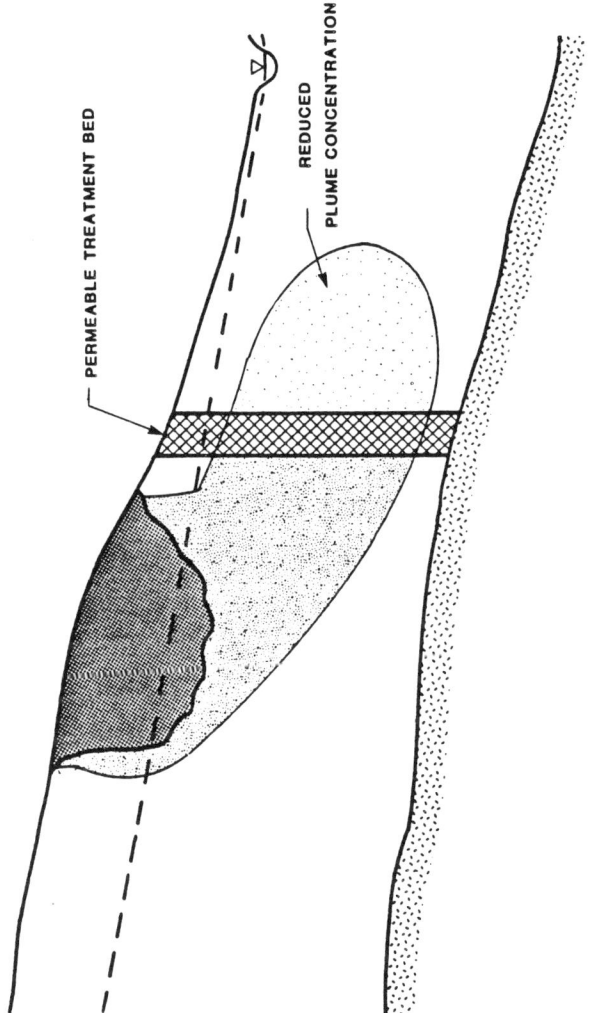

Figure 4.10 Permeable treatment bed or chemical/microbe injection.

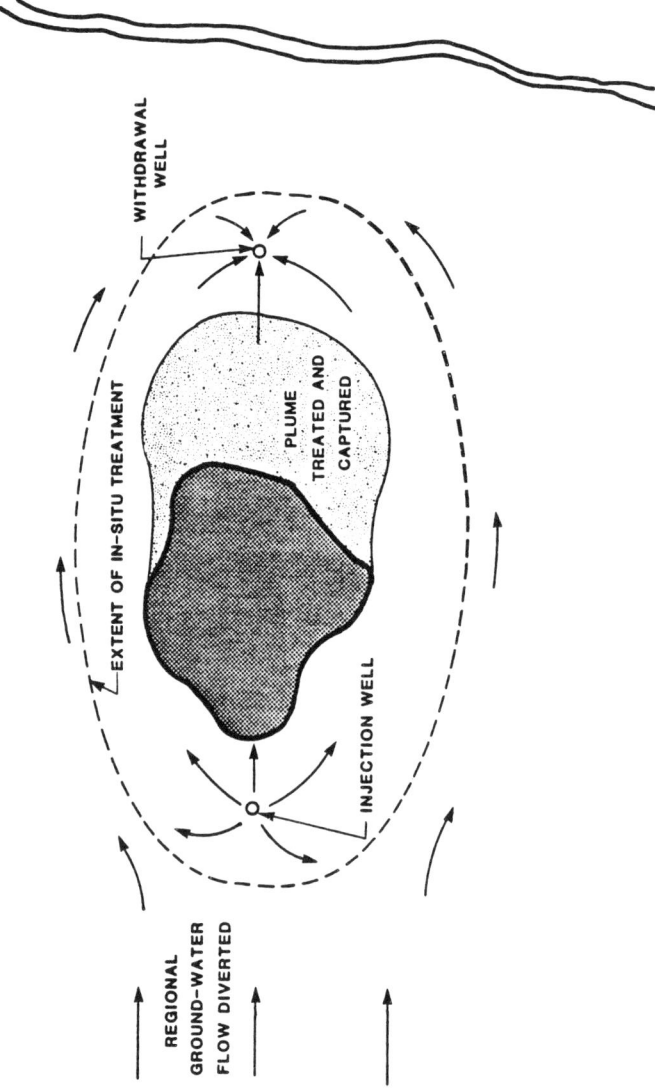

Figure 4.11 Bioreclamation and chemical injection using an injection/withdrawal doublet (plan view).

extent of treatment for a chemical injection system composed of an injection/withdrawal doublet.

### 4.4.4 Solution Mining (Extraction)

Solution mining is similar to chemical injection in that both methods chemically alter the pollutant. However, solution mining involves the injection into a landfill of a chemical solvent, which desorbs or frees the pollutant so that it may be mobilized in a larger leachate flow. The leachate is then collected by interceptors and/or well points (see Figure 4.12). The objective is to <u>increase</u> the mobility of the contaminant. Adequate confinement and collection of the resultant leachate is necessary to prevent increased aquifer contamination. Contaminant movement is also increased by solution injection and collection.

### 4.4.5 Excavation and Hydraulic Dredging

Excavation and hydraulic dredging involve the removal of the waste source itself. Hydraulic dredging may be used to remove liquids and/or sludges from lagoons or surface impoundments. After the waste area has been excavated or dredged, it may be backfilled and capped to control infiltration.

Depending on the permeability of the backfill material and other site restoration actions, infiltration to the unsaturated zone may increase or decrease. This will in turn lead to a decrease or increase in water percolation in the unsaturated zone and drainage to the saturated zone. Since the measure leads to the removal of waste materials, there should be a major decrease in leachate generation and plume size.

228  *Modeling Remedial Actions at Uncontrolled Hazardous Waste Sites*

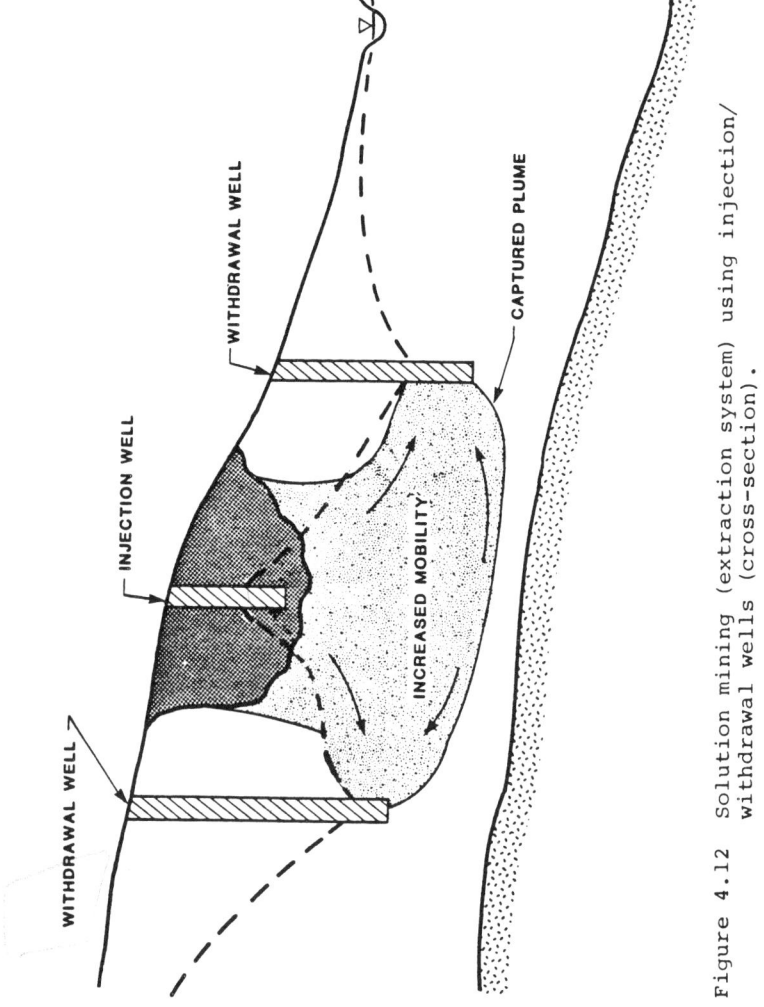

Figure 4.12  Solution mining (extraction system) using injection/withdrawal wells (cross-section).

# 5. Numerical Model Application Guidelines

## 5.1 OVERVIEW

Numerical models of the subsurface and surface environments provide capabilities which both complement and exceed those of field data analysis and simplified methods. In contrast to simplified methods, numerical models approximate process equations using finite difference or finite element solution techniques that make it possible to represent important spatial and temporal variations in site conditions. Along with this benefit, however, comes the cost of gathering the field data required to describe key variations. Consequently, a trade-off must be made between the ease of solution, computational accuracy, limited resolution and limited applicability of simplified methods and greater resolution, more general applicability, increased complexity and increased costs for numerical models. Key attributes of numerical models can be summarized as follows:

1. Fewer simplifying assumptions are required, although the simplicity and computational efficiency of the solution algorithm depend, in part, on assumptions made.

2. Values of key quantities (e.g., velocity and chemical concentration) are computed at discrete space and time intervals selected by the user. These intervals can be adjusted to achieve the accuracy and specificity required by site conditions and the problem being addressed.

3. Multiple independent variables (e.g., velocity, temperature, and chemical concentration) can be simulated simultaneously, including interactions between these variables.

4. Numerical solutions to the governing equations are approximate and subject to computational errors due to truncation, roundoff and numerical dispersion.

Choice of solution scheme can have a substantial effect on these errors.

5. Resources required to implement numerical models depend on the dimensionality, resolution, number of independent variables being predicted, and solution scheme. Required resources include: user expertise in modeling, field data, personnel time, and computer facilities. It is reasonable to expect that needed resources will be two to ten times those required for simplified model applications.

A number of authors provide overviews of numerical models and their use in the analysis of surface and subsurface problems, including Mercer and Faust (1981), Javandel et al., (1984), Bachmat et al., (1978), Orlob (1971) and Donigian (1981). The reader is refered to these sources for more information on model theory, structure, implementation and use.

Numerical models, as a result of the above attributes, are most appropriately used for the analysis of physical and chemical processes and site conditions which cannot be adequately represented with simplified methods. Situations which may justify numerical models include:

1. Local and/or off-site media properties which vary significantly with location or direction causing complex flow and transport conditions;

2. Highly variable, discontinuous or geometrically complex boundary conditions (e.g., mixed flow and no-flow boundaries) which require detailed representation;

3. Time varying sources, sinks, or boundary conditions (e.g., seasonal fluctuations in river water levels or infiltration rates) which strongly influence flow and transport; and

4. Remedial actions which, when implemented, result in one or more of the conditions listed previously (e.g., impermeable barriers).

Volume 1 of this report presents a methodology for determining when numerical models should be used for analysis of surface and subsurface remedial actions.

## 5.2 NUMERICAL MODEL CAPABILITIES

A broad spectrum of numerical models, potentially applicable to remedial action assessment, have been developed. For ease of discussion, models, are often classified by the types of problems they can solve and the solution techniques used. Categories include: solution domain or zone, independent variables considered and numerical solution technique.

As was noted in Section 3, the environment in the immediate vicinity of a waste site can be divided into four zones: 1) atmospheric, 2) surface, 3) unsaturated, and 4) saturated; surface water bodies (e.g., lakes and rivers) are considered to be a separate zone. Water and contaminant movement in each of these zones is controlled by different processes; the governing equations are enough different so that separate solution schemes are usually required. The capabilities of models for the atmospheric zone are outside the scope of this volume and models for surface water bodies are discussed in Volume 4. Models for the remaining zones or solution domains are discussed below.

Independent variables can be grouped as flow-related and transport-related. Flow models solve the applicable momentum, continuity and pressure equations to yield estimates of fluid movement and storage. Transport models use estimates of fluid movement to predict chemical migration and fate. Flow and transport calculations are often performed separately and in sequence. Some code designers have chosen to have entirely separate codes for the two computations so that each code is as simple and efficient as possible. This latter approach assumes that chemicals and other constituents being transported will not affect fluid flow. Situations involving chemicals that are denser or less dense than water, usually cannot be simulated in this way.

Numerical solution procedures fall into two general categories: finite difference methods (FDM) and finite element methods (FEM). Other methods, such as integrated finite difference and method of characteristics combine attributes of FDM and FEM, but are generally referred to as finite difference methods because of the way the solution domain is represented. Mercer and Faust (1981) provide a brief discussion of all of these techniques, including references to in-depth treatments of each technique.

### 5.2.1 Surface Zone Models

Section 3 of this volume provides a description of key processes controlling water and chemical movement in the surface zone. In essence, the surface domain extends from the surface of a hazardous waste site to the root zone and downslope to a receiving water body (see Figure 3.1).

Most readily available surface zone models represent the surface zone with single or multiple "land segments," each having uniform properties (e.g., slope, surface roughness, and vegetative cover). The most comprehensive surface zone models also simulate soil moisture in the unsaturated and saturated zones to obtain improved estimates of infiltration rates. Less comprehensive models use empirical relationships for antecedent soil moisture. Numerical solution schemes for these models use simple finite difference techniques. The majority of surface zone models are event-based. That is, they simulate one hydrologic event at a time. Models which provide continuous simulation of runoff and soil moisture, as well as water quality, require efficient data manipulation and storage routines due to the large number of parameters and the frequent time steps needed to represent runoff and infiltration processes. Several models provide some data management capability but only one, the Hydrologic Simulation Program - FORTRAN (HSPF) (Johanson et al., 1981), provides comprehensive data manipulation and storage capabilities. Donigian (1981) discusses the evolution of surface zone models and existing model capabilities and limitations in more detail.

Volume 1 presents one approach to the selection of models for remedial action evaluation. Examples of representative surface runoff models which are potentially suitable for remedial action evaluation are discussed below as a starting point for those interested in applying such models. More detailed listings of models and a discussion of model attributes can be found in Onishi et al., (1983), and Donigian (1981). Table 5.1 summarizes the capabilities of five surface zone models, fifteen saturated and seven unsaturated zone models. The characteristics of three particularly versatile codes are shown in Table 5.2. These three models are described briefly below.

HELP (Schroeder et al., 1984a and 1984b) estimates daily water movement on the surface and through a landfill by partitioning precipitation (and runoff entering the site) into runoff, evapotranspiration, infiltration, and lateral drainage. The SCS Curve Number method is used to estimate runoff on a continuous (daily) basis, using a soil moisture accounting procedure to determine infiltration. The landfill is divided into discrete layers and moisture is routed vertically from one layer to the next using Darcy's Law. Although the original version of HELP does not simulate leachate quality, Bicknell (1984) has modified HELP to simulate chemical losses from a landfill. Both leaching and volatilization losses can be estimated. HELP has been used in the analysis of existing landfills and the design of new sites.

TABLE 5.1  GENERAL CAPABILITIES OF SELECTED SATURATED, SURFACE AND UNSATURATED ZONE MODELS

| MODEL NAME (References) | LAND SEGMENT (Single/Multiple) | SURFACE ZONE | | | | | UNSATURATED ZONE | | | | | SATURATED ZONE | | | | MAJOR CODE LIMITATIONS |
|---|---|---|---|---|---|---|---|---|---|---|---|---|---|---|---|---|
| | | Sorption | Degradation | Erosion | Dimension (1,2,3) | | Dispersion | Sorption | Degradation | Dimension (1,2,3) | | Dispersion | Sorption | Degradation | Documentation | |
| **SURFACE ZONE MODEL** | | | | | | | | | | | | | | | | |
| HSPF (EPA) | (M) | X | X | X | 1 | | | | | | | | | | C | • Single segment and no sediment-pollutant adsorption |
| SWMM (EPA) | (M) | X | X | X | 1 | | | | | | | | | | C | |
| CREAMS (USDA / Corps of Engrs.) | (S) | X | X | X | 1 | | | | | | | | | | C | • Single segment and does not consider erosion |
| SEASOIL (A.D. Little, Inc.) | (S) | X | X | X | 1 | | | | | | | | | | C | |
| HELP (EPA / Corps of Engrs.) | (S) | | | | | | | | | | | | | | C | |
| **UNSATURATED ZONE MODEL** | | | | | | | | | | | | | | | | |
| FEMWATER / FEMWASTE (ORML) | | | | | | | X | X | X | 2 | | | | | C | |
| TRUST / MILTRAN (LBL / Battelle) | | | | | | | | X | X | 2 | | | | | I | |
| COLUMN TRANSPORT WITH SORPTION (Kipp, Kenneth L.; England) | | | | | | | X | X | X | 1 | | | | | | • 1 dimensional |
| ODMOD (Argonne National Lab) | | | | | | | | X | X | 1 | | | | | | • 1 dimensional, no dispersion and degradation |
| NMODEL (Univ. of Florida) | | | | | | | X | | | 1 | | | | | | • 1 dimensional, no adsorption and degradation |
| PERCOL (Battelle) | | | | | | | | X | X | 1 | | | X | | | • 1 dimensional, steady state, no dispersion and adsorption |
| PRZM (EPA / Athens) | (S) | X | | X | 1 | | X | X | X | 1 | | | | | | • 1 dimensional |

Legend: (M) = Multiple land segments
(S) = Single land segment
X = Considered
C = Complete documentation
I = Incomplete documentation or user's guide
? = Unknown

(continued)

TABLE 5.1 (continued)

| SATURATED ZONE MODEL | LAND SEGMENT (Single/Multiple) | SURFACE ZONE | | | | | | UNSATURATED ZONE | | | | SATURATED ZONE | | | | DOCUMENTATION | MAJOR CODE LIMITATIONS |
|---|---|---|---|---|---|---|---|---|---|---|---|---|---|---|---|---|---|
| | | Adsorption | Degradation | Erosion | Dimension (1,2,3) | | | Dispersion | Adsorption | Degradation | Dimension (1,2,3) | Dispersion | Adsorption | Degradation | Dimension (1,2,3) | | |
| FEWA / FEMA (ORNL) | | | | | | | | | | | | X | X | | 2 | C | • Being analytical the model has limited spatial resolution |
| SWIFT (Intera) | | | | | | | | | | | | X | X | | 3 | C | |
| MCTM (Intera) | | | | | | | | | | | | X | X | | 3 | C | |
| FE3DGW / CFEST (Battelle) | | | | | | | | | | | | X | X | | 3 | C | |
| AT123D (ORNL) | | | | | | | | | | | | X | X | X | 3 | I | • 1 dimensional unsteady state or 2 dimensional steady state |
| PLASM (Prickett & Lonnquist) | | | | | | | | | | | | X | | | 1 | C | • No adsorption and degradation |
| WASTE (Analytical Science Corp.) | | | | | | | | | | | | X | X | X | 2 | C | • No adsorption and degradation |
| GWSIM-II (Texas Dept. of Water Resources) | | | | | | | | | | | | | | | 2 | | |
| MOC (Konikow & Bredehoeft, USGS) | | | | | | | | | | | | X | | | 2 | C | • No adsorption and degradation |
| GROUNDWATER COMPUTER PACKAGE (Marlon-Lambert, J.; Canada) | | | | | | | | | | | | X | X | | 2 | C | • No documentation |
| PATHS (Battelle) | | | | | | | | | | | | X | X | | 2 | C | • Being analytical the model has limited spatial resolution |
| TRANSCOL / FRACSOL (Prickens, J.F.; Canada) | | | | | | | | | | | | | | | 1 | | • 1 dimensional, no dispersion and degradation |
| GETOUT (Burkholder, et al) | | | | | | | | | | | | | X | | 1 | C | • 1 dimensional |
| NEWSAM (Ledoux, E.; France) | | | | | | | | | | | | | X | | 1 | C | • 1 dimensional, no dispersion and degradation |
| VTT (Battelle) | | | | | | | | | | | | | | | 2 | C | • 2 dimensional, no pollutant transport sub-module |

Legend: 
(M) = Multiple land segments
(S) = Single land segment
X = Considered
C = Complete documentation
I = Incomplete documentation or user's guide
? = Unknown

TABLE 5.2 DETAILED CAPABILITIES OF SELECTED SURFACE, UNSATURATED AND SATURATED ZONE MODELS

| CODE NAME (Reference) | LAND SEGMENT (Single/Multiple) | SURFACE ZONE |  |  |  |  |  | UNSATURATED ZONE |  |  |  |  |  | SATURATED ZONE |  |  |  |  | SPECIAL CONSIDERATIONS |  |  |
|---|---|---|---|---|---|---|---|---|---|---|---|---|---|---|---|---|---|---|---|---|---|
| | | Sorption | Degradation | Runoff | Evapo-transpiration | Erosion | INTER-ZONE TRANSFER Infiltration DIMENSION (1,2,3) | Percolation/Leaching | Dispersion | Retardation | Degradation | INTER-ZONE TRANSFER Drainage DIMENSION (1,2,3) | Groundwater Movement | Dispersion | Retardation | Degradation | Variable Spatial Resolution | Case Studies | Documentation | User's Guide |
| **SURFACE CODE** | | | | | | | | | | | | | | | | | | | | | |
| HSPF (EPA) | (M) | x | x | x | x | x | I | | | | | | | | | | | x | x | x | |
| CREAMS (USDA / Corps of Engrs.) | (S) | x | x | x | x | x | I | | | | | | | | | | | x | x | x | |
| HELP (EPA) / Corps of Engrs.) | (S) | | | x | x | | 1 | x | | | | 1 | | | | | | x | x | x | |
| **UNSATURATED ZONE CODE** | | | | | | | | | | | | | | | | | | | | | |
| FEMWATER / FEMWASTE[1] (ORNL) | | | | | | | I/S | x | x | x | | 2 | | | | | | | | | |
| TRUST / MLTRAN (LBL / Battelle) | | | | | | | I/S | x | x | x | | 2 | x | x | x | x | | | | | |
| **SATURATED ZONE CODE** | | | | | | | | | | | | | | | | | | | | | |
| FEMA / FEMA (ORNL) | | | | | | | | | | | | | | x | x | x | | x | O | x | x |
| SWIFT (Intera) | | | | | | | | | | | | | | x | x | x | x | x | x | x | x |
| MCTM (Intera) | | | | | | | | | | | | | | x | x | x | x | x | x | x | x |
| FE3DGW / CFEST (Battelle) | | | | | | | | | | | | | | x | x | x | x | x | x | x | x |
| PLASM (Prickett & Lonnquist) | | | | | | | | | | | | | | x | | | | x | x | x | x |

Footnote:
1. Flow Model / Transport Model

Legend:
(M) = Multiple Land Segment
(S) = Single Land Segment
I = Infiltration
S = Seepage (handling of seepage pond)

x = Considered
O = Case studies - unpublished
    Documentation - only for flow model
                    not for transport code
    User's Guide - only for flow model
                    not for transport code

CREAMS (Knisel, 1980) simulates surface hydrologic processes, either continuously using the Green and Ampt formulation or for discrete events using the SCS Curve Number approach. Like HELP, it provides for only a single land segment and cannot represent spatial variations in hydrologic conditions. It simulates most of the important processes, including sediment production and transport.

HSPF (Johanson et al. 1981) is the most recent version of a family of watershed hydrology and quality models which have the Stanford Watershed model as a base. HSPF simulates surface and subsurface processes for multiple land segments and is capable of representing complex hydrologic and chemical transport conditions. Additional modules simulate transport in surface water bodies and the interactions between surface water and subsurface water and chemical movement. A sophisticated data base management system is included as part of this model.

### 5.2.2 Unsaturated Zone Models

For our purposes, the unsaturated zone begins at the base of the root zone and extends to the water table (or capillary fringe, if considered). Because moisture content is less than porosity, the properties influencing water movement in this zone (moisture content and hydraulic conductivity) depend upon pressure head. Water movement is predominantly vertical. Soil heterogeneities can result in lateral migration of water and contaminants around clay layers and other discontinuities.

Available unsaturated zone models vary widely in their capabilities and characterisitcs. While two-dimensional, finite element codes appear to be the most common, finite difference codes are also readily available. Separate codes for flow and transport calculations are common, due to the complexity of water movement. A number of codes can simulate both unsaturated and saturated conditions and may be potentially useful where fluctuating water table elevations or perched saturated conditions are important. The most complex models also simulate multi-phase flow and/or heat transfer and may be appropriate if detailed modeling of multi-phase transport is required.

The number and diversity of unsaturated zone codes often makes selection difficult. Brief discriptions of several codes are given below as a starting point. Surveys and critiques of available codes can be found in Kincaid et al. 1984, Nelson et al. (1982), and Oster (1982). The International Ground Water Modeling Center (IGWMC) operated by Holcomb Research

Institute, Butler University in Indianapolis, Indiana provides a clearinghouse for information on the capabilities of a number of different unsaturated zone models. Table 5.1 summarizes the capabilities of seven codes while the characteristics of two particularly versatile codes are presented in Table 5.2.

TRUST and MLTRAN (Narasimhan and Witherspoon, 1976, Reisenauer et al. 1981 and 1982) are companion flow transport codes for variably saturated media. TRUST is a two-dimensional integrated finite difference code. TRUST considers soil processes such as deformation, as well as the wetting-drying front problem. TRUST output is formatted for direct input to MLTRAN. MLTRAN was originally developed for the evaluation of low level radioactive wastes and computes travel path and travel time for water and a chemical. Retardation and degradation of constituents are simulated, but dispersion is not included.

FEMWATER and FEMWASTE (Yeh and Ward, 1979 and 1981) are companion flow and transport codes that simulate two-dimensional unsaturated/saturated ground-water systems. Boundary condition options allow representation of seepage from ponds, as well as surface infiltration. The simulated plane can be vertical (x-z) or horizontal (x-y), allowing simulation of lateral drainage. FEMWASTE represents all important contaminant transport processes, including dispersion. Heterogeneous soil properties, including the effects of remedial actions, can be represented using a variable finite element grid.

### 5.2.3 Saturated Zone Codes

The saturated zone extends from the water table downward to underlying basement rock. Fluid and contaminant flow are controlled by pressure head and hydraulic conductivity, and are fundamentally three-dimensional.

A broad spectrum of saturated zone models are currently available, varying from one-dimensional finite difference flow codes to three-dimensional finite element codes that include multiple phases, temperature effects, transport in fractured media and geochemical and biological reactions. Numerous surveys and critiques of saturated zone models are available, including, Kincaid et al. (1984), Thomas et al., (1982), Javandel et al. (1984), Gelhar (1977), Bachmat et al. (1978), van Genuchten (1978a), Anderson (1979), Grove and Kipp (1980), Knox and Canter (1980), Lappalla (1980), Moiser et al. (1980), SAI (1981) and Koines (1982). The first two references provide detailed reviews of a limited number of models, while the others are more comprehensive inventories. Again, IGWMC provides information on a number of saturated zone models. Brief descriptions of five saturated zone models (see Table 5.2) are provided below as examples of codes potentially suitable for remedial action evaluation. Table 5.1 gives general characteristics of 15 potentially applicable codes.

PLASM, or the "Random-Walk" Solute Transport Model, developed by Prickett and Lonnquist (1981) is a two-dimensional (x-y), transient model. It considers all important saturated zone processes and inter-zone transfer processes. Judgement is needed to arrive at an acceptable solution, since improper discretization may cause the predicted concentrations to be greater than the initial concentration. The "lumpy" character of output (expressing concentration in terms of number of particles) requires computer plotting and smoothing routines to draw meaningful results. Such subroutines had not been incorporated into the computer code at the time of this review.

FE3DGW/CFEST (Gupta et al., 1979 and 1982) are two finite element models which can simulate two-dimensional or three-dimensional systems which are complex and multi-layered. Flexible boundary conditions, an easily defined and modified finite element structure, and the capability to model point sources and sinks make this model both powerful and adaptable. CFEST now simulates both retardation and degradation.

HCTM, or the Hydrologic Contaminant Transport Model, developed by Intera, Inc., considers all the required saturated zone processes such as adsorption, degradation and dispersion, as well as inter-zone transfer processes. It handles heterogeneous soil properties and provides variable spatial resolution. It is a proprietary model and is not available to the public, except by purchase.

SWIFT is generically related to HCTM. It was developed by Intera and Sandia National Laboratories for the Nuclear Regulatory Commission from the earlier USGS model SWIP (predecessor of the DWDM - Deep Well Disposal Model). The model, is more complex and costs more to run than the HCTM code (Lantz, R., personal communication) as it couples a heat transport sub-module to the original fluid and contaminant transport codes. Unlike HCTM, SWIFT is not proprietary and iswell documented with a user's manual and self teaching guide (Dillion, et. al., 1978; Finley and Reeves, 1968; Reeves and Cranwell, 1981).

FEWA/FEMA has been developed by Oak Ridge National Laboratory (Yeh, G., unpublished draft). It is designed to be compatible with FEMWATER/FEMWASTE. Like HCTM and SWIFT, it considers all important saturated zone processes and inter-zone transfer processes. It also handles heterogeneous soil properties and provides variable spatial resolution. Unlike HCTM and SWIFT, FEWA/FEMA is not three-dimensional but two-dimensional, and is less complex. It has a user's guide but complete documentation was not available at the time of the review.

## 5.3 INTERACTIONS BETWEEN MODELS

Most hazardous waste sites hydrologically and chemically, influence more than one zone. While most remedial actions typically focus on a specific zone, they almost always change water and chemical movement in other zones. Consequently, more than one model will often be required to adequately represent certain remedial actions. Table 4.1 lists the inter-zone transfer processes (i.e., infiltration, percolation, drainage, and pumping) affected by different actions.

All numerical models have a limited solution domain because differences in physics and, to a lesser extent, chemistry between zones require substantially different governing equations and solution techniques. Furthermore, ease of use dictates that numerical codes be limited in size and complexity, often yielding separate codes for flow and contaminant transport calculations. When a complex hazardous waste site must be modeled or the effects of certain remedial actions predicted, several codes may be required. Because of the interactions between zones, the codes must communicate with one another.

Inter-code communication or linkage can be provided in one of three ways:

1. Transfer of data between models by hand,

2. Integration of governing equations and solution techniques into a separate computer code (hard linkage), and

3. The use of external data management programs to indirectly link the programs (soft linkages).

Hand transfer of data between codes is the most common, least efficient and least reliable method for linking models. It requires little advance preparation and no new software, but can be very labor intensive if the number and extent of model interactions are large. Hard linkage integrates the separate computer codes so that all equations are solved simultaneously and information is passed between the models during each computation cycle (i.e., time step). Hard linkage requires, in a practical sense, that the codes be merged into a single code. This type of undertaking is ambitious and can result in a comprehensive, but complex, code. Soft linkage allows the codes to remain separate and retain their original data structures, computational sequences, and input/output structures. Linkage is implemented via an external data management program, often referred to as a "bridge program," which accepts output from one code, makes necessary modifications, and inputs data to another code. Data transfers typically occur only in one direction; consequently, "feedback" from the second code to the first cannot occur. The codes are run sequentially, with computations in the first code proceeding independently from any conditions or results in the second code. Of these three linkage techniques the soft linkage or "bridge program" is most commonly used. The next two sections discuss some of the considerations and design procedures typically needed to use a soft linkage.

### 5.3.1 Soft Linkage of Codes

A typical soft linkage of surface, unsaturated and saturated zone codes for the assessment of remedial action performance is shown schematically in Figure 5.1. Site processes are represented by the unidirectional movement of water and waste constituents between zones. This requires that the surface zone code includes the plant root zone where transpiration can be removed from infiltration, leaving "net infiltration" for input to the unsaturated zone code. It also requires that the position of the water table remain fairly constant since there is no feedback between the saturated and unsaturated zone codes.

Remedial actions, such as subsurface drains and ground-water pumpings can potentially cause feedback problems if water and contaminants are withdrawn from one zone are re-introduced to another zone through land application or seepage basins. To account for this feedback, flow quantities and chemical concentrations, including the affects of treatment, must initially be estimated, checked, and possibly adjusted through an iterative procedure.

For the type of linkage shown in Figure 5.1, the following simulation steps would need to be performed to represent an entire site, including remedial actions:

1. Input chemical/biological conditions and meteorologic, hydrologic, and hydraulic conditions to the surface zone code and run the code over a selected simulation period. If remedial actions include land application, estimate an application rate and waste

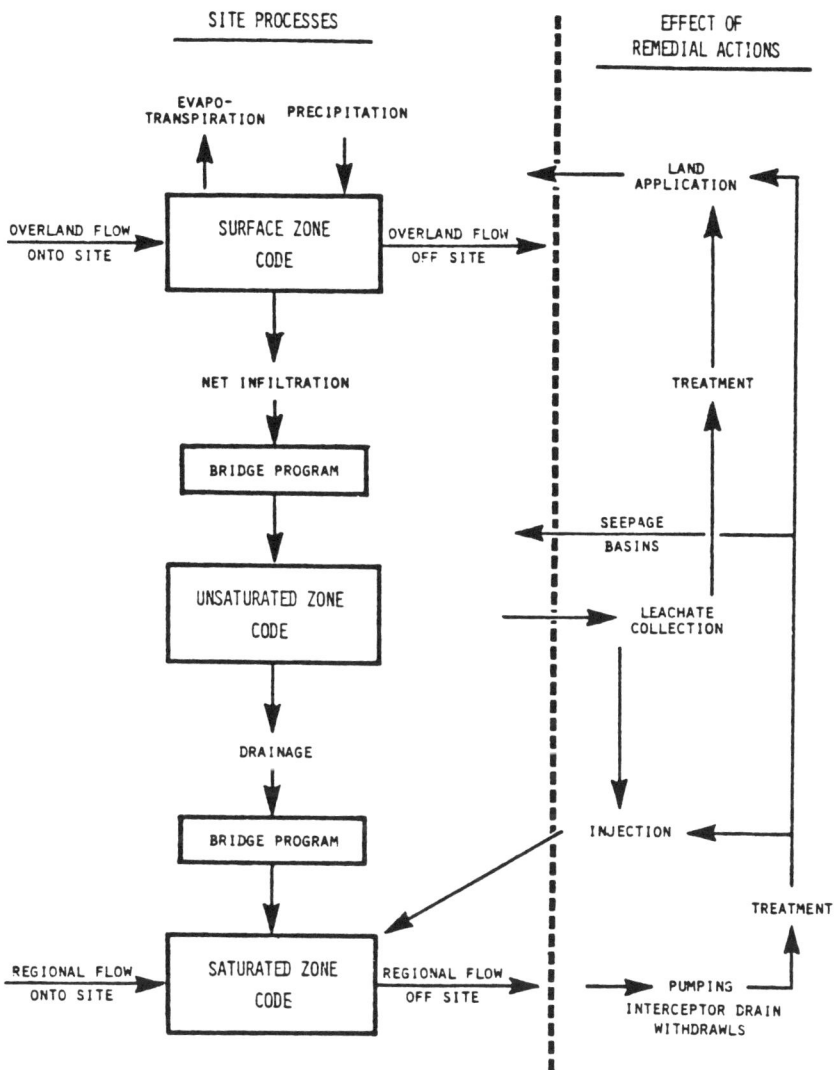

Figure 5.1  Typical soft linkage of surface, unsaturated and saturated zone codes.

constituent concentration.

2. Transfer the predicted net infiltration rates for the simulation period to the bridge program for processing and then to the unsaturated zone code.

3. Run the unsaturated zone code over the simulation period using net infiltration inputs, as well as any leachate collection rates and estimated seepage basin water surface elevations or seepage rates.

4. Transfer the drainage rates for the simulation period to the bridge program and then the saturated flow code, making any necessary conversions.

5. Run the saturated zone code over the simulation period using drainage rate inputs and any pumping/injection rates and interceptor drain withdrawals.

6. Compare estimated land application rates and basin flows and associated contaminant concentrations assumed in Steps 1 and 3 with the model results in steps 3 and 5. If the estimates are inappropriate, adjust and rerun the models.

Since the soft linkage does not allow feedback between the codes, an iterative procedure will often be necessary to properly simulate transfers of water and waste constituents associated with certain remedial actions.

### 5.3.2 Generic Bridge Program Design

The design of bridge programs to link codes basically involves identifying the specific model results that need to be transferred between codes. In general, these results will be in the form of time series (i.e., a chronologically ordered series of values). The design process also involves determining whether any unit conversions are required. Finally, the need to aggregate or disaggregate time series to account for differences in model time step requirements and the need to combine or separate time series to account for differences in spatial discretization have to be considered.

The time series that must be transferred between the surface and unsaturated zone codes are net infiltration and any associated contaminant concentrations. The time series that must be transferred between the unsaturated and saturated zone codes are drainage of water and associated contaminant concentrations.

Due to the difference in time scale for processes in the three zones, time stepping will differ between codes. Typical time steps are minutes to hours, hours to days, and days to months for the surface, unsaturated and saturated zone codes, respectively.

Spatial discretization or computational element size will also typically be different between models. Most surface zone codes use relatively large single or multiple land segments. Unsaturated zone codes may need to represent vertical and horizontal variations in soil properties due to waste site conditions and remedial actions. As a result, relatively small, variable element sizes are often used. Saturated zone code element sizes will vary with aquifer geometry and type of remedial action, but will often be larger than the unsaturated zone elements. In addition to differences in land segment and element sizes between codes, different dimensionalities are typically used. The surface zone is always represented in one-dimension, while the unsaturated zone is usually represented in either one (vertical or z) or two dimensions (longitudinal-vertical or x-z). If a two-dimensional representation is used, consideration must be given to how the surface zone code results will be "mapped" onto the two-dimensional unsaturated zone grid. A similar situation arises when unsaturated zone code results in two-dimensions (x-z) have to be mapped onto the y dimension of a two- or three-dimensional saturated zone grid (see Figure 5.2). The combination and separation of time series to account for differences in element sizes and dimensionalities will be specific to the codes selected and site being assessed.

The operation of multiple codes as a single system requires that certain consistency checks be made to ensure accurate results. The most important of these is conservation of mass. Linkage procedures need to be checked to ensure that the total mass of water and contaminant output from one code is input _exactly_ into the next code. This is often complicated by the spatial and temporal differences between codes, as discussed above. An input vs output mass balance should be computed within each bridge program.

## 5.4 MODEL APPLICATION PROCESS

The process of "setting up" a computer code to simulate the key processes controlling water and waste constituent movement at a specific site is called the "model application" process. It involves combining one's understanding of how a _code_ represents individual processes with one's understanding of their actual occurrence in the field to obtain a _model_ of the site. Here, a code refers to the computer program that solves

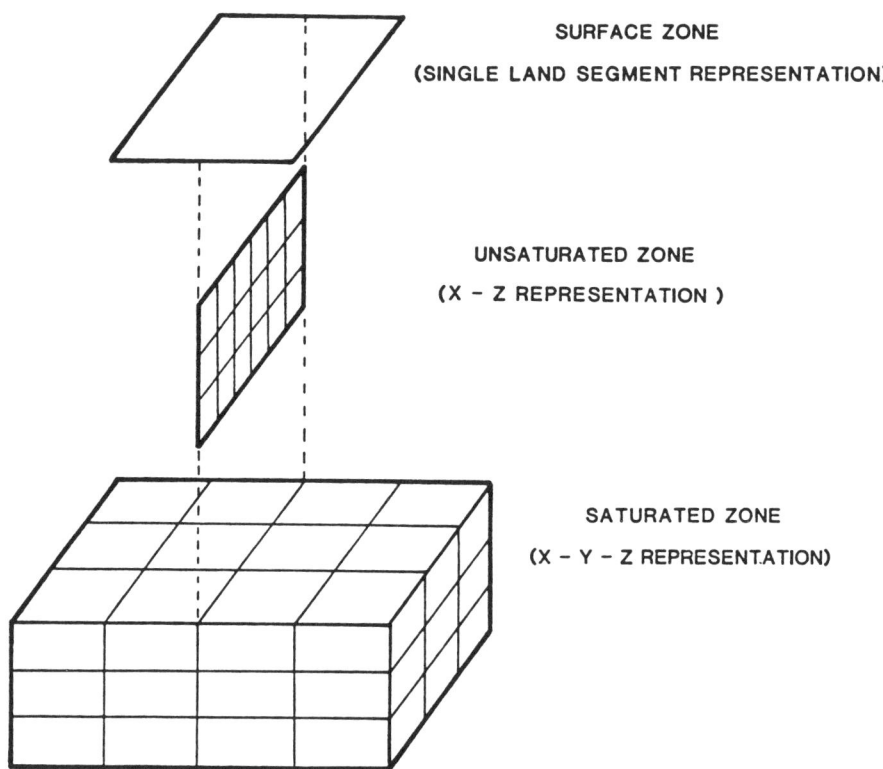

Figure 5.2  Typical dimensionalities used to represent surface, unsaturated and saturated zones.

a set of equations. A model can either be generic or site-specific. A generic model is the representation of a generic physical system by a set of equations, conditions and assumptions. A site-specific model is obtained by applying a generic model to a particular site. The latter is based on available site data and past experience. Application involves using the model to analyze target situations, in this case the performance of potential remedial actions. Mercer and Faust (1981) discuss ground-water model development and application, including data requirements, sources of error and possibilities for misuse. Figure 5.3, taken from their article, shows the steps in the model development and application process. Once the need for numerical modeling has been determined and appropriate models selected, the following steps may be taken:

1. The conceptual understanding of site conditions is further defined and quantified through the collection and analysis of site data. This "conceptual model" may also include approximate effects of potentially feasible remedial actions.

2. The conceptual model is then used to define the numeric model structure required for each zone, the types of outputs needed, and the required spatial (i.e., dimensionality) and temporal resolution.

3. Linkages between codes for each zone can also be specified by the conceptual model. The design of these linkages will depend on the structure of each code and the required interactions between zones.

4. Individual codes are installed on an appropriate computer and the site model implemented by creating an appropriate structure (i.e., grid configuration, boundary conditions, and sink and source node locations).

5. Values for individual model parameters are estimated from field data and then verified by comparing model predictions with available site data (i.e., calibration or history matching).

6. Appropriate linkages between zone models may be implemented to form a complete model which represents all important aspects of the site. In this way, the inter-zone movement of water and contaminants can be simulated.

7. Adjustments to model parameters and model structure can then be made to represent the effect of alternative remedial actions on water and constituent

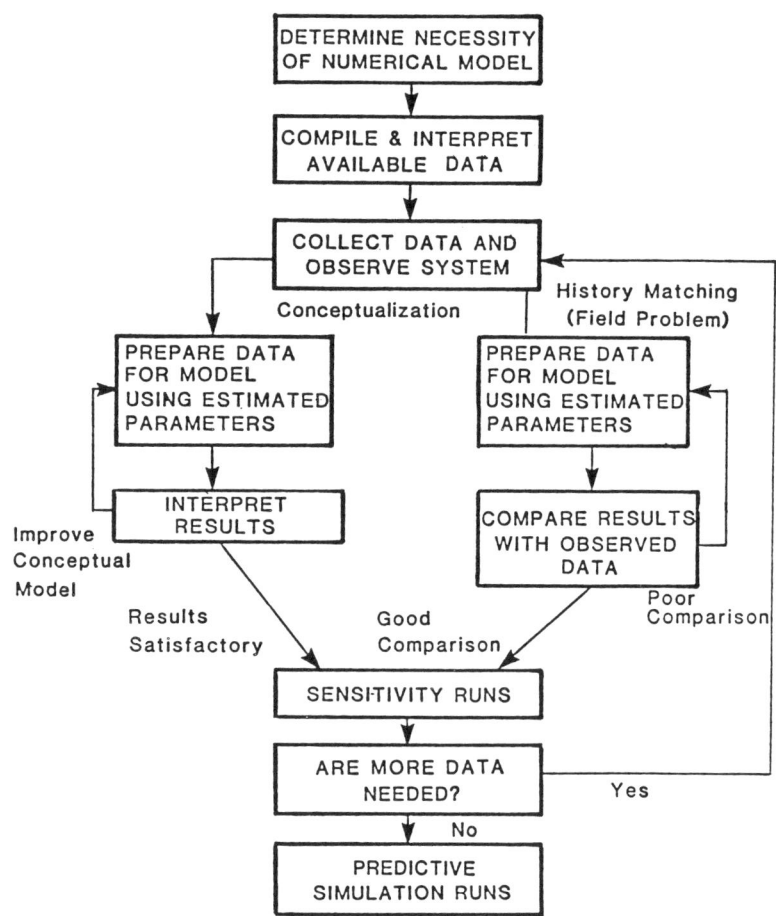

Figure 5.3  Model application process (from Mercer and Faust, 1981). Copyrighted by National Water Well Association.

movement. Model parameter adjustments required to represent specific remedial action alternatives are discussed in detail in Section 6. The simulation of certain actions (e.g., bioreclamation) may require the adjustment of selected parameters with time. Codes with a "restart" capability are particularly well suited to this type of analysis. The restart capability simply allows the user to stop a simulation run, adjust one or more parameters, and then start the simulation again.

8. The models (either individual or linked) can now be run to predict future conditions with and without remedial actions. Various combinations of actions can be explored. Where data uncertainties exist, sensitivity analyses can be used to estimate the range of outcomes.

The development of a conceptual model for a site and the collection of key site data to be used in models is discussed by several authors, including Mercer and Faust (1981) and Javandel et al. (1984). Model verification and parameter adjustment is discussed in the user's guides for most codes, and in numerous reports and papers--see the discussions and references in Bedient et al. (1981) and Knox and Canter (1980).

A number of important issues can be addressed when applying numerical models, including:

1. Existing exposure routes and levels of exposure for specific chemicals

2. Future exposures if no action is taken

3. Effects of alternative remedial actions on conditions at and near the site

4. Future exposures during and after the implementation of alternative remedial actions

Most of these questions will need to be answered during the screening and analysis of alternatives. While screening may require simplified methods, numerical models will find use in the analysis of alternatives where complex site conditions exist or complex remedial actions are anticipated.

During the remedial investigation, site characterization data are collected. Site characterization could also include the use of numerical models to specify chemical sources, chemical migration pathways, and potential receptors.

## 5.5 USER EXPERTISE AND RESOURCE REQUIREMENTS

The application of numerical models requires a level of expertise that goes beyond that needed for the simplified methods discussed in Volume 2. This is largely because both computers and numerical methods are required to efficiently solve practical problems, whereas simplified methods can often be solved by hand or through the use of programmable calculators and micro-computers. The following four basic areas of expertise are required:

1. <u>Hydrology/hydrogeology</u> - Model users should have the ability to conceptualize hydrologic systems and identify key processes controlling water movement at a site. Since both are largely derived from available site characterization data, an understanding of the limitations associated with different field sampling methods is required.

2. <u>Environmental Chemistry</u> - Model users should be able to identify important chemical migration and fate processes, including the estimation of physical-chemical properties, transfer coefficients, and rate constants. The need to consider multi-phase transport, density driven transport and interactions that occur in complex mixtures is also required. Again, since site characterization and literature data provide much of the basis for parameter estimation the user should have an understanding of sampling procedures, analytical methods and chemical property estimation methods.

3. <u>Numerical Analysis</u> - Numerical methods are used in all numerical models, and even some simplified methods, to solve basic driving equations. Errors can be introduced in simulation results, unless the user clearly understands the limitations associated with different methods. These limitations can include grid spacing (i.e., spatial discretization) and size of the time step needed to obtain a stable, accurate solution. A related area of expertise is in the linkage of models. Users must understand how to correctly transfer model results and map them onto grids with different dimensionalities.

4. <u>Computer Operations</u> - At a minimum, numerical models should be solved on a micro-computer. The application of multi-dimensional models to large problems will generally require a mini-computer or a main-frame system. The efficient use of numerical

models requires expertise in code implementation, programming, data management, data processing and computer graphics. Computer operations expertise is especially important if models require linkage through bridge programs.

Clearly, few individuals have all of the above expertise. For this reason, it is common that a team approach will be followed. It is important that the team members not only have training in the above areas, but also considerable *experience*. In many cases, the level of previous experience with similar site conditions and a similar, or the same, model will determine the success and quality of a modeling effort.

As with user expertise, more resources are generally required to apply numerical models. Here resources refers to:

1. Computer facilities - As was stated above, access to at least a micro-computer is required. Generally, the user must have access to a mini-computer or mainframe system.

2. Data - One of the major benefits of numerical models is that spatial and temporal variations in site conditions can be considered. To take advantage of this benefit, data must be available to describe variations in key parameters. Considerable data are also required for model testing (i.e., calibration/verification or history matching).

3. Time/manpower - The collection/reduction of site characterization data, the development of a conceptual understanding of important processes, and model calibration/verification are the three most time consuming steps in applying a numerical model; relatively little time is required to analyze remedial action performance once these steps are completed. While it is difficult to specify the exact time required for each step, a complete numerical modeling study can easily require between 3 and 6 months of calender time and at least twice this amount in manpower.

It is important to recognize the need to be able to commit these levels of resources prior to initiating a numerical modeling study.

## 5.6 ANALYSIS OF REMEDIAL ACTION PERFORMANCE USING NUMERICAL MODELS

The evaluation of remedial action performance initially involves screening out those actions that clearly will not meet site clean-up goals. Best engineering judgement supported by the use of simplified methods, like those discussed in Volume 2, are usually sufficient to determine which general technologies are likely to work. This screening effort is followed by a more detailed analysis of the remaining actions in terms of technical feasibility and environmental, public health, institutional and cost concerns. The technical feasibility of an action relates to the degree to which design objectives are achieved (i.e., effectiveness) and the length of time that effectiveness is maintained (i.e., durability). It also relates to the ease of implementation of a remedial action and possible concerns associated with risk of failure. Environmental concerns are the incremental impacts and benefits associated with the implementation of a remedial action, while public health concerns relate to reductions in human and environmental exposure levels. Institutional concerns are related to relevant local, state and Federal regulation. Cost concerns are the capital and operating and maintenance costs for a given action.

Based on the results of the detailed analysis, one or more actions are then selected for conceptual design. This effort involves determining the optimal location, size and configuration of a remedial action alternative.

Recent applications of numerical models (e.g., Silka and Mercer, 1982; Cole et al., 1983; Mercer et al., 1983; Cohen and Mercer 1984; and Anderson et al., 1984;) have shown how they can be used to support the analysis of: 1) reductions in exposure levels, 2) uncertainty regarding remedial action performance, 3) optimization of remedial action designs, and 4) design life and impacts of failure.

### 5.6.1 Assessment of Reductions in Future Exposure Levels

Narrowing the large number of possible remedial actions down to a set of technically feasible actions may be difficult if one depends only on best engineering judgement. While it may be relatively easy to determine that a subsurface control measure is needed to clean-up a contaminated aquifer, site conditions may make it difficult to determine whether a pumping/injection system, up-gradient cut off wall, downgradient cutoff wall, interceptor drains or combination of these actions will be most effective in terms of providing the greatest reduction in exposure levels or which actions can meet established site clean-up goals. Similarly, it is difficult to determine what level of reductions are achievable where specific clean-up goals are subjective or not established. One of the benefits associated with using numerical models is that environmental concentrations useful for exposure and risk assessment can be estimated for a number of locations of interest, including drinking water wells, site boundaries or nearby surface water bodies.

Figure 5.4 shows the results of a model-based evaluation of remedial action performance for the La Bounty Landfill in Charles City, Iowa (Cole et al., 1984). This figure shows predicted concentrations of arsenic levels in a river (the Cedar River) adjacent to the landfill under low flow conditions. The pre-restoration curve shows the predicted build-up of arsenic concentrations from 1967 to 1983. The curve labelled clay cap shows how concentrations were predicted to change after the installation of a clay cap; this curve represents the base case. All of the other curves are for potential remedial actions proposed for implementation in conjunction with the clay cap. Given a primary drinking water standard of 50 ppb for arsenic as an example of a site clean-up goal, the model results in Figure 5.4 can be used to identify those actions that will lead to the greatest overall reduction in exposure levels.

Figure 5.4 shows the importance of considering time when comparing the performance of remedial actions. The pump and treat and downgradient cut off wall alternatives reduce arsenic concentrations within a few years of their implementation, whereas the limited bottom lining, stabilization and limited excavation alternatives take a number of years to achieve the same reduction. However, these three alternatives ultimately lead to the greatest reductions. This point is more clearly demonstrated in Table 5.3. It shows how the relative ranking of alternatives, in terms of reductions in Cedar River contamination levels, changes depending upon which point in time is chosen to evaluate performance.

### 5.6.2 Uncertainty Regarding Remedial Action Performance

Field data are frequently insufficient to accurately characterize site conditions. This is especially true for the unsaturated and saturated zones. Additionally, the actual performance of a remedial action may not be known until it has been implemented and tested. Numerical models can be particularly efficient and insightful tools for studying potential uncertainties. Sensitivity analyses can be

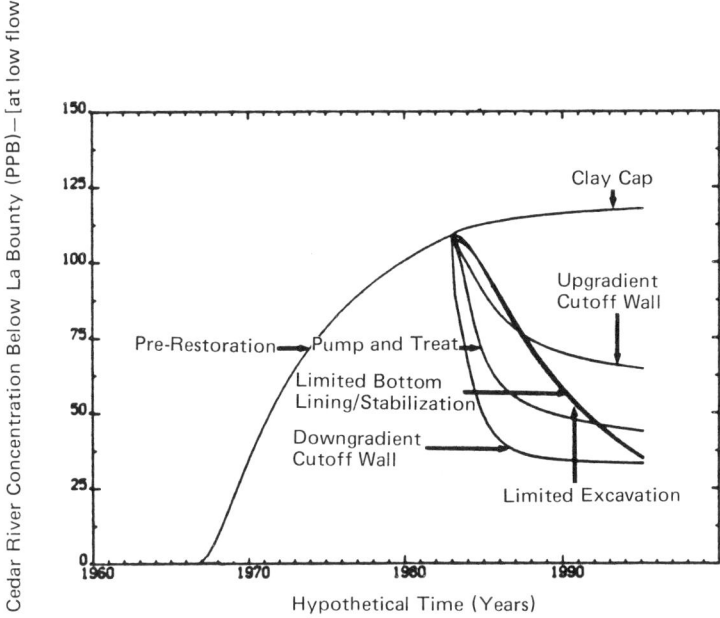

Figure 5.4 Predicted performance of different remedial action alternatives in reducing arsenic concentrations in the Cedar River under low flow conditions (after Cole et al., 1984).

TABLE 5.3 RELATIVE RANKING OF POTENTIAL ALTERNATIVE LA BOUNTY LANDFILL REMEDIAL ACTIONS AT DIFFERENT POINTS IN TIME USING LEVEL OF CONTAMINATION REDUCTION IN THE CEDAR RIVER AS A MEASURE OF PERFORMANCE (Taken from the results by Cole et al., 1984)

| Remedial Action | Years After Implementation | | |
|---|---|---|---|
| | 2 | 6 | 12 |
| Downgradient cutoff wall | 1 | 1 | 1 |
| Pump and Treat | 2 | 2 | 4 |
| Upgradient cutoff wall | 3 | 5 | 5 |
| Limited excavation | 5 | 4 | 3 |
| Limited bottom lining stabilization | 4 | 3 | 2 |

1 = Largest reduction in contamination levels
5 = Smallest reduction in contamination levels

performed by varying uncertain parameters, making runs and observing the changes in model outputs. Such changes include water levels, flow directions and rates, and chemical concentrations.

Silka and Mercer (1982) used sensitivity analyses to investigate the probable effects of installing a subsurface drain at Love Canal, NY. Parameters evaluated included hydraulic conductivity, effective porosity and recharge. Figure 5.5 shows the effect of changes in shallow aquifer hydraulic conductivity on water table elevations, as simulated by a two-dimensional saturated zone model. Substantial differences in elevation near the drain are predicted, indicating that accurate specification of hydraulic conductivity is important. Through comparison of model predictions of drain flux and water table elevation with field measurements, the authors were able to estimate average or bulk hydraulic conductivities for the shallow aquifer.

Mercer et al. (1983) conducted both sensitivity analyses and more rigorous uncertainty analysis in a later remedial action evaluation at Love Canal. Their sensitivity analyses considered conditions along two of the model boundaries, aquifer transmissivity, confining bed hydraulic conductivity and shallow system water levels. Their uncertainty analysis considered the effects of uncertainties in hydraulic conductivity and porosity on contaminant travel times. A Monte Carlo technique was used to select conductivity and porosity values from estimated frequency distributions.

### 5.6.3 Optimization of Remedial Action Design

Once remedial action alternatives have been identified, their design may be refined as part of the development of a conceptual design. Optimizing a design involves evaluating alternative locations, pumping rates and remedial action configurations to identify which specific combination will be most effective. Modeling is ideally suited to this type of analysis because a number of alternative designs can be evaluated rapidly and quantitatively.

Cole et al. (1984) evaluated several alternative designs for a proposed upgradient cut off wall at the La Bounty Landfill. They showed that by changing the location of the cut off wall and by lowering the head in the subsurface drain located on the upgradient side of the wall, arsenic loadings to the Cedar River could be reduced by about 30 percent. This reduction was sufficient to make the upgradient cut off wall a feasible alternative.

Anderson et al. (1984) analyzed alternative remedial action

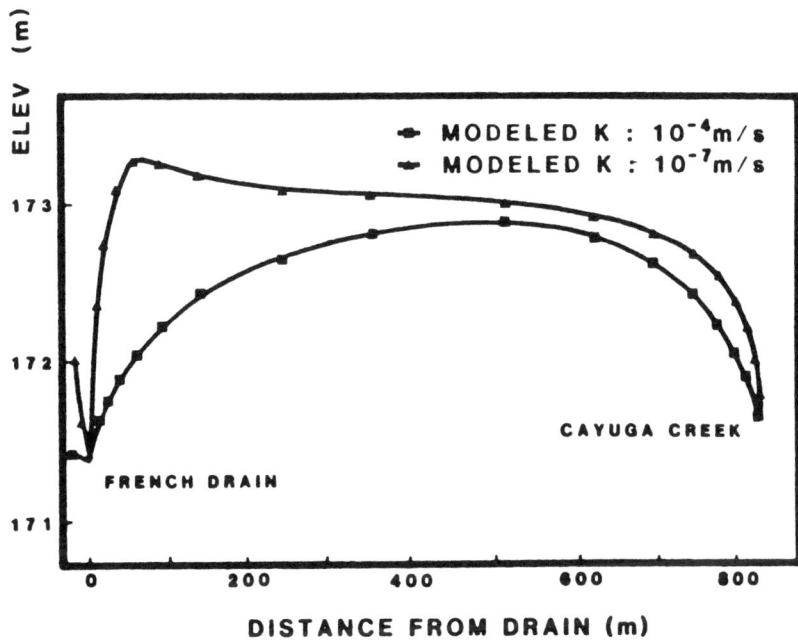

Figure 5.5  Predicted effects of two values of hydraulic conductivity on the shape of the water table with installed French Drain  (from Silka and Mercer, 1982).

designs for the Lipari landfill in New Jersey using a two-dimensional (x-y) finite difference model of the saturated zone. Actions simulated included slurry walls, drains and clay caps, alone and in combination. The effect of drain depths, a partial or full clay cap and a slurry barrier wall on discharge of ground water to seeps and to drains over time were estimated. Figure 5.6 shows predicted variations in drain discharge with time for different drain depths.

Cohen and Mercer (1984) used a two-dimensional (x-y) model to evaluate proposed additional remedial measures for Love Canal. They analyzed the effectiveness of different designs that included a concrete cut off wall and a synthetic cover. The simulation results showed that the cut off wall would provide only a minor reduction in drainage to the French Drain compared to the cover. As Cohen and Mercer note, based in part on the model results, the State of New York decided not to construct the proposed cut off wall.

Optimization of remedial action design can and has been taken one step further in sophistication. Mathematical programming such as linear or quadratic programming can be used in conjunction with numerical modeling to directly optimize pumping rates and well locations, eliminating a tedious trial and error search. This technique involves defining an objective function (such as minimizing the costs of pumping) and a set of constraints that might require that certain hydraulic gradient or head conditions are met. The mathematical algorithm then finds the optimal solution for the given problem. Atwood (1984) used linear programming to optimize well selection and pumping rates in a hydraulic containment and extraction design. The optimal design called for an outer set of wells to initially stabilize contaminated ground water. As the plume diminishes in size, an inner set of wells is determined to be more efficient. An optimal schedule of extraction and injection rates for the sixteen year clean-up period was also determined by the program. Another example of combining optimization techniques with numerical modeling to design extraction well systems is discussed by Shafer (1984). Gorelick (1983) reviews the state-of-the-art research on these management techniques for water quality and water allocation problems. A more recent article by Gorelick et al. (1984) introduces the use of non linear programming techniques.

## 5.6.4. Assessment of Design Life and Impacts of Failure

All remedial actions have a finite life that needs to be considered when evaluating their performance. Covers erode, drainage systems clog and wells collapse. Remedial actions

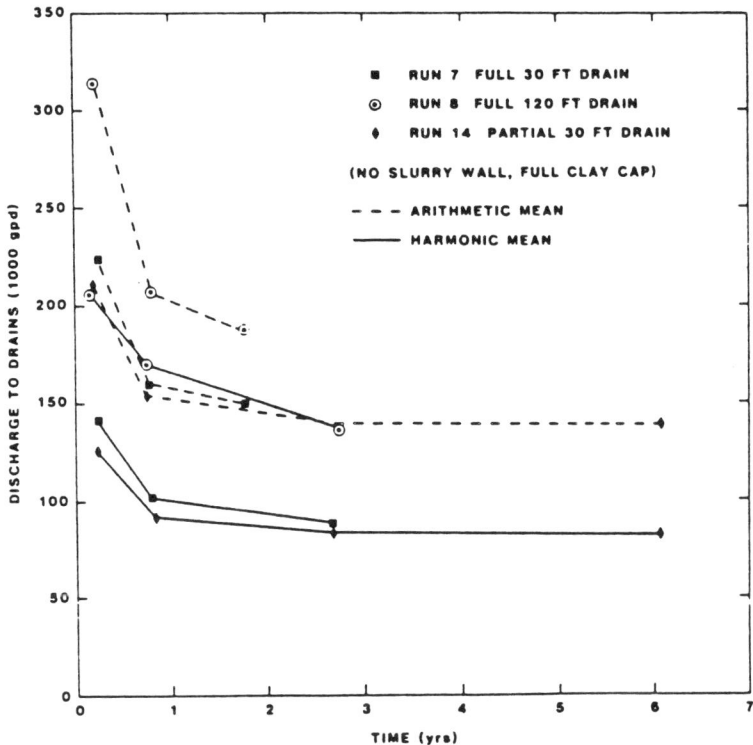

Figure 5.6 Discharge to drains at Lipari Landfill for different drain depths (from Anderson et al., 1984).

can also simply fail, either catastrophically or progressively. Synthetic liners for example, should be effective for a number of years. However, if they are not properly selected, installed and used, they can fail in relatively short periods of time. Recent research has shown that the permeability of clay liner materials can change over time if exposed to hydrophobic pollutants (Green et al., 1983). The permeability of bentonite slurry materials used in impermeable barriers has also been found to increase in the presence of certain organic and inorganic compounds (Spooner et al., 1983).

Numerical modeling has not been used to any large extent to evaluate the impact of these types of failures when assessing remedial action performance. In at least one situation, however, design life and failure mode considerations were incorporated into a model-based analysis of new waste disposal facilities. A multi-disiplinary team was assembled by the EPA Office of Solid Waste to investigate the influence of site conditions, disposal facility design, and failure mode on leachate migration. As described by Brown et al., (1984), three numerical models were used. HELP (Schroeder et al., 1984a and 1984b) was used to estimate leachate generation, vertical movement through the facility and release through the liner into the unsaturated zone. The Pesticide Root Zone Model (Carsel et al., 1984) was used to predict the transport of leachate vertically through the unsaturated zone. The Combined Fluid, Energy, Solute Transport (CFEST) model (Gupta et al., 1979) was used to estimate chemical movement in the saturated zone.

Examples of typical model results for one of the facility design/failure mode scenarios are given in Figures 5.7 and 5.8. Figure 5.7 shows the calculated leachate loading from the base of a landfill with a single clay cover and a leachate collection system. The progressive increase in leachate loading over the first 20 years shows the impact of increasing facility size by opening new waste cells. The slight decrease in loading is due to the installation of a cap after the facility is closed. The rapid increase in loading after about 50 years is due to the failure of the leachate collection system. In this scenario, the facility is sited in a humid location with high intensity rainfall. Figure 5.7 also shows the predicted mass loading of leachate to the saturated zone over a 200 year time frame.

Figure 5.8 shows predicted time histories of leachate concentration at a monitoring well 100 m downgradient from the facility and a ground-water discharge point (i.e., stream) 300 m downgradient. The concentrations shown in this figure are "relative concentrations." That is, Figure 5.8 shows predicted ground-water concentrations relative to the initial

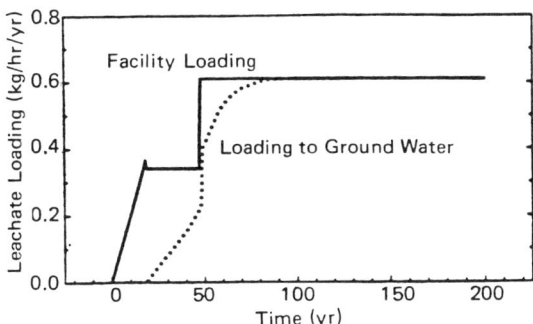

Figure 5.7  Facility leachate loading and loading to ground water.

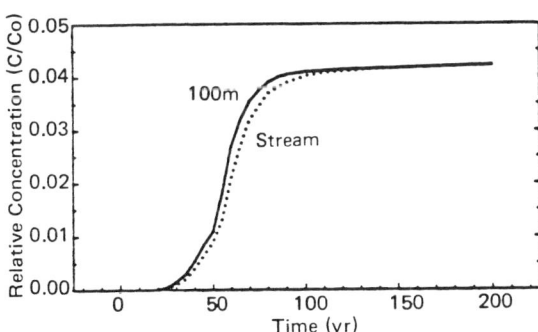

Figure 5.8  Relative leachate concentration at monitoring well (100 m) and stream.

leachate concentration (Co). Thus, for this scenario the maximum relative concentration (C/Co) in the saturated zone is 0.04 or 4 percent of the original concentration.

Results such as those shown in Figures 5.7 and 5.8 were used to evaluate the impacts of a number of facility designs and modes of failure on ground-water quality.

# 6. Remedial Action Modeling Requirements

## 6.1 OVERVIEW

In using numerical models to evaluate remedial action performance it is important to recognize that different remedial actions can have substantially different modeling requirements. The type of model(s) (i.e., surface, unsaturated or saturated zone) required to simulate the effects of an action can vary, as can the dimensionality and grid configuration. In addition, the model parameters that must be adjusted to represent the effects of implementing different actions can vary. As Volume 1 notes, all of these modeling requirements must be considered, hopefully early enough in the Feasibility Study/Remedial Investigation process to have an impact on the specific model(s) selected for use in remedial action evaluation.

Section 6 seeks to define numerical modeling requirements for specific remedial actions and groups of actions. Here, modeling requirements refer to: 1) the type of model(s) that are required, 2) dimensionality and grid configuration considerations, and 3) model parameter adjustments. Guidance is provided on sources of information and available techniques for parameter estimation for situations where field data are not available. The modeling requirements defined herein were, in large part, taken from previous remedial action modeling studies (e.g., Cole et al., 1984; Mercer and Silka, 1981; Mercer et al., 1983; Anderson et al., 1984; and Cohen and Mercer, 1984).

As was noted in Section 4, similarities in design objectives and remedial action configuration made it possible to condense the large number of available technologies into fourteen remedial measures under the general categories of surface control, subsurface control, and waste control. These fourteen measures can be condensed further due to similarities in modeling requirements. An example would be the grouping together of bio-reclamation and chemical injection. Both of these measures can be modeled in a similar fashion: injection and extraction wells are used and the degradation rates assigned to different elements or blocks in the model grid are adjusted to represent the enhanced degradation of the chemical in the treated zone. The fourteen remedial measures were also re-grouped because they tend to be used conjunctively. For example, the remedial actions of capping, grading, and revegetation were grouped together because they are often implemented as a group to control infiltration and runoff. Given the above, the fourteen remedial measures discussed in Section 4, were reduced to the nine remedial action groups shown in Table 6.1.

Prior to presenting modeling requirements for each group of remedial measures, several key points need to be addressed.

1. Only those modeling requirements associated with a given group of remedial measures are discussed. Requirements associated with the use of numerical models for site characterization and assessment are not presented. Thus, the guidance presented herein is <u>in addition</u> to that needed to develop a model of the site.

2. Certain model parameter adjustments are highly site-specific. Thus, it is difficult to provide guidance on their estimation.

3. Data on certain model parameters are, on the whole, quite sparse due to a lack of field data on the performance of some remedial measures. In many cases, only laboratory or pilot scale data or parameter values from previous modeling studies are available.

## 6.2 MODELING REQUIREMENTS

The modeling requirements for each group of measures are presented in terms of the following:

1. Model Type - Model type refers to whether a surface, unsaturated or saturated zone model, or some combination of the three, is required.

2. Dimensionality and Grid Configuration - Dimensionality refers to the directions (i.e., x, y, and z) of water and chemical movement that can be simulated; grid configuration refers to the spatial discretization needed to represent a site and the remedial action.

3. Parameter Adjustments - Parameter adjustments refer to

TABLE 6.1   REMEDIAL MEASURES

Capping, Grading and Revegetation
Surface Water Diversion and Collection
Ground-Water Pumping and Interceptor Trenches
Impermeable Barriers
Subsurface Drains and Solution Mining
Excavation
Hydraulic Dredging and Seepage Basins
Bioreclamation and Chemical Injection
Permeable Treatment Beds

the model inputs that must be modified to represent a remedial measure.

Table 6.2 summarizes the modeling requirements for each measure. The following discussion provides more detailed guidance.

### 6.2.1 Capping, Grading, and Revegetation

Capping, grading, and revegetation are often used to reduce infiltration and control erosive runoff. Since these three remedial actions are commonly implemented together, they were grouped into one remedial measure. The purpose of modeling is to: 1) estimate reductions in chemical loadings to adjacent surface water bodies and 2) estimate reductions in infiltration into the waste site and associated leachate generation.

#### Model Type

Figure 6.1 shows two typical cap designs composed of vegetative, barrier, gas channel, filter, and buffer layers overlying waste materials.

Two types of models may be required to evaluate this measure: a surface zone model and an unsaturated zone model. Typically, the surface zone model is applied to only the upper portion of the cap. The vegetative layer would constitute the surface zone for the designs shown in Figure 6.1. Time series of rainfall, potential evapotranspiration and possibly other meterological conditions are input to the surface zone model to generate time series of net infiltration into the layer below the vegetative layer and time series of runoff, erosion, and contaminant loadings from the site.

The remainder of the cap and the waste itself would be analyzed with the unsaturated zone model. The net infiltration time series generated by the surface zone model can be used as a flux boundary condition in the unsaturated zone model. This boundary condition is applied to the first compartment representing the interface between the surface and unsaturated zones.

#### Dimensionality and Grid Configuration

Either a single or a multiple land segment configuration can be used to represent the disposal site with the surface zone model. If runoff, erosion and chemical loadings from the site itself are of concern, a single land segment with uniform properties (e.g., slope, roughness and infiltration capacity)

TABLE 6.2 SUMMARY OF MODELING REQUIREMENTS FOR EACH REMEDIAL MEASURE

| Remedial Measure | Model Type | Dimensionality Grid Configuration | Parameter Adjustments | Comments |
|---|---|---|---|---|
| Capping, Grading and Revegetation | SF | S, M | SR, ER, ET, IN, | Number of land segments depends on site conditions |
| | UZ | 1D/z | MC, HC, PO, DS, BD | - |
| Surface Water Diversion and Collection | SF | M | SR, ER | Channel segments will also be required |
| Ground-Water Pumping and Interceptor Trenches | SZ | 2D/x-y | NW, NC | 3D model may be needed for partially-penetrating wells/drains |
| Impermeable Barriers | SZ | 2D/x-y, 2D/x-z, 3D | HC | Dimensionality dependant upon barrier design |
| Subsurface Drains and | UZ | 2D/x-z | NW, NC, AD | Model type dependant on site conditions |
| Solution Mining | SZ | 2D/x-z | AD | - |
| Excavation | UZ | 1D/z | MC, HC, NC, PO, BD | Model type dependant on site conditions |
| | SZ | 2D/x-y | HC, NC | - |
| Hydraulic Dredging | UZ | 1D/z | NW, NC | - |
| and Seepage Basins | SZ | 2D/x-y | - | Saturated zone model not required if mounding not of concern |
| Bioreclamation and Chemical Injection | SZ | 2D/x-y | NW, NC, DG | Estimation of extent of treated zone must be estimated prior to degradation rate |
| Permeable Treatment Beds | SZ | 2D/x-z | AD, PO, HC, BD | Hydraulic conductivity adjustment dependant on materials in treatment bed |

(continued)

TABLE 6.2 (continued)

LEGEND:
- SF  Surface zone model
- UZ  Unsaturated zone model
- SZ  Saturated zone model

- S   Single land segment
- M   Multiple land segment

- 1D  One-dimensional
- 2D  Two-dimensional
- 3D  Three-dimensional

- x   Longitudinal direction
- y   Lateral direction
- z   Vertical direction

- SR  Surface roughness
- ER  Soil erodibility
- ET  Evapotranspiration
- IN  Interception
- IF  Infiltration

- MC  Moisture content
- HC  Hydraulic conductivity
- DG  Degradation
- DS  Dispersivity
- PO  Porosity
- BD  Bulk density
- AD  Sorption

- NW  Nodal water flux or held head
- NC  Nodal chemical flux or held concentration

Remedial Action Modeling Requirements 267

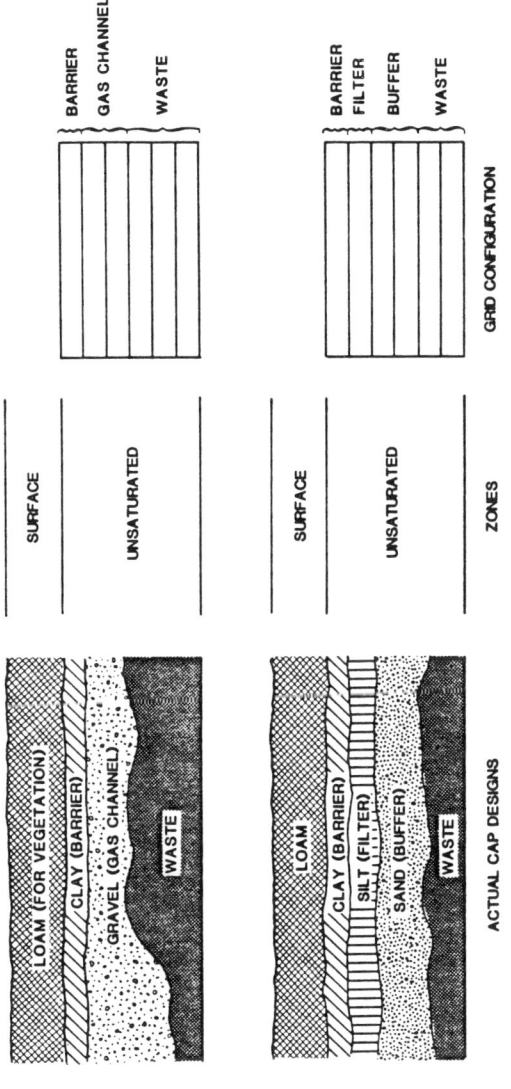

Figure 6.1 Two typical cap designs showing layers in each zone (after CRB Associates, 1982).

can be used. In cases, however, where runoff from areas surrounding the site or loadings to a nearby surface water are of concern, multiple land segments may be required. As the areal extent of the surface zone increases, care must be exercised in selecting the number of land segments and their characteristics. In addition, users should recognize the possible need to represent channel processes should the drainage area encompass well-defined surface drainage features.

The minimum dimension of the unsaturated zone model should be one, in the vertical or z-direction. As Figure 6.1 shows, the cap and disposal site can be represented as a series of compartments of equal thickness corresponding to the layers below the vegetated layer and the waste materials. Each layer can be assigned varying properties (e.g., hydraulic conductivities and porosities), depending upon the site conditions and choice of materials in the cap design. The thickness of the cap and drainage layers can range from 0.5 to 1.0 meters (JRB Associates, 1982; Mercer and Silka, 1982).

Parameter Adjustments

The general surface zone model parameters that need to be adjusted to represent the effects of capping, grading and revegetation are:

- o interception storage
- o surface roughness
- o infiltration capacity
- o evapotranspiration rate
- o soil erodibility

The first four parameters largely affect runoff and infiltration, while the remainder affect soil erosion. Subsection 6.3 provides guidance on the estimation of these parameters. The simulation of this measure will not require the adjustment of those parameters affecting chemical migration and fate. If a prior modeling study was not conducted during site characterization, sorption coefficients and degradation rates will have to be estimated.

Unsaturated zone model parameters that need to be adjusted include:

- o moisture content characteristics
- o hydraulic conductivity
- o porosity

They include those parameters related to the hydraulic properties of the individual layers used in the cap. In their analysis of the clay cap at Love Canal, Silka and Mercer

(1982) used a hydraulic conductivity of $10^{-9}$ m/sec. Cole et al., (1984) used a conductivity of $3.5 \times 10^{-10}$ m/sec for the Charles City clay cap.

In situations where a synthetic material is used as a cover, a common assumption in modeling the unsaturated zone is to use a zero infiltration rate; this is the assumption Cohen and Mercer (1984) made in their analysis of a synthetic cover extension for Love Canal. A similar asssumption can also be made for more regional analyses of clay or synthetic covers; Anderson et al., (1984) made this assumption in their analysis of the Lipari Landfill.

Again, if an unsaturated zone model was not used for site characterization, parameters related to the hydraulic properties of the waste materials will need to be estimated, as will those related to chemical transport and fate. The latter include sorption coefficients, degradation rates, dispersivities and bulk densities.

### 6.2.2 Surface Water Diversion and Collection

Surface water diversion and collection actions are designed primarily to route runoff away from a hazardous waste site. Reductions in runoff, erosion, infiltration and off-site transport of waste constituents are the primary changes that need to be analyzed when using models to evaluate this measure.

#### Model Type

The evaluation of this remedial measure can be accomplished with a surface zone model. As with the capping, grading, and revegetation measure, time series of meteorological conditions are input to the model to generate time series of runoff losses, erosion losses and chemical loadings.

#### Dimensionality and Grid Configuration

A model that is capable of considering multiple land and channel segments is required. At least one land segment is required for the waste site, the others are required to represent areas adjacent to the site and channels collecting diverted runoff. Runoff from land segments upgradient from the site can be used as input to channel segments to represent the diversion of runoff around the site.

Model Parameter Adjustments

The parameters that must be adjusted to represent the effect of surface water diversion and collection are those related to changes in the topography of the land surface and those related to the addition of drainage structures. These parameters include:

- o  surface/channel roughness
- o  soil erodibility

Parameters related to surface hydrology (e.g., infiltration rate and interception storage) and waste constituent transport (e.g., sorption coefficients and degradation rates) will need to be estimated, if they are not available from an earlier modeling study.

### 6.2.3  Ground-Water Pumping and Interceptor Trenches

Numerical models can be used to evaluate a number of different changes induced by the implementation of a ground-water pumping system or interceptor trenches. Changes in heads, directions of water and contaminant migration, and rates of plume withdrawal can all be evaluated.

Model Type

A saturated zone model is required to evaluate this measure. Pumping wells are represented by assigning heads or fluxes to nodes in the grid; injection wells are represented in a similar manner except contaminant concentrations also have to be assigned if any residual contamination will be reinjected following treatment. Trenches are normally represented by assigning heads to a line of nodes. The performance of such a measure can be assessed by modifying the number, placement, and withdrawal rates of the wells or trenches.

Dimensionality and Grid Configuration

At a minimum, a two-dimensional (x-y) simulation is required to represent mounding and depression of the water table. It is important to note, however, that a two-dimensional (x-y) representation inherently assumes that the wells or trenches fully penetrate the saturated zone. If field conditions dictate that withdrawal/injection occur over specified depth intervals or the trenches be partially penetrating, a more rigorous three-dimensional representation may need to be used. A two-dimensional (x-z) representation is rarely used for wells because they must be represented as a trench or line of closely spaced wells in the y dimension. This can create a problem when specifying pumping rates, since water is not withdrawn from the y-direction. In addition, this representation makes it impossible to examine the potential for plume excursion around or between pumping wells. An x-z representation can be used for an interceptor trench, however. Such a representation is only reasonable however, near the middle of the trench where flows are mainly in the x and z directions.

In designing a grid for a ground-water pumping system or interceptor trench both regular and irregular grid spacings can be used. The key constraint is that a node be positioned near the proposed location of each well/trench. If several wells are close together, their discharges may be combined and assigned to a single node. The grid should be designed to accommodate a number of different well/trench locations to avoid having to restructure the grid for each alternative. Depending on the level of analysis, the size of the grid blocks or elements may be reduced near the wells/trenches to obtain greater spatial resolution of predicted heads and ground-water flow directions. Figure 6.2 shows the grid configuration used by Silka and Mercer (1982) to represent the French Drain at Love Canal. Note the change in grid spacing near the french drain.

Parameter Adjustments

The parameter adjustments for this measure are relatively straight forward. Heads or fluxes for the nodes representing the wells or trenches need to be specified. The heads or fluxes can be constant or time varying. The only other required parameter adjustment is to assign contaminant concentrations to those nodes representing injection wells. These concentrations will have to be estimated based on the concentration of waste constituents in the aquifer and the efficiency of the on-site treatment system.

6.2.4 Impermeable Barriers

As is noted in Volume 2, simplified methods can be used to analyze only a few of the many impermeable barrier design objectives and possible barrier configurations. For this reason, numerical models will often be used. Of primary interest are the extent to which a barrier will prevent contaminated ground water from migrating away from the site or divert uncontaminated ground water around a site.

Model Type

A saturated zone model is needed to evaluate impermeable

272  *Modeling Remedial Actions at Uncontrolled Hazardous Waste Sites*

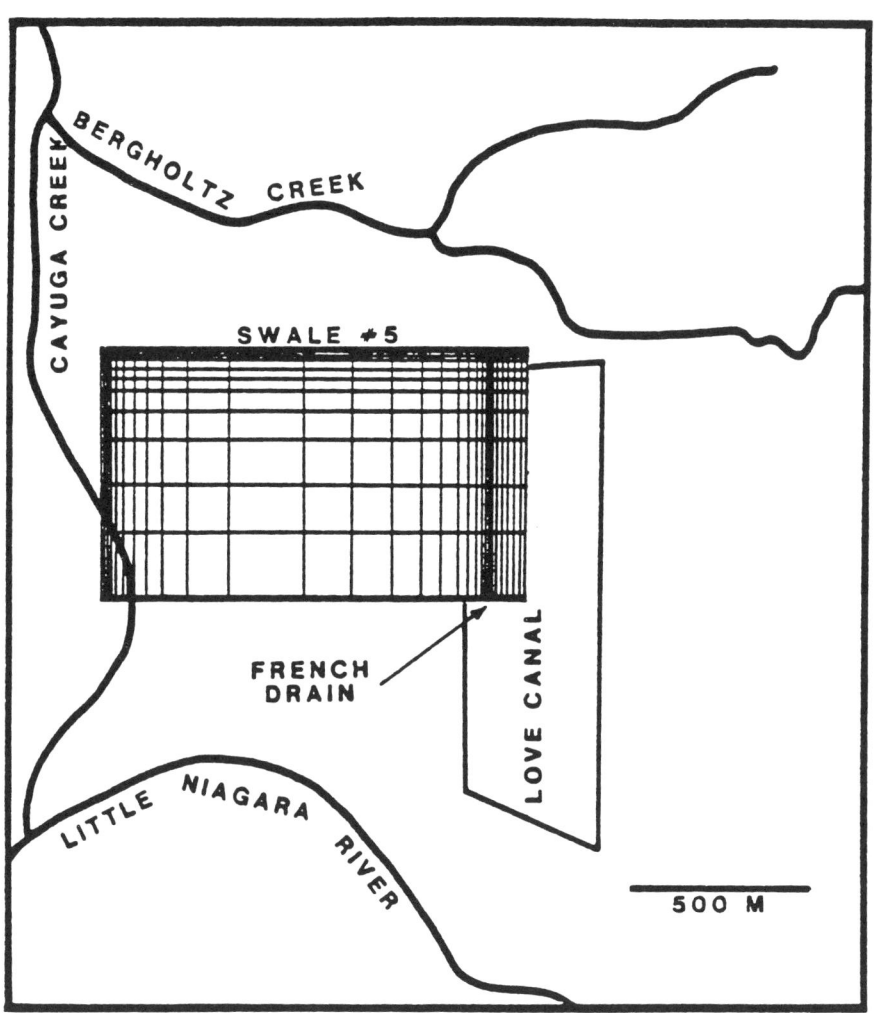

Figure 6.2  Example x-y representation and grid used to evaluate the French Drains at Love Canal (taken from Silka and Mercer, 1982).

barriers.

## Dimensionality and Grid Configuration

The required dimensionality of the saturated zone model is highy dependent upon the design of the barrier. Barrier designs can include: 1) an upgradient barrier keyed into an impermeable layer, with an optional drainage system upgradient of the wall to reduce mounding; 2) a partially penetrating, upgradient barrier that is keyed into natural impervious boundaries at each end; 3) a fully-penetrating downgradient barrier with or without a drainage system; 4) both upgradient and downgradient barriers; or 5) a fully-penetrating barrier that completely surrounds a site.

A two-dimensional (x-y) model can be used to evaluate all of designs where the barrier is keyed in at the bottom; as with ground-water pumping, the use of an x-y representation inherently assumes that the barrier fully penetrates the aquifer and that there is no flow under the barrier. Designs involving partially penetrating barriers or barriers keyed into leaky formations require at least a two-dimensional (x-z) representation. Figure 6.3 shows how an x-z representation was used by Cohen and Mercer (1984) to evaluate the benefits of a new cut-off wall at Love Canal. They modeled a cross-section of the site and assumed symmetry along the centerline of Love Canal. Thus, the grid in Figure 6.3 is only for one-half of the total cross-section. Using an x-z representation, however, assumes that the barrier is infinitely long in the y-direction. Thus, flow conditions near the end of a barrier cannot be analyzed. In relatively complex situations a three-dimensional representation can be used to evaluate the potential for contaminant movement both under and around the ends of a barrier.

Simulation of the effects of an impermeable barrier involves designing a grid with elements or blocks in the approximate location of the barrier. Element or block widths are a function of the barrier design; the usual thickness is around 1 meter (JRB Associates, 1982). Generally an irregular grid spacing is used particularly in the immediate vicinity of the barrier where directions of water movement change rapidly. Figure 6.4 shows the grid configuration used by Anderson et al. (1984) for the Lipari Landfill. This grid is for a two-dimensional (x-y) representation. The grid blocks representing the barrier are shown. Figure 6.3 shows the x-z grid configuration used by Cohen and Mercer (1984). Notice the variable grid spacing around the partially-penetrating barrier and drain.

If a drainage system is used in conjunction with a barrier, it may be represented by a set of nodes with fixed heads

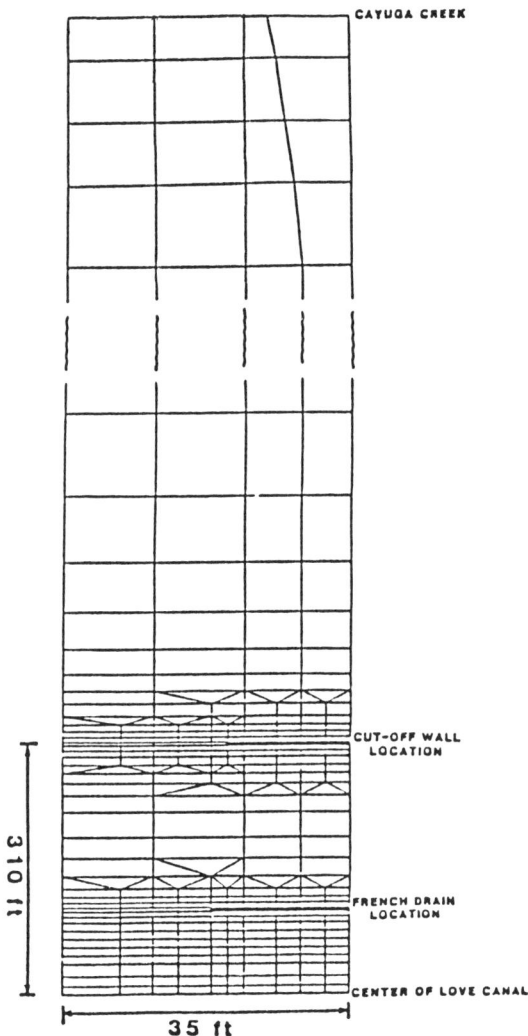

Figure 6.3 Two-dimensional (x-z) grid configuration used by Cohen and Mercer to evaluate a proposed cut-off wall at Love Canal. Copyrighted by National Water Well Association.

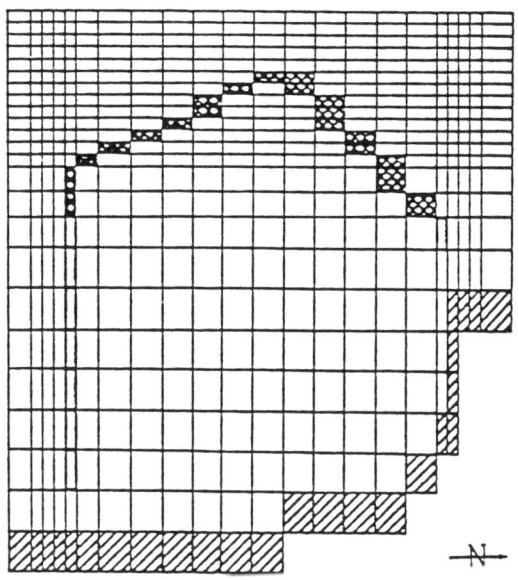

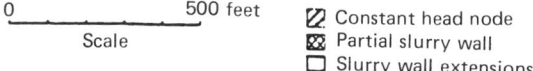

Figure 6.4  Two-dimensional (x-y) grid configuration used by Anderson et al. (1984) to evaluate a proposed slurry wall at the Lipari Landfill.

corresponding to the estimated water elevation in the drain.

Parameter Adjustments

The only parameter adjustment required to represent an impermeable barrier is the hydraulic conductivity assigned to those grid blocks or elements representing the barrier. The actual conductivity values will depend on the material used for the barrier and the construction method. As will be discussed later in this section, there are some data available for the hydraulic conductivity of soil-bentonite and cement-bentonite slurry materials. No data were found for other grout materials or sheet pile barriers.

### 6.2.5 Subsurface Drains and Solution Mining

Subsurface drains and solution mining are grouped together because of similarities in required model type, dimensionality and parameter adjustments. In analyzing these actions, the primary use of a model is to determine the extent to which leachate generation can be controlled.

Model Type

The type of model required to evaluate subsurface drains or solution mining will depend upon site conditions. In situations where the wastes are disposed above the water table, an unsaturated zone model should be used. When the wastes are inundated by ground water both unsaturated and saturated zone models may be required. To properly represent the effects of these actions on water movement and contaminant migration, the unsaturated and saturated zone models may have to be linked or, if possible, a combined unsaturated/saturated zone model can be used.

Dimensionality and Grid Configuration

In modeling the effectiveness of these actions the primary focus will usually be on changes in leachate generation rather than reductions in ground water concentrations some distance from the site. For this reason, the site can be represented with a typical cross-section. Thus, a two-dimensional (x-z) model can be used for both the unsaturated and saturated zones.

A single node or group of nodes can be used to represent a subsurface drain. Fluxes or heads must be assigned to each node to obtain proper withdrawal rates. A similar approach can be used to represent drains or well points used to extract leachate generated as a result of solution mining. The injection of the chemical solvent used to mobilize contaminants can be simulated by assigning fluxes or held concentrations to nodes in the waste layer.

The grid spacing for either action can be regular or irregular. Often the size of the grid will be reduced near the nodes simulating the drains or well points. Figure 6.5 shows an example x-z grid configuration for the analysis of a subsurface drain (Nelson et al., 1983). This grid was used in an evaluation of leachate migration from a uranium mill tailings disposal site.

Parameter Adjustments

No additional parameter adjustments other than assigning heads or fluxes to selected nodes are required to evaluate subsurface drains.

For solution mining, however, the sorption coefficient or retardation factor must be adjusted for those elements or grid blocks receiving the injected chemical solvent. Either parameter needs to be reduced to reflect the effect of increased mobility. The amount of reduction is waste and solvent specific. No data are available on possible parameter ranges, largely because this technology has received limited use in the field.

### 6.2.6 Excavation

In the evaluation of excavation actions, models can be used to estimate reductions in leachate quality associated with the removal of waste materials.

Model Type

As with the previous measure, the required model type depends on site conditions. If the wastes are disposed of above the water table, an unsaturated zone model can be used. If the wastes extend into the saturated zone, the use of a saturated zone model is required.

Dimensionality and Grid Configurations

The minimum dimensionality for the unsaturated zone is a one-dimensional (z) representation. If there are lateral heterogenities in the waste materials or subsoils, a two-dimensional (x-z) representation should be used. For the saturated zone, a two-dimensional (x-z) representation is appropriate for near field analyses. A two-dimensional (x-y) representation is more appropriate for a more regional analysis.

### Parameter Adjustments

Since limited excavation involves the replacement of waste materials with other clean soils, those parameters related to material properties need to be adjusted. These parameters include moisture content characteristics, hydraulic conductivity, bulk density, and porosity. The back fill materials will probably be taken from a locally available source. Values for a range of different material types are presented in Subsection 6.3.

### 6.2.7 Hydraulic Dredging and Seepage Basins

These remedial actions are represented in a group because of similar simulation requirements. Hydraulic dredging is used to remove liquids and/or sludge from lagoons or surface impoundments. Seepage basins are used to recharge treated water back into the ground. Such water may originate from pumping wells or surface water diversion structures.

### Model Type

An unsaturated zone model can be used for both actions. If the extent of mounding caused by a seepage basin is of concern, a saturated zone model can be linked to the unsaturated zone model.

### Dimensionality and Grid Configuration

A one-dimensional (z), representation would be the minimum for the unsaturated zone. The grid configuration would be a series of compartments or elements representing the soils below the base of the pond or basin. Vertical variations in soil characteristics can be represented by assigning different properties to the compartments/elements. In cases where lateral variations are important or seepage from the sides of the pond/basin need to be considered, a two-dimensional (x-z) representation is needed.

The dimensionality for the saturated zone model would be x-y. Such a representation would make it possible to predict changes in water table elevations (i.e., mounding) produced by recharge from a seepage basin.

### Parameter Adjustments

The required parameter adjustments are limited to changing

Remedial Action Modeling Requirements 279

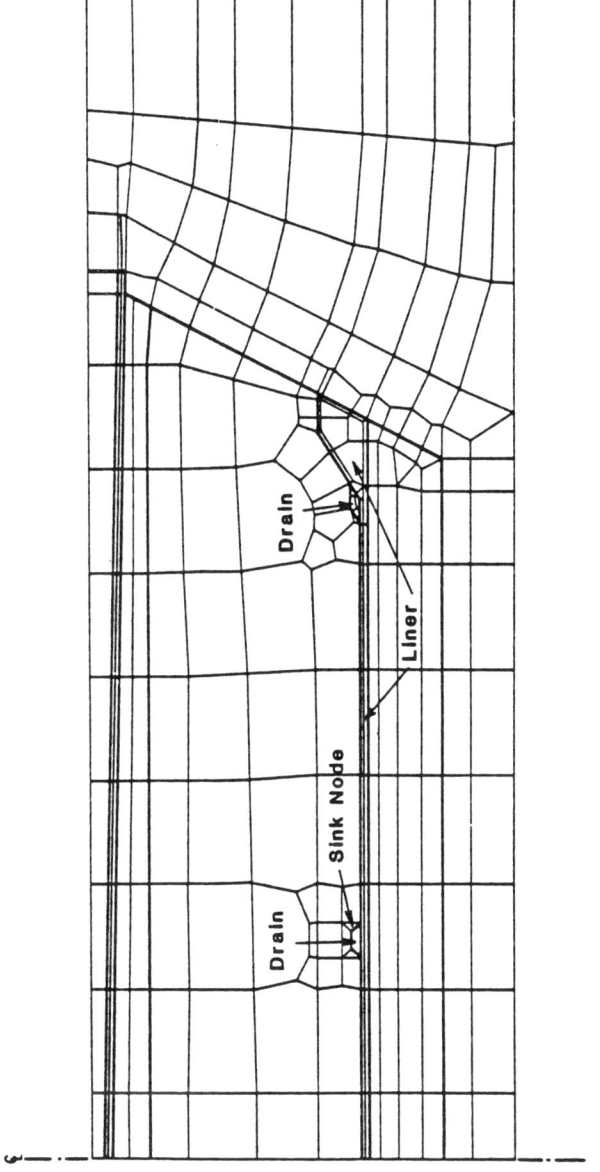

Figure 6.5 Example representation and grid for a drain system used to evaluate Uranium mill tailings seepage into the unsaturated zone (after Nelson et al., 1983).

heads or fluxes to represent the removal of water and waste materials as in the case of hydraulic dredging or to represent the ponding of water as in the case of seepage basins.

### 6.2.8 Bioreclamation and Chemical Injection

The simulation requirements for these actions are similar to those for ground-water pumping. The major difference is in the parameter adjustments required to simulate the in-situ treatment of waste constituents.

#### Dimensionality and Grid Configuration

The dimensionality and grid configuration requirements are basically the same as for a ground-water pumping action: two-dimensional (x-y) with either regular or irregular grid spacing. The analysis of partially-penetrating injection/withdrawal wells may require a three-dimensional model.

#### Parameter Adjustment

Heads or fluxes and held concentrations must be specified for those nodes representing injection wells. The held concentrations will depend upon the efficiency of the on-site treatment system. Withdrawal wells are represented by assigning nodes or fluxes to appropriate nodes.

The effects of chemical injection and bioreclamation require that the degradation rate assigned to some of the grid blocks or elements be adjusted. Such an adjustment is complicated by the fact that the elements requiring adjustment will change with time as the chemical or bacteria migrate away from the injection wells. Few, if any, saturated zone models can handle such changes in parameter values unless they offer a restart capability. Therefore, the only way to evaluate the performance of these technologies is through a steady-state analysis. Such an analysis would initially involve using a flow model in a steady-state mode, to identify the region of the flow field influenced by the injection and extraction wells (i.e., the treated zone). All water within this region is influenced by the wells, while all water outside the region is influenced by the regional ground-water flow system. Once the region has been identified, the steady-state flow field can be input to a transport model. The degradation rate for those elements encompassing the region can be assigned values typical of those for the action. The degradation rates for the other elements would remain unchanged.

Limited data are currently available on ranges of degradation rates for bioreclamation and chemical injection.

### 6.2.9 Permeable Treatment Beds

The purpose of modeling a permeable treatment bed is to determine the extent to which plume concentrations are reduced as a result of in-situ treatment.

#### Model Type

A saturated zone model is appropriate for most analyses.

#### Dimensionality and Grid Configuration

In most situations, a two-dimensional x-z representation can be used. In using a x-z representation it is impossible to evaluate the possibility for plume excursion around the ends of the treatment bed. An x-y representation would be required if plume excursion is of concern. Such a representation, however, presumes the treatment bed is fully-penetrating.

The only major consideration in designing the grid is to ensure that grid blocks or elements are established to represent the treatment bed. Since the treatment beds are designed to have permeabilities similar to the surrounding media, they generally will not alter ground-water flow patterns significantly (JRB Associates, 1982). For this reason, there is no need to modify grid spacing near the treatment bed.

#### Parameter Adjustments

Assuming the permeability of the material selected for the treatment bed is similar to that for the surrounding media only the retardation factor assigned to those grid blocks or elements representing the treatment bed need to be adjusted. In some models, the retardation factor is actually adjusted. In others, the sorption coefficient and/or porosity are adjusted so that the internally calculated retardation factor is correct. Only limited data are available on ranges of parameter values for treatment bed materials. Roberts (1982) reported that equilibrium partition coefficients for activated carbon can range between 0.005 to 0.1 ml/gm.

If the permeability of the bed materials is different, the material properties assigned to the elements representing the bed also need to be adjusted. No data were found on the in-situ permeability of activated carbon or crushed limestone materials. Subsection 6.3 provides parameter estimation guidance for natural aquifer materials.

## 6.3 PARAMETER ESTIMATION GUIDANCE

The parameters requiring adjustment to simulate the remedial measures discussed in the previous section can be grouped as follows: 1) surface zone modeling parameters, and 2) unsaturated and saturated zone modeling parameters. This section seeks to provide sources of data and techniques for the estimation of selected model parameters.

The guidance presented herein is only meant to be used in _support_ of, rather than in _place_ of on-site field measurements, sampling and laboratory studies. To the extent possible, values for model parameters should be determined as part of the site characterization phase of the Remedial Investigation process. This process is meant to fill limitations in the existing data base for a site and provide the data required to evaluate remedial action alternatives (EPA, ~1984). Hopefully, this section can be used to more fully understand those data required for remedial action modeling and, in the absence of site specific data, aid in parameter estimation.

Where available, data sources and estimation techniques pertinent to remedial action specific parameters are provided. Both are extemely limited, however. For this reason, more general data sources and estimation techniques are discussed to provide a basis for at least the initial determination of appropriate parameter values.

### 6.3.1 Surface Zone Model Parameter Guidance

The key surface zone model parameters requiring adjustment are those related to: 1) channel/surface roughness, 2) evapotranspiration, 3) interception, 4) infiltration, and 5) soil erodibility. Available guidance on the estimation of remedial action performance is provided below.

#### 6.3.1.1 Channel/Surface Roughness--

In most surface zone models, the roughness of land or channel segments is defined in terms of a parameter known as the roughness coefficent, the most common being Manning's "n." Donigian et al., (1983) note that most of the published values for Manning's "n" are for channel rather than overland flow.

Most standard open channel flow references provide ranges of values for different channel types. Table 6.3 lists values for lined and unlined channels typical of those that might be

TABLE 6.3  CHANNEL AND LAND SURFACE MANNING's 'n'
VALUES APPLICABLE TO REMEDIAL ACTION
MODELING

| Channel Type* | Manning's 'n' Value |
|---|---|
| Smooth concrete | 0.012 |
| Ordinary concrete | 0.013 |
| Shot concrete, untroweled and earth channels in good condition | 0.017 |
| Straight unlined earth channels in good condition | 0.020 |
| Grass covered waterways | 0.02-0.4 |

| Land Surface Condition** | |
|---|---|
| Smooth fallow | 0.15-0.20 |
| Rough fallow, cultivated | 0.20-0.30 |
| Light turf | 0.25-0.35 |
| Heavy turf | 0.30-0.40 |

---

\*     Values taken from Chow (1964)

\*\*    Values recommended by Donigian et al. (1983)

constructed at a hazardous waste site. As this table shows, Manning's "n" values for grass covered waterways can be highly variable. The actual value depends upon the vegetal retardance, flow velocity and hydraulic radius of the channel. Chow (1964) provides guidance on the estimation of appropriate values given the design of the waterway.

Table 6.3 also lists ranges of values recommended by Donigian et al., (1983) for different land surface conditions. Specific published values include those by Ree et al. (1977). They calculated values of 0.25 to 0.62 for grass cover conditions. Ross et al. (1977) used values of 0.35 to 0.40 for agricultural areas and 0.30 for forested areas.

6.3.1.2 Evapotranspiration--

Evapotranspiration is the process by which water is carried from the soil by either direct vaporization from the soil or by transpiration of plants. The maximum rate of evapotranspiration (potential ET) depends on the demand from the atmosphere and the nature of the evaporating surface be it soil or plant. The actual rate depends on the moisture available to evaporate from the surface and the soil. Linsley et al. (1982) discuss methods for calculating ET based on water and energy budget methods, meteorological data and pan evaporation data.

One common method is a two-phased approach. First, the potential ET is calculated using pan evaporation data. Most U.S. weather stations provide pan evaporation data along with other standard meteorological data. The pan factor is then used to convert the daily pan evaporation data into daily potential ET. The second phase involves calculating the actual ET from the surface and soil, based on the available water and potential ET. Models using this method only require the appropriate pan evaporation factor along with pan evaporation data from a nearby weather station. Figure 6.6 provides pan factors for the entire contiguous United States.

6.3.1.3 Interception--

The interception parameter in surface zone models represents a storage depth or volume for precipitation that is trapped on the surface of vegetation. Precipitation in excess of the interception storage is assumed to reach the soil surface. Interception storage is related directly to the density of the vegetation cover. Several publications provide ranges of values for different agricultural crops (Woolhiser, 1976; Donigian et al., 1983; Knisel, 1980; and Carsel et al., 1984). Typical values range from 0.0 to 0.25 cm. Table 6.4 lists general ranges of values for different vegetation densities.

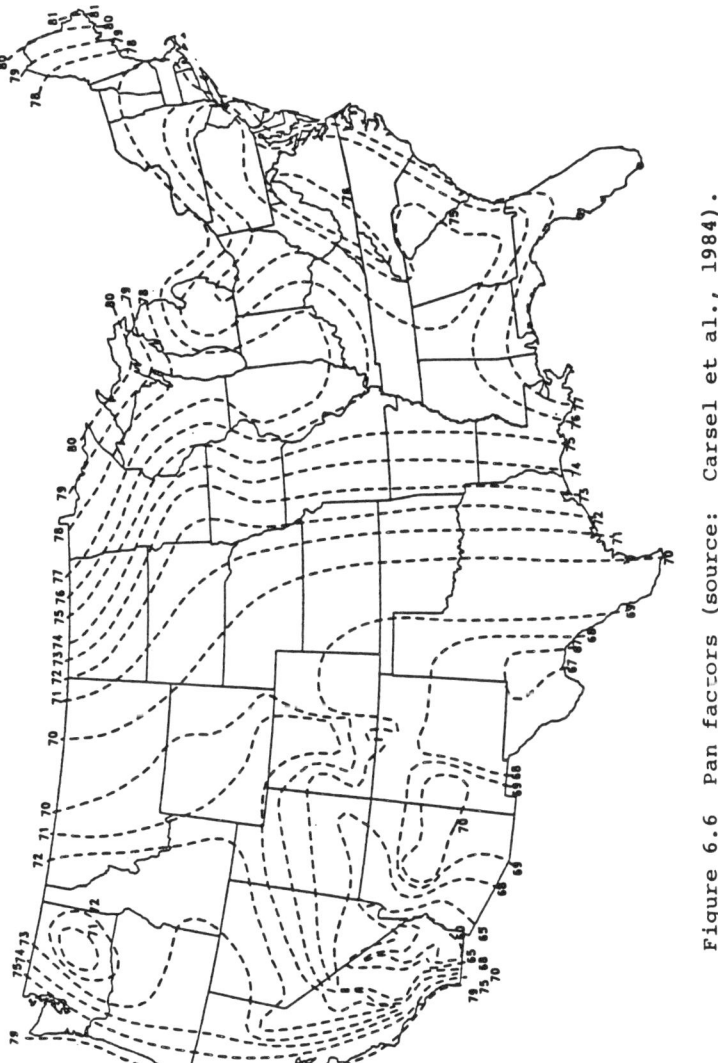

Figure 6.6 Pan factors (source: Carsel et al., 1984).

TABLE 6.4  INTERCEPTION STORAGE FOR DIFFERENT VEGETATIVE DENSITIES

| Density | Interception Storage (cm) |
|---|---|
| Light | 0.0 - 0.15 |
| Moderate | 0.20 - 0.30 |
| Heavy | 0.30 - 0.45 |

### 6.3.1.4 Infiltration--

Infiltration is a complex process that depends on many physical factors: 1) the soil type, 2) antecedent moisture, 3) organic matter, 4) vegetative cover, and 5) rate of water supply to surface. The sophistication with which surface zone models handle infiltration varies. Often the infiltration rate is calculated within the model and does not require any special parameters. Two examples of infiltration estimations are presented.

One of the simplest approaches is the calculation of the average infiltration rate, the W index (see Linsley et al., 1982):

$$W = \frac{1}{t} (P - Q_s - S) \qquad (6.1)$$

where: $W$ = average infiltration rate, L/T
$t$ = duration of precipitation, T
$P$ = total precipitation during time, L
$Q_s$ = surface runoff, L
$S$ = effective surface retention, L.

Another approach developed by Holtan et al., (1975) incorporates the effects of vegetative cover in the calculation of the maximum infiltration rate, f:

$$f = a (S_c - S)^{1.4} + f_c \qquad (6.2)$$

where $f$ = infiltration rate, L/T
$a$ = vegetative parameter, L/T (see Table 6.5)
$S_c$ = soil water capacity exceeding wilting point, $L^3/L^3$
$S$ = soil water in excess of wilting point, $L^3/L^3$
$f_c$ = minimum infiltration rate after prolonged wetting, L/T

The maximum infiltration capacity as suggested by the equation above depends on the antecedent moisture content. The minimum infiltration rate, $f_c$, is the saturated hydraulic conductivity. Once the profile is saturated, the infiltration rate is limited by the speed at which water can move in the soil represented by the saturated hydraulic conductivity.

### 6.3.1.5 Soil Erodibility--

A range of algorithms are used in surface zone models to simulate the process of soil erosion. Some are based on more

TABLE 6.5  VALUES OF 'a' FOR EQUATION (6.2)
(From Holtan et al., 1975)

|  | 'a'<br>(in./hr per in.**1.4 of available storage) | |
|---|---|---|
| Cover | Poor Condition | Good Condition |
| Fallow | 0.10 | 0.30 |
| Row crops | 0.10 | 0.20 |
| Small grains | 0.20 | 0.30 |
| Hay | | |
|    Legumes | 0.20 | 0.40 |
|    Sod | 0.40 | 0.60 |
| Pasture | | |
|    Bunchgrass | 0.20 | 0.40 |
|    Temporary (sod) | 0.40 | 0.60 |
|    Permanent (sod) | 0.80 | 1.00 |
| Woods and forests | 0.80 | 1.00 |

mechanistic descriptions while others are strictly empirical. Thus, it is difficult to provide guidance on the adjustment of specific parameters because they are often model-dependent. Some general guidance can be provided, however, since most of the commonly used algorithms were derived to take advantage of the wealth of information generated by the Soil Conservation Service in their development of the Universal Soil Loss Equation (USLE):

$$Y(s) = A\ (R \cdot K \cdot LS \cdot C \cdot P)\ S_d \qquad (6.3)$$

where  $Y(s)$ = sediment loading from surface erosion, tons/year
$A$ = drainage area, acres
$R$ = rainfall factor, expressing the erosion potential of average annual rainfall
$K$ = soil erodibility factor, expressed in tons per acre per R unit
$LS$ = topographic factor, a combination of the slope-length and slope-steepness, dimensionless
$C$ = Cover management factor, representing the degree of soil disturbance and vegetative cover density, dimensionless
$P$ = erosion control practice factor, accounting for practices that act to reduce erosion, dimensionless
$S_d$ = sediment delivery ratio, dimensionless

Parameter values for most of the factors are well documented for agricultural areas. However, values have also been derived for construction and mining conditions that would be applicable to hazardous waste sites. Detailed guidance and estimation methods are available for each factor. Rather than repeat it herein, it is suggested that the following sources be consulted: Wischmeier and Smith, 1978; EPA, 1975; and Mills et al., 1982.

### 6.3.2 Subsurface Modeling Parameters

The subsurface modeling parameters are divided into two categories: 1) flow-related parameters and 2) transport-related parameters. Flow parameters affect the hydraulic flow field and the general velocity of the ground water. The transport-related parameters affect the migration and fate of the contaminant.

6.3.2.1 Flow-related Parameters--

Three key parameters that affect the flow of water are moisture content characteristics, hydraulic conductivity and, for transient flow problems, the storage coefficient. All three of these parameters vary over a large range and are highly specific to the materials being considered.

6.3.2.1.1. Moisture Content Characteristics--In the unsaturated zone, where all the pores are not filled with water, the soil moisture content is an important physical property which affects plasticity, strength, microbial activity and the chemical state of the soil. The negative pressure head (suction) determines the degree with which water is held in the soil matrix. This is the amount of suction that is required to remove the water. At zero pressure all the pores are filled with water and the soil is saturated. As the pressure decreases (suction increases) the water content decreases.

The relation between moisture content and pressure head is described by the soil moisture characteristic curve. The shape and intercepts of this curve depends on the physical properties of the soil matrix: the pore size distribution, grain size, and mean pore diameter. Figure 6.7 illustrates the influence of soil type on the characteristic curve. Empirical measurements in the lab or field must be made for each site to accurately determine the characteristic curve. Hillel (1982), discusses methods for measuring water content and pressure head.

In the absence of any laboratory or field data several methods have been developed to determine the moisture content at given pressure heads. One method, developed by Rawls and Brakensiek (1982), uses bulk density and organic matter content, as well as soil texture. They developed the following regression equation to estimate the water content at given negative pressure heads (suction):

$$\Theta = a + (b \times sand\%) + (c \times clay\%) + (d \times organic\ matter) + (e \times bulk\ density\ (gr/cm^3)) \tag{6.4}$$

where $\Theta$ = water content, $L^3/L^3$
a-e = regression coefficients

Table 6.6 shows the values of the regression coefficients to to be used at selected pressure heads.

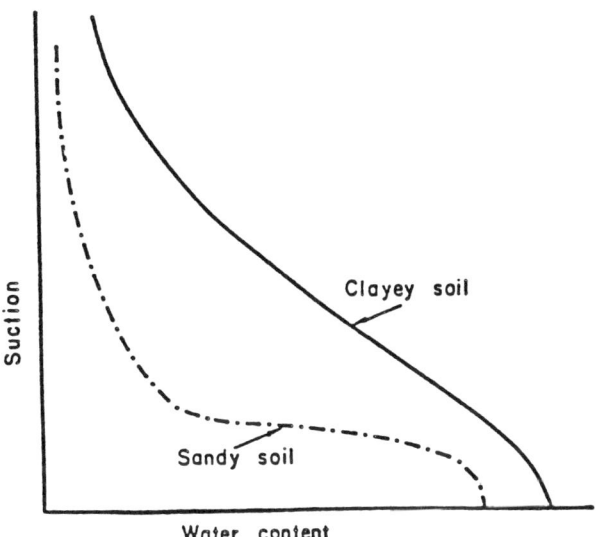

Figure 6.7  The effect of soil type on soil-water retention (source: Hillel, 1982). Copyrighted by Academic Press.

TABLE 6.6  COEFFICIENTS FOR LINEAR REGRESSION EQUATIONS FOR PREDICTION OF SOIL WATER CONTENTS AT SPECIFIC MATRIC POTENTIALS (source: Carsel et al., 1984)

| Matric Coefficient | Intercept a | Sand (%) b | Clay (%) c | Organic Matter (%) d | Bulk Density (g cm$^{-3}$) e | $R^2$ |
|---|---|---|---|---|---|---|
| -0.20 | 0.4180 | -0.0021 | 0.0035 | 0.0232 | -0.0859 | 0.75 |
| -0.33 | 0.3486 | -0.0018 | 0.0039 | 0.0228 | -0.0738 | 0.78 |
| -0.60 | 0.2819 | -0.0014 | 0.0042 | 0.0216 | -0.0612 | 0.78 |
| -1.0 | 0.2352 | -0.0012 | 0.0043 | 0.0202 | -0.0517 | 0.76 |
| -2.0 | 0.1837 | -0.0009 | 0.0044 | 0.0181 | -0.0407 | 0.74 |
| -4.0 | 0.1426 | -0.0007 | 0.0045 | 0.0160 | -0.0315 | 0.71 |
| -7.0 | 0.1155 | -0.0005 | 0.0045 | 0.0143 | -0.0253 | 0.69 |
| -10.0 | 0.1005 | -0.0004 | 0.0044 | 0.0133 | -0.0218 | 0.67 |
| -15.0 | 0.0854 | -0.0004 | 0.0044 | 0.0122 | -0.0182 | 0.66 |

6.3.2.1.2 Hydraulic Conductivity--The hydraulic conductivity is a measure of the ease with which a fluid is transmitted through a porous medium. It is one of the most important and most variable physical properties governing flow in both the saturated and unsaturated zones. For remedial action modeling, the correct assessment of hydraulic conductivities for example in a slurry trench is critical for the accuracy of a modeling effort. Both the fluid properties and the media properties contribute to the hydraulic conductivity:

$$K = \frac{k \rho g}{\eta} \qquad (6.5)$$

where  $K$ = hydraulic conductivity, L/T
$k$ = intrinsic permeability of the porous medium, $L^2$
$\eta$ = viscosity of fluid, M/TL
$g$ = gravitational constant, $L/T^2$
$\rho$ = fluid density, $M/L^3$

For most studies, the fluid of concern is water. However when a dissolved contaminant affects the density and viscosity of the ground water, these factors should be considered. Viscosity accounts for the fluid's internal resistance to flow. Density compensates for the effects of gravity. The tables (and equations) that follow assume water is the fluid.

The effects of the media on hydraulic conductivity are encompassed in the intrinsic permeability. Attempts have been made to quantify the media effects based on porosity, pore size distribution, and surface area without much success. The Kozeny-Carman theory estimates the hydraulic conductivity of the well graded sands based on a pore shape factor, porosity, specific surface area and tortuosity. Its application however is limited by the difficulty to measure the dependent variables.

The most reliable method is to directly measure the hydraulic conductivity in the field or laboratory. Roberts (1984) describes a variety of measurement techniques for hydraulic conductivity. In the absence of any field or laboratory measurement, Table 6.7 lists order of magnitude estimates of hydraulic conductivity for selected materials that could be encountered in remedial action modeling. For more detailed guidance on hydraulic conductivities relating to slurry walls and grout curtains, see JRB Associates (1984) and Shafer et al. (1984).

In a previous modeling study Cohen and Mercer (1984) used 3.5 x $10^{-12}$ m/sec as a conductivity for a proposed concrete cut-off wall at Love Canal. Anderson et al. (1984) used conducivity of $10^{-6}$ times lower than that of the surrounding aquifer materials in their analysis of a slurry wall at the Lipari

TABLE 6.7 RANGES OF HYDRAULIC CONDUCTIVITIES FOR DIFFERENT MATERIALS (adapted from Spooner et al., 1983, Freeze and Cherry, 1979, and Morris and Johnson, 1967)

| Material | Hydraulic Conductivity (m/s) |
|---|---|
| Clay | $10^{-10} - 10^{-8}$ |
| Soil Bentonite | $5 \times 10^{-10} - 10^{-7}$ |
| Cement Bentonite | $10^{-8}$ |
| Silt/Loess | $10^{-9} - 10^{-5}$ |
| Sand | |
| fine | $10^{-5} - 10^{-4}$ |
| medium | $10^{-4} - 10^{-3}$ |
| course | $10^{-3} - 10^{-2}$ |
| Gravel | $10^{-3} - 10^{-1}$ |

Landfill. Silka and Mercer (1982) used conductivities of $10^{-5}$ m/sec and $10^{-6}$ m/sec for barrier drain materials and a clay cap in a analysis of Love Canal.

In the unsaturated zone the hydraulic conductivity depends on soil moisture and the pressure head, as well as fluid and media properties. The hydraulic conductivity in the soil can vary over several orders of magnitude simply depending on the soil moisture. The physical properties of the soil influence the relation between hydraulic conductivity and suction as shown by the curve in Figure 6.8.

As for the soil-moisture characteristic curve, it is best to determine the relationship between hydraulic conductivity and pressure head in the field or laboratory. Alternately, the unsaturated hydraulic conductivity can be determined by knowing the saturated hydraulic conductivity and the soil-moisture curve. van Genuchten (1978b) has developed a closed form analytical solution for unsaturated hydraulic conductivity based on the soil-moisture characteristic curve.

6.3.2.1.3  Storage Coefficient--The storage coefficient is the volume of water released from storage per unit surface area of aquifer for a unit decline in the water table elevation or piezometric surface in an unconfined or confined aquifer. This parameter is necessary when simulating transient flow conditions.

In unconfined aquifers the storage coefficient more commonly known as the specific yield, is much higher in magnitude than the storativity of a confined aquifer. Specific yield generally ranges from 0.01 to 0.30 (Freeze and Cherry, 1979). Table 6.9 in the section on porosity shows typical values for selected materials (porosity and effective porosity).

In confined aquifers storativity ranges from .005 to .00005 (Freeze and Cherry, 1979). The higher values correspond to aquifers with more easily compressed materials. The water released from storage comes from the compaction of the aquifer material and the expansion of water under lower-pressure. In an unconfined aquifer the water released from storage actually drains from the aquifer. These two processes account for the difference in magnitude.

6.3.2.2 Transport-Related Parameters--

The important transport parameters that require adjustment for remedial action modeling are: 1) dispersivity, 2) porosity, 3) bulk density, 4) sorption coefficient, and 5) degradation rate. All five of these parameters affect the movement of hazardous wastes constituents and must be correctly adjusted to represent the effects of different

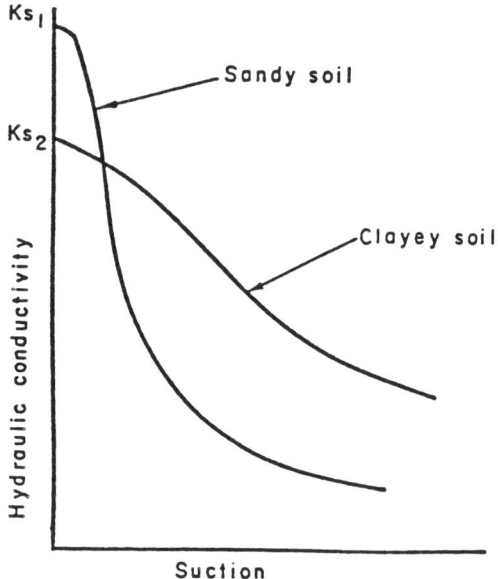

Figure 6.8 Dependence of conductivity on suction in soils of different texture--log-log scale (source: Hillel, 1982). Copyrighted by Academic Press.

remedial actions.

6.3.2.2.1 Dispersivity--Dispersion is the spreading of a contaminant due to two processes: 1) velocity- and flow-related mechanical dispersion and 2) molecular diffusion. Mathematically, the dispersion coefficent is estimated by:

$$D = \alpha \overline{V} + D_m \qquad (6.6)$$

where  $D$ = dispersion coefficient, $L^2/T$
$\alpha$ = dispersivity, L
$\overline{V}$ = mean ground-water velocity, L/T
$D_m$ = molecular diffusion, $L^2/T$

In most modeling studies the molecular diffusion is considered negligible. Mechanical dispersion is the product of dispersivity and ground-water velocity. It is the result of velocity variations in an individual pore space and in pores of differing sizes and because of the tortuous flow path that water must take around the grains in the porous medium.

Experimental and theoretical work has shown that dispersivity is scale dependent. Lab experiments report small dispersivities on the order of centimeters. Field scale experiments have dispersivities ranging from a few meters up to hundreds of meters (see Figure 6.9). However, there has been little success in developing techniques for accurate estimation. For the most part dispersivity has been used as a model calibration parameter, not necessarily reflecting a physically meaningful number (see Anderson, 1984 for more detailed discussion). Table 6.8 shows dispersivity values that have been measured and used in other modeling studies.

6.3.2.2.2 Porosity--Porosity is defined as the percent void volume in a representative volume of the porous medium. In the saturated zone this entire volume is filled by water. In the unsaturated zone the pore space is filled by both water and air. In modeling contaminant transport, porosity is necessary to determine the average ground-water velocities and associated contaminant velocities. For a given average flux rate (specific discharge), a porous medium with high porosity will have a slower pore water velocity than material with a low porosity. Low porosity material has fewer voids for water to flow through so a higher velocity is necessary for the specific discharge to equal that of a high porosity material.

Porosity depends on particle size, particle size distribution and degree of lithification. For a single particle size class, in general, the larger the particle size the higher the porosity. However, if the aquifer material can fill in the

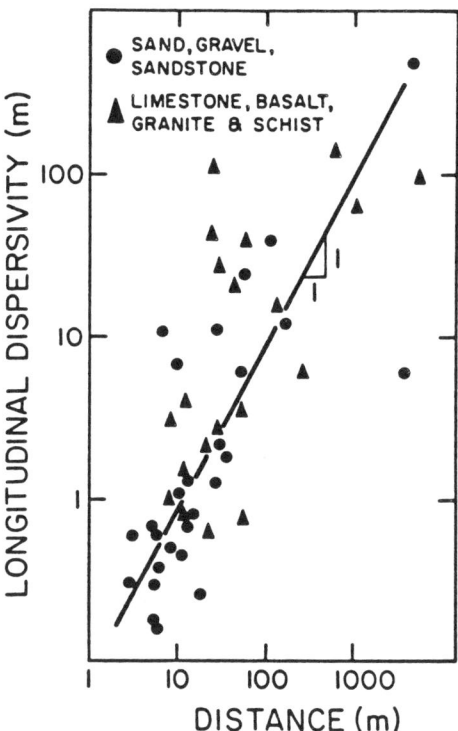

Figure 6.9 Variation of dispersivity with distance (source: Anderson 1984; adapted from Lallemand-Barres and Peaudecerf, 1978).

TABLE 6.8  SMALL SCALE AND REGIONAL DISPERSIVITY VALUES
(adapted from Anderson, 1984)

Localized Scale

| Aquifer Type | Location | Distance Between Wells | $\alpha x$ | $\alpha x/\alpha y$ |
|---|---|---|---|---|
| Alluvial | Chalk River, Ontario | | 0.34 – 1 | |
| | Lyons, France | | 5.0 | |
| | | | 12.0 | .9 – 3.9 |
| | | | 8.0 | .8 – 530 |
| | | | 5.0 | .34 – 34.5 |
| | | | 7.0 | 7.0 – 780 |
| | Barstow, CA | 6.4 | 12.0 | 3.0 |
| | Tucson, AZ | 80 | 15.2 | |
| | | | 15.2 | |
| Fractured Dolomite | Carlsbad, NM | 38.1 – 54.9 | 38.1 | |
| Fractured Chalk | Dorset, England | 8 | 3.1 | |
| Chalk | Dorset, England | 8 | 1 | |

Regional Scale

| Aquifer Type | Location | Distance Between Wells | $\alpha x$ | $\alpha x/\alpha y$ |
|---|---|---|---|---|
| Alluvial | Rocky Mt. Arsenal, CO | | 30.5 | 1 |
| | Colorado | | 30.5 | 3.3 |
| | California | | 30.5 | 3.3 |
| | Sutter Basin, CA | | 80 – 2000 | 10 |
| | Alsace, France | | 15 | 15 |
| Limestone | Cutler, FL | | 22 | 10.0 |
| | Brunswick, GA | | 61 | 3.3 |
| Glacial Deposits | Long Island, NY | | 21.3 | 5 |
| Glacial Till | Alberta, CA | | 3–6 | 5 |

space between large particles, the porosity will decrease. Lithification decreases porosity by compacting the sediments and eliminating the pore space.

Porosity and effective porosity should be distinguished. Most transport equations use effective porosity which does not include dead-end and unconnected pores. Effective porosity approximately equals the specific yield, which is the amount of water that will drain from a given saturated soil sample. Table 6.9 lists ranges of porosity and effective porosity for selected materials that may be used in different remedial actions.

6.3.2.2.3 Bulk Density--Bulk density is the dry particle mass per unit volume of soil. It is a basic property in the estimation of the retardation factor. Table 6.10 and Figure 6.10 can be used to estimate values for different materials.

6.3.2.2.4 Sorption Coefficients--The most common sorption parameter in unsaturated and saturated zone models is the partition coefficient or $K_d$. In using this parameter to simulate sorption, it is assumed that the process is linear, completely reversible and rapid relative to the time step in the model. Where these assumptions are not valid, other descriptions must be used (e.g., Freundlich and Langmuir isotherms). Cherry et al. (1984) discuss the assumptions and important limitations inherent in the use of different isotherms to describe the sorption process, particularly with respect to inorganic pollutants whose mobility is controlled by precipitation/dissolution, oxidation/reduction and chemical speciation reactions. Both Cherry et al. (1984) and Rai et al. (1984a) discuss the role these reactions play in controlling the chemical mobility.

There are three basic approaches for estimating sorption parameter values. The most preferred is to conduct in-situ tracer experiments or laboratory batch or column experiments using soil and ground-water samples from the site. In many cases, however, time and resource limitations preclude the use of such procedures.

A second approach is to use literature data derived from field or laboratory experiments. There are a growing number of useful compilations of sorption data for both organic and inorganic pollutants,. Dawson et al. (1980) reviewed literature data on 250 chemicals commonly found in hazardous waste streams. Available determinations of $K_d$ values are presented for each chemical. Another useful source is the Fate of Organic Compounds in Soils (FOCIS) database currently being developed by Battelle, Pacific Northwest Laboratories for the EPA Environmental Research Laboratory in Ada, Oklahoma. This computerized database contains literature data

TABLE 6.9  RANGES OF POROSITY AND EFFECTIVE POROSITY VALUES FOR SELECTED MATERIALS (sources: Morris and Johnson, 1967 and Davis and DeWeist, 1966)

|          |        | Porosity |         | Effective Porosity Specific Yield | |
|----------|--------|---------|---------|-------|---------|
| Material |        | Range   | Average | Range | Average |
| Clay     |        | 33-65   | 42      | 0-18  | 5       |
| Silt     |        | 33-61   | 46      | 1-39  | 15      |
| Sand     | fine   | 25-53   | 43      | 1-46  | 30      |
|          | medium | 27-49   | 39      | 16-46 | 32      |
|          | coarse | 30-46   | 39      | 18-42 | 29      |
| Gravel   | fine   | 25-40   | 34      | 12-40 | 28      |
|          | medium | 24-44   | 32      | 17-43 | 24      |
|          | coarse | 24-35   | 28      | 12-26 | 22      |

TABLE 6.10  RANGE OF BULK DENSITY (gm/cm$^3$) FOR DIFFERENT MATERIALS (source: Morris and Johnson, 1967 and Baes and Sharp, 1983)

| Material | | Range | Mean |
|---|---|---|---|
| Sandstone | fine | 1.34 - 2.32 | 1.76 |
| | medium | 1.50 - 1.86 | 1.68 |
| Siltstone | | 2.52 - 2.89 | 2.65 |
| Claystone | | 2.50 - 2.76 | 2.66 |
| Shale | | 2.47 - 2.83 | 2.69 |
| Sand | fine | 1.13 - 1.99 | 1.55 |
| | medium | 1.27 - 1.93 | 1.69 |
| | coarse | 1.42 - 1.94 | 1.73 |
| Gravel | fine | 1.60 - 1.99 | 1.76 |
| | medium | 1.47 - 2.09 | 1.85 |
| | coarse | 1.69 - 2.08 | 1.93 |
| Silt | | 1.01 - 1.79 | 1.38 |
| Clay | | 1.18 - 1.72 | 1.49 |
| Silt Loams | | 0.86 - 1.67 | 1.32 |
| Clay and Clay loams | | 0.94 - 1.54 | 1.30 |
| Sandy Loams | | 1.25 - 1.76 | 1.49 |
| Silt Loams | | 1.02 - 1.58 | 1.22 |
| Loams | | 1.16 - 1.58 | 1.42 |
| All Soils | | 0.86 - 1.76 | 1.35 |

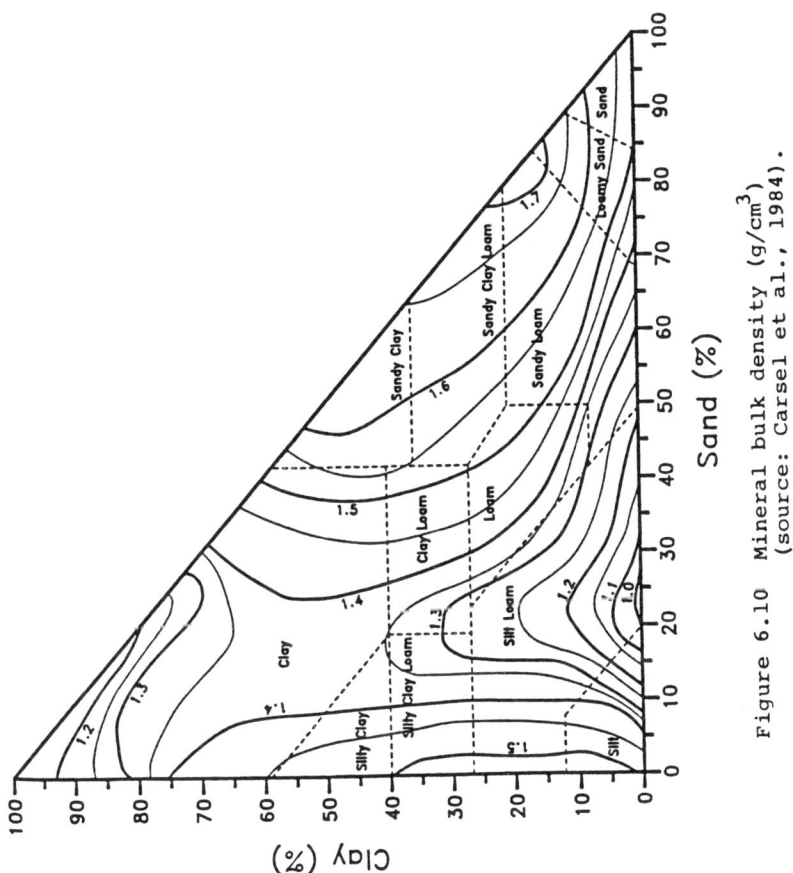

Figure 6.10 Mineral bulk density (g/cm³) (source: Carsel et al., 1984).

on sorption parameters, including details on experimental conditions (i.e., soil properties, water chemistry and laboratory methods). Published data on the sorption of inorganic constituents commonly found in utility solid wastes can be found in a recent publication by Rai et al. (1984a and 1984b). The first volume provides summaries of sorption parameter values for each constituent, while the latter is an annotated bibliography of over 350 publications. In using any of the above databases, it is critical to remember that sorption parameters are <u>only</u> applicable to the experimental conditions under which they were measured. The application of parameter values to other conditions should be approached with great care.

The final approach is to use one of the many available empirical estimation techniques that provide relationships between the sorption coefficient and other basic physical-chemical properties (e.g., solubility and octanol water partition coefficient). One useful summary of available relationships can be found in Lyman et al., (1982); this summary is given in Table 6.11. Most of the available relationships give a $K_{oc}$ or organic carbon partition coefficient. A $K_{oc}$ can be converted to a $K_d$ by:

$$K_d = K_{oc} \cdot f_{oc} \qquad (6.7)$$

where $f_{oc}$ = weight percent of the solid phase composed of organic carbon.

Typical values of $f_{oc}$ range from 0.4 to 10.0 percent (Brady, 1974). Values of solubility or octonal-water partition coefficients can be found in a number of compilations of chemical properties (Dawson et al., 1980; Mabey et al., 1982; Leo et al., 1971; Hansch and Leo, 1979; Verschueren, 1977; and Sax, 1979). Lyman et al., (1982) provide techniques for estimating solubility and octonal-water partition coefficients when literature data are not available.

In using the relationships in Table 6.11, or other similar relationships, it is important to recognize that sorption is assumed to be "keyed" solely to the organic carbon content of the soil/sediment (Cherry et al., 1984). McCarty et al., (1981) note that at low organic contents (say <1%) typical of those found in deep aquifer materials, the inorganic composition of a soil can have a larger affect on sorption. McCarty et al., (1981) provide a relationship determining the critical organic fraction:

TABLE 6.11  REGRESSION EQUATIONS FOR THE ESTIMATION OF $K_{oc}$ (source: Lyman et al., 1982.) Copyrighted by McGraw-Hill

### Regression Equations for the Estimation of $K_{oc}$

| Eq. No. | Equation[a] | No.[b] | $r^2$[c] | Chemical Classes Represented | Ref. |
|---|---|---|---|---|---|
| 4.5 | log $K_{oc}$ = –0.55 log S + 3.64* | 106 | 0.71 | Wide variety, mostly pesticides | 26 |
| 4.6 | log $K_{oc}$ = –0.54 log S + 0.44** | 10 | 0.94 | Mostly aromatic or polynuclear aromatics; two chlorinated | 25 |
| 4.7[d] | log $K_{oc}$ = –0.557 log S + 4.277*** | 15 | 0.99 | Chlorinated hydrocarbons | 11 |
| 4.8 | log $K_{oc}$ = 0.544 log $K_{ow}$ + 1.377 | 45 | 0.74 | Wide variety, mostly pesticides | 26 |
| 4.9 | log $K_{oc}$ = 0.937 log $K_{ow}$ – 0.006 | 19 | 0.95 | Aromatics, polynuclear aromatics, triazines and dinitroaniline herbicides | 9 |
| 4.10 | log $K_{oc}$ = 1.00 log $K_{ow}$ – 0.21 | 10 | 1.00 | Mostly aromatic or polynuclear aromatics; two chlorinated | 25 |
| 4.11 | log $K_{oc}$ = 0.94 $K_{ow}$ + 0.02 | 9 | e | s-Triazines and dinitroaniline herbicides | 7 |
| 4.12 | log $K_{oc}$ = 1.029 log $K_{ow}$ – 0.18 | 13 | 0.91 | Variety of insecticides, herbicides and fungicides | 36 |
| 4.13[d] | log $K_{oc}$ = 0.524 log $K_{ow}$ + 0.855 | 30 | 0.84 | Substituted phenylureas and alkyl-N-phenylcarbamates | 5 |
| 4.14[d,f] | log $K_{oc}$ = 0.0067 (P – 45N) + 0.237 | 29 | 0.69 | Aromatic compounds: ureas, 1,3,5-triazines, carbamates, and uracils | 10 |
| 4.15 | log $K_{oc}$ = 0.681 log BCF(f) + 1.963 | 13 | 0.76 | Wide variety, mostly pesticides | 26 |
| 4.16 | log $K_{oc}$ = 0.681 log BCF(t) + 1.886 | 22 | 0.83 | Wide variety, mostly pesticides | 26 |

*S in mg/ℓ.        **S in mol fraction.        ***S in μmol/ℓ.

[a] $K_{oc}$ = soil (or sediment) adsorption coefficient; S = water solubility; $K_{ow}$ = octanol-water partition coefficient; BCF(f) = bioconcentration factor from flowing water tests; BCF(t) = bioconcentration factor from model ecosystems; P = parachor; N = number of sites in molecule which can participate in the formation of a hydrogen bond.
[b] No. = number of chemicals used to obtain regression equation.
[c] $r^2$ = correlation coefficient for regression equation.
[d] Equation originally given in terms of $K_{om}$. The relationship $K_{om} = K_{oc}/1.724$ was used to rewrite the equation in terms of $K_{oc}$.
[e] Not available.
[f] Specific chemicals used to obtain regression equation not specified.

$$f_{oc} = \frac{S}{200} \frac{1}{K_{ow}^{0.89}} \qquad (6.8)$$

where  S = silica - specific surface area, $m^2/gm$

6.3.2.2.5  Degradation  Rate--The degradation of organic pollutants can occur as a result of chemical and biological reactions. The most important chemical reactions are oxidation, hydrolysis and reduction. Oxidation requires the presence of an oxidant (e.g., gaseous oxygen or free radicals like OH or peroxy RO2). Hydrolysis involves the introduction of a hydroxyl group into a compound. Organics susceptible to hydrolysis can be found in Callahan et al. (1979). Reduction involves the removal of a halogen atom through oxidation-reduction. This process is generally only of importance in low-redox state ground waters (Cherry et al., 1984).

Biological reactions are generally enzymatic reactions induced by bacteria. Until recently, biodegradation in ground water was neglected because microbial activity was assumed to be limited. Research by Wilson et al. (1982) and others has shown that significant microbial activity can occur in the saturated zone and that indigenous species are capable of degrading selected organics. Kobayashi and Rittman (1982) note that only broad guidelines can be given regarding the susceptibility of compounds to biodegradation.

It is important to note that only limited data exist on changes in biodegradation rates associated with the implementation of in-situ treatment remedial actions. This is in part due to the proprietary nature of certain treatment schemes and a general lack of data on the performance of the limited number of treatment systems that have been implemented.

As with sorption parameters, three basic approaches exist for estimating degradation rates for remedial action assessment. Field and/or laboratory determinations are clearly preferred whenever possible.

Literature data provide a second source. Table 6.12 lists the limited data that were found for bioreclamation actions. There are a number of useful compilations of measured rate constants for both chemical and biological degradation reactions (Callahan et al., 1979; Mills et al., 1982; Lyman et al., 1982; and Dawson et al., 1980). The FOCIS database being developed by Battelle, Pacific Northwest Laboratories for

TABLE 6.12 BIORECLAMATION DEGRADATION RATES FOR SELECTED WASTE CONSTITUENTS (Source: Personal communition with Mr. John Zikopoulos, Polybac, Inc., Allentown, Pennsylvania, April, 1983)

| Waste Constituent | Degradation Rate (/day) |
|---|---|
| Polyvinyl alcohol | 0.63-2.5 |
| Benzoic acid | 0.076-1.0 |
| Chloropropham | 0.01-0.03 |

EPA-Ada will also contain literature data on degradation rates, including field or laboratory experimental conditions. In using any of these sources it is again important to recognize the difficulties inherent in extrapolating rate constants to different site-specific conditions, particularly since much of the published data are for surface water rather than ground-water systems.

The third approach is to use estimation procedures such as those given in Lyman et al. (1982). Mills (1980) notes that oxidation and hydrolysis rates can be estimated within a factor of 3-5 and 2-3, respectively. The predictions of biodegradation rates is all but impossible according to Cherry et al. (1984).

# References

Anderson, M.P. 1979. Using Models to Simulate the Movement of Contaminants Through Ground-Water Flow Systems, In: CRC Critical Reviews in Environmental Control, Vol 9, No. 2.

Anderson, M. 1984. "Movement of Contaminants in Ground Water Transport - Advection and Dispersion," In: Ground Water Contamination, Studies in Geophysics, National Research Council, National Academy Press, Washington, D.C.

Anderson, P.F., C.R. Faust and J.W. Mercer. 1984. "Analysis of Conceptual Designs for Remedial Measures at Lipari Landfill, New Jersey," Ground Water, Vol. 22., No. 2.

Atwood, D.F. 1984. Management of Contaminated Ground Water with Aquifer Simulation and Linear Programming: The Development of a Hydraulic Control Procedure, MS Thesis, Stanford University, Stanford, CA.

Bachmat, Y., B. Andrews, D. Holtz, and S. Sebastian. 1978. Utilization of Numerical Ground Water Models for Water Resource Management, EPA 600/8-78-012, U.S. Environmental Protection Agency, Environmental Research Laboratory, Ada, OK.

Baes, C.F., III and R.D. Sharp. 1983. "A Proposal for Estimation of Soil Leaching Constants for Use in Assessment Models," J. Environmental Quality, Vol. 12, No. 1.

Barnwell, T.O., Jr. and R.C. Johanson. 1981. HSPF: A Comprehensive Package For Simulation of Watershed Hydrology and Water Quality. In: Nonpoint Pollution Control: Tools and Techniques for the Future. Interstate Commission on the Potomac River Basin Rockville, MD.

Bedient, P.B., N.K. Springer, C. J. Cook and M. B. Thompson 1982. "Modeling Chemical Reactions and Transport in Groundwater Systems: A Review," In: Modeling the Fate of Chemicals in the Aquatic Environment, K. L. Dickson, A. W. Maki and J. Cairns, Jr. eds., Ann Arbor Science Publishers, Ann Arbor, MI.

Bicknell, B.R. 1984. Modeling Chemical Emissions from Lagoons and Landfills, Final Report, U.S. Environmental Protection Agency, Environmental Research Laboratory, Athens, GA.

Brady, N.C. 1974. The Nature and Properties of Soils. 8th E. McMillan Publishing Co., New York, N.Y.

Brown, S.M., A.S. Donigian, Jr., S.B. Yabusaki and J.T. Bachmaier. 1984. "Locational Factors Affecting Leachate Migration," Proceedings of the 1984 Specialty Conference on Environmental Engineering, ASCE, New York, N.Y.

Callahan, M.A., M.W. Slimak, N.W. Gable, I.P. May, C.F. Fowler, J.R. Freed, P. Jennings, R.L. Durfee, F.C Whitmore, B. Maestri, W.R. Mabey, B.R. Holt, and C. Gould. 1979. Water Related Environmental Fate of 129 Priority Pollutants, Volumes 1 and 2, EPA 600/3-82-023ab, U.S. Environmental Protection Agency, Office of Water Planning and Standards, Washington, D.C.

Carsel, R.F., C.N. Smith, L.A. Mulkey, J.D. Dean, and P. Jowise. 1984. User's Manual for the Pesticide Root Zone Model (PRZM), U.S. Environmental Protection Agency, Environmental Research Laboratory, Athens, GA.

Cherry, J.R., R.W. Gillham and J. F. Barker. 1984. "Contaminants in Ground Water: Chemical Processes," In: Ground Water Contamination, Studies in Geophysics, National Research Council, National Academy Press, Washington, DC.

Chow, V.T. 1964. Handbook of Applied Hydrology, McGraw-Hill Company, New York, N.Y.

Cohen, R.M. and J.W. Mercer. 1984. "Estimation of a Proposed Synthetic Cap and Concrete Cut-Off Wall at Love Canal Using a Cross-Sectional Model," Proceediungs of the Fourth National Symposium and Exposition on Aquifer Restoration and Ground Water Monitoring, National Water Well Association, Washington, D.C.

Cole, C.R., F.W. Bond, S.M. Brown and G.W. Dawson. 1984. Demonstration/Application of Ground-Water Modeling Technology for Evaluation of Remedial Action Alternatives, Draft Final, U.S. Environmental Protection Agency, Municipal Environmental Research Laboratory, Cincinnati, OH.

Davis, S.N. and R.J.M. DeWiest. 1966. Hydrogeology, John Wiley and Sons, New York, N.Y. 463 p.

Dawson, G.W., C.J. English and S.E. Petty. 1980. Physical Chemical Properties of Hazardous Waste Constituents, U.S. Environmental Protection Agency, Environmental Research Laboratory, Athens, GA.

Dillion, R.T., R.B. Lantz and S.B. Pahwa. 1978. Risk Methodology for Geologic Disposal of Radioactive Waste: The Sandia Waste Isolation Flow and Transport (SWIFT) Model, SAND78-1267.

Donigian, Jr., A.S., J.C. Imhoff, B.R. Bicknell and J.L. Kittle, Jr. 1984. Guide to the Application of the Hydrological Simulation Program - FORTRAN (HSPF). EPA 600/3-84-065, U.S. Environmental Protection Agency, Athens, GA.

Donigian, Jr., A.S., T.Y.R. Lo and E.W. Shanahan. 1983. Rapid Assessment of Potential Ground-Water Contamination under Emergency Response Conditions, EPA 600/8-83-030, U.S. Environmental Protection Agency, Washington, D.C.

Donigian, Jr., A.S. 1981. "Water Quality Modeling in Relation to Watershed Hydrology," In: Modeling Components of Hydrologic Cycle, ed. V.P. Singh, Water Resource Publications, Littleton, CO.

Donigian, Jr., A.S., J.L. Baker, D.A. Haith and M.F. Walter. 1983. HSPF Parameter Adjustments to Evaluate the Effects of Agricultural Best Management Practices, EPA PB83-247-171, U.S. Environmental Protection Agency, Environmental Research Laboratory, Athens, GA.

Duguid, J.O. and M. Reeves. 1976. Material Transport in Porous Media: A Finite-Element Galerkin Model, ORNL-4928, Oak Ridge National Library, Oak Ridge, TN.

Edwards, A.L. 1972. TRUMP: A Computer Program for Transient and Steady-State Temperature Distributions in Multidimensional Systems, Report UCRL-14754, NTIS, Springfield, VA.

Environmental Protection Agency. 1975. Control of Water Pollution from Cropland, Vol. I, A Manual for Guideline Development, EPA-600/2-75-026a, U.S. Environmental Protection Agency, Environmental Research Laboratory, Athens, GA.

Finley, N.C. and M. Reeves. 1968. SWIFT - Self Teaching Curriculum, NUREG/CR -1968, SAND81-0410.

Freeze, R. and J. Cherry. 1979. Groundwater, Prentice-Hall Press, N.Y.

Gelhar, L.W. 1977. A Comparison of Ground Water Quality Modeling Techniques, Paper presented at Water Resource Center Conference, September 6-8, 1976, Berkshire, U.I.C.

Gorelick, S.M., C.I. Voss, P.E. Gill, W. Murray, M.A. Saunder and M.H. Wright. 1984. Aquifer Reclamation Design: The Use of Contaminant Transport Simulation Combined with Nonlinear Programming", Water Resources Res., Vol. 20, No. 4.

Gorelick S.M. 1983. "A Review of Distributed Parameter Groundwater Management Modeling Mehtods," Water Resources Research, Vol. 19, No. 2.

Green, W.J., G.F. Lee, R.A. Jones and T. Palit. 1983. "Interaction of Clay Soils with Water and Organic Solvents: Implications for the Disposal of Hazardous Wastes," Environmental Science and Technology, Vol. 17, No. 5.

Grove, D.B. and K.L. Kipp. 1980. Modeling Contaminant Transport in Porous Media in Relation to Nuclear Waste Disposal, A Review, In: Modeling and Low-Level Waste Management, An Interagency Workshop, ORD-821, Oak Ridge National Laboratory, Oak Ridge, TN.

Gupta, S. K., C. R. Cole and F. W. Bond. 1979. Finite-Element Three-Dimensional Ground-Water (FE3DGW) Flow Model - Formulation, Program Listings and User's Manual, PNL-2939, Battelle, Pacific Northwest Laboratory, Richland, WA.

Gupta, S.K., C.T. Kincaid, P.R. Meyer, C.A. Newbill and C.R. Cole. 1982. A Multi-Dimensional Finite Element Code for the Analysis of Coupled Fluid, Energy and Solute Transport (CFEST), PNL 4260, Battelle, Pacific Northwest Laboratory, Richland, WA.

Hansch, C. and A. Leo. 1979. Substituent Constants for Correlation Analysis in Chemistry and Biology, Wiley, New York, NY.

Hillel, D. 1982. Introduction to Soil Physics, Academic Press, New York, N.Y.

Holton, H.N., G.J. Stiltner, W.H. Menson and N.C. Lopez. 1975. USDAHL-74, Revised Model of Watershed Hydrology, Tech. Bull. 1518, USDA, Washington, D.C.

Javandel, I., C. Doughty and C.F. Tsang. 1984. Groundwater Transport: Handbook of Mathematical Models, American Geophysical Union, Water Resources Monograph 10, Washington, D.C.

Johanson, R., H. Davis, J. Imhoff, J.L. Kittle, and A.S. Donigian, Jr. 1981. User's Manual for the Hydrological Simulation Program - FORTRAN (HSPF), Release 7.0, U.S. Environmental Protection Agency, Environmental Research Laboratory, Athens, GA.

JRB Associates. 1982. Handbook for Remedial Action at Waste Disposal Sites, U.S. Environmental Protection Agency, Cincinnati, OH. EPA 625/6-82-006

JRB Associates, 1984. Slurry Trench Construction for Pollution Migration Control, EPA 540/2-84-001, U.S. Environmental Protection Agency, Office of Emergency and Remedial Response, Cincinnati, OH.

Kincaid, C.T., J.R. Morrey and J.E. Rogers. 1984. Geohydrochemical Models for Solute Migration, Volume 1: Process Description and Computer Code Selection, EA-3417, Electric Power Research Institute, Palo Alto, CA.

Knisel, W.G. 1980. CREAMS: A Field Scale Model for Chemical Runoff, and Erosion from Agricultural Management Systems, United States Department of Agriculture, Conservation Research Report Number 26.

Knox, R.C. and L.W. Canter. 1980. Summary of Ground Water Modeling in the United States, Report No. NCGWR 80-20, National Center for Ground Water Research, Norman, OK.

Kobayashi, H. and B.E. Rittman. 1982. "Microbial Removal of Hazardous Organic Compounds," Environmental Science and Technology, Vol. 16, No. 3.

Koines A. 1982. Technical Review of Ground-Water Models, Contract No. 68-01-6404, Draft Report, U.S. Environmental Protection Agency, Office of Solid Waste, Washington, D.C.

Lappala, E.G. 1980. Modeling of Water and Solute Transport Under Variable Saturated Conditions: State-of-the-Art, In: Modeling and Low-Level Waste Management, An Interagency Workshop, ORD-821, Oak Ridge National Laboratory, Oak Ridge, TN.

Leo, A., C. Hansch and D. Elkins. 1971. "Partition Coefficeints and Their Use," Chem. Rev., Vol. 71.

Linsley, R.K., M.A. Kohler and J.H.L. Paulhus. 1982. Hydrology for Engineers, McGraw-Hill, New York, N.Y.

Lyman, W.J., W.F. Reehl and D.H. Rosenblatt. 1982. Handbook of Chemical Property Estimation Methods. McGraw-Hill, New York, N.Y.

Mabey, W.R., J.H. Smith, R.T. Podoll. H.L. Johnson, T. Mills, T.W. Chou, J. Gates, I. Waight Partridge and D. Vandenburg. 1982. Aquatic Fate Process Data for Organic Priority Pollutants, U.S. Environmental Protection Agency, Office of Water Regulation and Standards, Washington, D.C.

McCarty, P.L., M. Reinhard and B.E. Rittman. 1981. "Trace Organics in Ground Water," Environmental Science and Technology, Vol. 15, No. 1.

Mercer, J.W. and C.R. Faust. 1981. Ground-Water Modeling, National Water Well Association, Worthington, OH.

Mercer, J.W., S.D. Thomas and B. Ross. 1982. Parameters and Variables Appearing in Repository Siting Models, NUREG/CR-3066, U.S. Nuclear Regulatory Commission, Washington, D.C.

Mills, W., J. Dean, D. Porcella, S. Gherini, R. Hudson, W. Frick, G. Rupp and G. Bowie. 1982. Water Quality Assessment: A Screening Procedure for Toxic and Conventional Pollutants, Vol. 1 and 2, EPA 600/6082-004ab, U.S. Environmental Protection Agency, Athens GA.

Moiser, J.E., J.R. Fowler, C.J. Barton, W.W. Tolbert, S.C. Meyers, J.E. Vancil, H.A. Price, M.J.R. Vasco, E.R. Rutz, T.X. Wendel and L.D. Rickertson. 1980. Low-Level Waste Management: A Complilation of Models and Monitoring Techniques, ORNL/SMB-79/13617/2, SAI/OR-565-2, Oak Ridge National Laboratory, Oak Ridge, TN.

Morris, D.A. and A.I. Johnson. 1967. Summary of Hydrologic and Physical Properties of Rock and Soil Materials as Analyzed by the Hydrologic Laboratory of the U.S. Geogiocal Survey 1948-60, USGS Water Supply Paper 1839-D.

Narasimhan, T.N. and P.A. Witherspoon. 1976. An Integrated Finite Difference Method for Analyzing Fluid Flow in Porous Media, Water Resources Research Vol. 12, No. 1.

Nelson, R.W., G.W. Gee and E.M. Arnold, Eds. 1982. "Proceedings of the Symposium of Unsaturated Flow and Transport Modeling," NUREG/CP-0030, PNL-SA-10325, Battelle, Pacific Northwest Laboratory, Richland, WA.

Nelson, R.W., P.R. Meyer, P.L. Oberlander, S.C. Sneider, D.W. Mayer and A.E. Reisenauer. 1983. Model Evaluation of Seepage from Uranium Tailings Above and Below the Water Table, NUREG/CR-3078, PNL-4461, Battelle, Pacific Northwest Laboratory, Richland, WA.

Onishi, Y., R.J. Serne, E.M. Arnold, C.E. Cowan and F.L. Thompson. 1981. Critical Review: Radionuclide Transport, Sediment Transport, and Water Quality Modeling; and Radionuclide Adsorption/Desorption Mechanisms, U.S. Nuclear Regulatory Commission, NUREG/CR-1322, Washington, D.C.

Orlob, G. 1971. "Mathematical Modeling of Estuarial Systems," International Symposium of Mathematical Modeling Techniques.

Oster, C.A. 1982. Review of Ground-water Flow and Transport Models in the Unsaturated Zone, NUREG/CR-2917, PNL-4427, Battelle, Pacific Northwest Laboratory, Richland, WA.

Prickett, T.A. and C.G. Lonnquist. 1971. Selected Digital Computer Techniques for Groundwater Resource Evaluation, Bulletin 55, Illinois State Water Survey, Champaign, IL.

Prickett, T.A., T.G. Naymik and C.G. Lonnquist. 1981. A "Random-Walk" Solute Transport Model for Selected Ground Water Quality Evaluations, Bulletin 65, Illinois State Water Survey, Champaign, IL.

Rai, D., J.M. Zachara, R.A. Schmidt and A.P. Schwab. 1984a. Chemical Attentuation Rates, Coefficients and Constants in Leachate Migration, Volume 1: A Critical Review, EA-3356, Volume 1, Electric Power Research Institute, Palo Alto, CA.

Rai, D., J.M. Zachara, R.A. Schmidt and A.P. Schwab. 1984b. Chemical Attenuation Rates, Coefficients and Constants in Leachate Migration, Volume 2: An Annotated Bibliography, EA-3356, Volume 2, Electric Power Research Institute, Palo Alto, CA.

Rawls, W.J. and D.L. Brakensiek. 1982. "Estimating Soil Water Retention from Soil Properties," Proc. ASCE, Vol 108, No. IR2.

Ree, W.U., F.L. Wimerley and R.W. Crow. 1977. "Manning's n and the Overland Flow Equation," ASAE Trans., Vol. 22, No. 1

Reeves, M. and R.M. Cranwell. 1981. User's Manual for the Sandia Waste Isolation Flow Transport Model (SWIFT), Release 4.81, NUREG/CR-2324, SAND81-2516, Washington, D.C.

Reeves, M. and J.O. Duguid. 1976. Water Movement Through Saturated Unsaturated Porous Media: A Finite-Element Galerkin Model, ORNL-4297, Oak Ridge National Laboratory Report, Oak Ridge, TN.

Reisenauer, A.E., S.K. Gupta, R.W. Nelson and C.A. Newbill. 1981. Advective Radionuclide Transport with Soil Interaction Under Variably Saturated Flow Conditions, PNL-3994, Battelle, Pacific Northwest Laboratory, Richland, WA.

Reisenauer, A.E., K.T. Key, T.N. Narasimhan and R.W. Nelson. 1982. TRUST: A Computer Program for Variably Saturated Flow in Multidimensional Media, NUREG/CR-2360, PNL-3975, Battelle, Pacific Northwest Laboratory, Richland, WA.

Roberts, P.V. 1982. "Performance of GAC for FOC Removal," Journal of American Water Works Association, Feb.

Roberts, D.W. 1984. Soil Properties, Classification, and Hydraulic Conductivity Testing, SW-925. Draft Technical Resource Document for Public Comments, U.S. Environmental Protection Agency, Municipal Environmental Research Laboratory, Cincinnati, OH.

Ross, B.B., D.N. Contractor, and V.O. Shanholtz. 1977. "Finite Element Simulation of Overland and Channel Flow," Trans. ASAE.

SAI. 1981. Tabulation of Waste Isolation Computer Models, ONWI-78, Prepared by Science Applications, Inc. for the Office of Nuclear Waste, Battelle, Memorial Institute, Columbus, OH.

Sax, N.I. 1979. Dangerous Properties of Industrial Materials 4th ed., Van Nostrand Reinhold, New York, N.Y.

Schroeder, P.R., J.M. Morgan, T.M. Walbki, and A.C. Gibson. 1984a. The Hydrologic Evaluation of Landfill Performance (HELP) Model, Volume I., User's Guide for Version 1, EPA/530-SW-84-009, U.S. Environmental Protection Agency, Office of Solid Waste and Emergency Response, Washington, D.C.

Schroeder, P.R., A.C. Gibson, and M.P. Smolen. 1984b. The Hydrolgic Ealuation of Landfill Performance (HELP) Model, Volume II., Documentaiton for Version I, EPA/530-SW-84-010, U.S. Environmental Protection Agency, Office of Solid Waste and Emergency Response, Washington, D.C.

SCS Engineers. 1982. Costs of Remedial Response Actions at Uncontrolled Hazardous Waste Sites, EPA 600/2-82-035, U.S. Environmental Protection Agency, Municipal Environmental Research Laboratory, Cincinnati, OH.,

Segol, G. 1976. A 3-D Galerkin Finite-Element Model for the Analysis of Contaminant Transport in Variably Saturated Porous Media, Proceeding of the 1st International Conference on Finite Elements in Water Resources, Princeton University, Pentech Press, London, England.

Shafer, J.M., P.L. Oberlander, and R.L. Skaggs. 1984. Mitigative Techniques and Analysis of Generic Site Conditions for Ground-Water Contamination Associated with Sewer Accidents, NUREG/CR-3681 PNL-5072. Prepared for U.S. Nuclear Regulatory Commission, Washington, D.C.

Shafer, J.M. 1984. "Determining Optimum Pumping Rates for Creation of Hydraulic Barriers to Ground Water Pollutant Migration," Fourth National Symposium and Exposition on Aquifer Restoration and Ground Water Monitoring, National Water Well Association, Washington, D.C.

Silka, L.R. and J.W. Mercer. 1982. "Evaluation of Remedial Actions for Groundwater Contamination at Love Canal, New York," Proceedings of National Conference on Management of Hazardous Materials Control Research Institute, Silver Spring, MD.

Thomas, S.D., B. Ross and J.W. Mercer. 1982. A Summary of Repository Siting Models, NUREG/CR-2782, U.S. Nuclear Regulatory Commission, Washington, D.C.

van Genuchten, R. 1978a. Simulation Models and Their Application to Landfill Disposal Siting: A Review of Current Technology, In: Land Disposal of Hazardous Waste, Proceedings of the Fourth Annual Research Symposium. EPA-600/9-78-016, San Antonio, TX.

van Genuchten, R. 1978b. Calculating the Unsaturated Hydraulic Conductivity with a new Closed-Form Analytical Model. Research Report 78-WR-08 Water Resources Program, Department of Civil Engineering, Princeton University, N.J.

Verschueren, K. 1977. Handbook of Environmental Data on Organic Chemicals, Van Nostrand Reinhold Co., N.Y.

Wilson, J.T., J.F. McNabb, D.L. Balkwill and W.C. Ghiorse. 1983. "Enumeration and Characterization of Bacteria Indigenous to a Shallow Water-Table Aquifer," Ground Water, Vol. 21, No. 2.

Wischmeier, W.H. and D.D. Smith. 1978. Predicting Rainfall Erosion Losses--A Guide to Conservation Planning, Agriculture Handbook No. 537, USDA, Agricultural Research Service.

Woolhiser, D.A. 1976. Hydrologic Aspects of Nonpoint Source Pollution, In: B.A. Stewart et al., Control of Water Pollution from Cropland, Vol. II, EPA-600/2-75-026b, U.S. Environmental Protection Agency, Washington, D.C.

Yeh, G.T. and D.S. Ward. 1979. FEMWATER: A Finite-Element Model of Water Flow Through Saturated-Unsaturated Porous Media, ORNL-5567, Oak Ridge National Laboratory, Oak Ridge, TN.

Yeh, G.T. and D.S. Ward. 1981. FEMWASTE: A Finite-Element Model of Waste Transport Through Saturated-Unsaturated Porous Media, ORNL-5601, Oak Ridge National Laboratory, Oak Ridge, TN.

Yeh, G.T. 1982a. Training Course No. 1: The Implementation of FEMWATER Computer Program--Final Report, NUREG/CR-2705, ORNL-5567, Oak Ridge National Laboratory, Oak Ridge, TN.

Yeh, G.T. 1982b. Training Course No. 2: The Implementation of FEMWATER Computer Program--Final Report, NUREG/CR-2706, ORNL-5601, Oak Ridge National Laboratory, Oak Ridge, TN.

# Appendix A: Supporting Information on HSPF, FEMWATER / FEMWASTE and FE3DGW / CFEST

At the request of EPA, surface, unsaturated and saturated zone models applicable to the assessment of a broad range of remedial actions were selected for implementation on the EPA-NCC computer system in Research Triangle Park, N.C. The intent was to provide a general capability for remedial action modeling. This appendix discusses the selection of the three codes; linkage considerations; their implementation on the NCC system; sources of code documentation and user assistance; and specific parameters requiring adjustment for each remedial measure discussed in Section 6.

A.1 CODE SELECTION

The process of reviewing and selecting models for remedial action assessment is described in detail in Volume 1. Section 5 of this volume discusses the model development and application process and evaluates a number of candidate codes. Starting with the candidate codes listed in Section 5.2, the following criteria were used to select one code for each zone.

1. <u>Dimensionality</u> - multiple land segment surface zone model, two - dimensional (x-z) unsaturated zone model, and three - dimensional saturated zone model.

2. <u>Time Frame</u> - continuous simulation with variable time step.

3. <u>Flow Processes</u> - advection, infiltration to the unsaturated zone, drainage to the saturated zone, and evapotranspiration.

4. <u>Transport Processes</u> - advection, dispersion, sorption, retardation, and degradation.

5. <u>Data Structure</u> - flexible input and output sequences, data management and storage capabilities.

6. <u>Ease of Use</u> - clear and complete documentation, program maintenance and user assistance available.

7. <u>Code Testing</u> - model tests under a variety of conditions, validation against field data.

Table A.1 summarizes these criteria for each zone.

The following codes were selected based on the above criteria: HSPF for the surface zone, FEMWATER/FEMWASTE for the unsaturated zone and FE3DGW/CFEST for the saturated zone. The principle reasons for selecting the codes are:

1. HSPF provides simulation of multiple land segments, which allows the representation of both the site and the drainage area surrounding the site. This is particularly important when modeling surface water diversion and collection type actions, which cause local changes in runoff.

2. The data base management software associated with HSPF is both flexible and powerful, and can provide a structure for easy linkage with the codes selected for the other zones.

3. FEMWATER/FEMWASTE provides all of the needed processes and also has been tested and applied to case studies. Documentation and ease of use are generally good.

4. FE3DGW/CFEST supports two-dimensional or three-dimensional simulation. Code tests have been conducted. Documentation of FE3DGW is reasonably complete; similar documentation is under prepration for CFEST. Both models were used for remedial action evaluation (Cole et al., 1983).

5. User support and code implementability can be major factors when selecting codes. HSPF is actively supported by the Water Quality Modeling Center at EPA-Athens (Contact: Mr. Thomas Barnwell) and through a maintenance contract that provides for training workshops and code updates. Individual support contracts can also be arranged. HSPF has been implemented on numerous computer systems, including the IBM at the National Computer Center. User support for FEMWATER/FEMWASTE is limited to that offered by ORNL. Both codes have been implemented on a number of computer systems. User support for FE3DGW and CFEST is offered by Battelle, Pacific Northwest Laboratory. The Department of Energy's, Office of Nuclear Waste Isolation has recently

Appendix A 321

TABLE A.1 CHARACTERISTICS OF CODES (BY ZONE) THAT WOULD SATISFY EPA'S NEED FOR A COMPREHENSIVE, MODEL-BASED REMEDIAL ACTION ASSESSMENT CAPABILITY

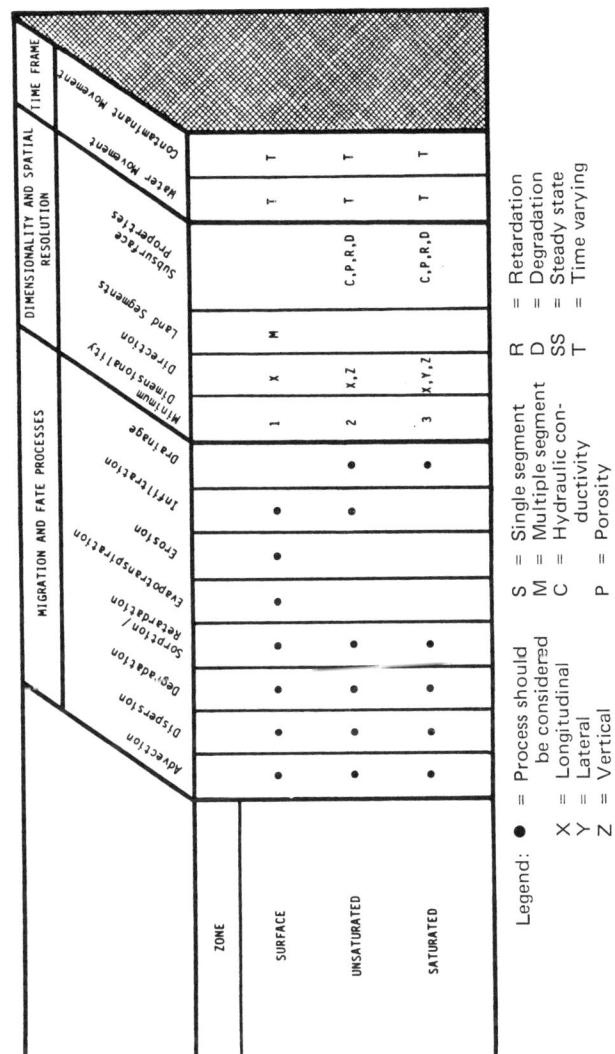

expanded the level of user support. CFEST has been implemented on several computers, including EPA's IBM system at the National Computer Center.

These models, as a group, provide most of the capabilities required for analysis of complex site conditions, including the following:

1. Representation of the surface hydrologic system, including precipitation, snow melt, evapotranspiration, runoff, and infiltration (HSPF).

2. Representation of sediment and sediment-related contaminant transport, including soil detachment, scour and deposition (HSPF).

3. Representation of percolation through the unsaturated zone, including soil wetting front movement, seepage from ponds, and lateral drainage (FEMWATER/FEMWASTE).

4. Representation of flow through heterogeneous aquifers and multi-aquifer systems with variable water table elevations (FE3DGW/CFEST).

5. Representation of all key chemical transport processes (advection, dispersion, retardation, and degradation) in the three zones.

6. Representation of complex boundary conditions caused by ponds, streams, aquicludes and basement rock, as well as by different remedial actions.

7. Representation of changes in most of the key processes affected by remedial actions.

## A.2 LINKAGE OF HSPF, FEMWATER/FEMWASTE AND CFEST

In cases where the use of two or more of these models is required to evaluate remedial action performance, linkage of the models may be required (see Section 5.3 of this volume). "Soft linkage" of the three codes is likely to be the most viable approach. No such linkage currently exists. The following discussion, however, presents a possible approach for linking the models.

Figure A.1 provides a schematic diagram for the proposed linkage. The linkage of HSPF and FEMWATER/FEMWASTE would take advantage of the data management utility routines already included in HSPF. TSMS, the Time Series Management System, is

*Appendix A* 323

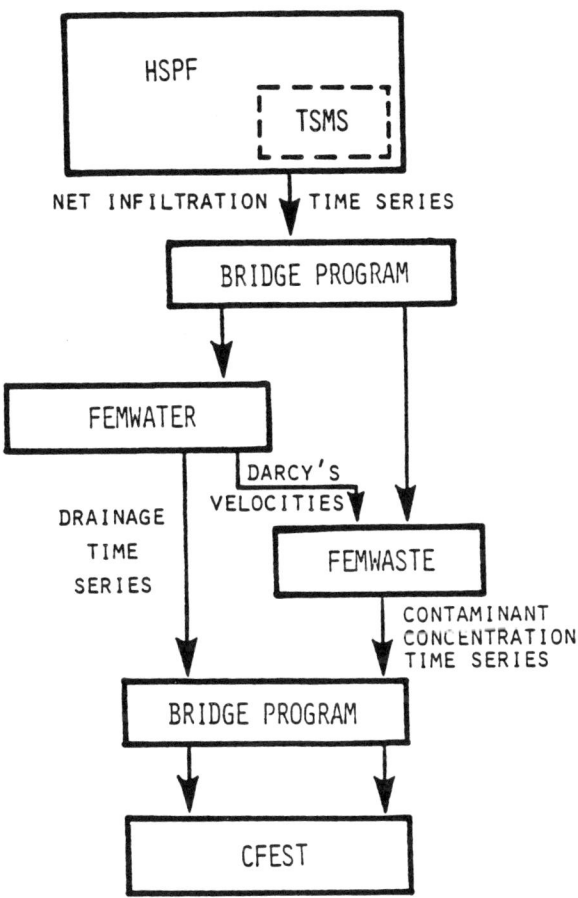

Figure A.1  Schematic diagram showing soft linkage of HSPF, FEMWATER/FEMWASTE and CFEST with bridge programs.

a series of routines that provide time series storage, transfer, conversion and mass balance checking capabilities. TSMS could be used aggregate or disaggregate and combine or separate net infiltration time series (from HSPF) for individual land segments to obtain net infiltration time series for input to individual nodes in the FEMWATER/FEMWASTE grid. Once these time series have been prepared with the TSMS routines, they could be transferred to a bridge program that simply reformats the time series for direct input to FEMWATER and FEMWASTE. Thus, the bridge program between these two codes would be relatively simple.

The bridge program between FEMWATER/FEMWASTE and FE3DGW/CFEST would be more complicated. It should provide for the aggregation and disaggregation of drainage and associated contaminant concentration time series, as well as the combination and separation of these time series. The former would make it possible to account for differences in time stepping, the latter would make it possible to account for differences in computational grids. The program also should provide for unit conversions, mass balance checks, and the reformatting of time series for input to CFEST.

### A.3 MODEL CODE IMPLEMENTATION

The codes as currently implemented are available on the NCC IBM computer system; a valid account is required for access to the codes. Existing program load modules are executed via input files containing the program input and required Job Control Language (JCL). JCL is used to execute load modules, create scratch files and output files, and pass control to subsequent modules.

The following procedure should be used for implementing the codes:

1. Contact the appropriate EPA official and request that all necessary files (i.e., program load modules, sample input/JCL files, and other necessary files) be copied to your account. Program documentation should also be obtained.

2. Using the sample input files and code documentation, develop input for your application.

3. Modify the JCL to reflect your account; and modify all file names.

4. Run the code.

Interested parties should contact Mr. Douglas Ammon for current information or program status and implementation of FE3DGW, CFEST, FEMWATER and FEMWASTE. Mr. Thomas O. Barnwell, Jr. should be contacted for information on HSPF.

## A.4 SOURCES OF CODE DOCUMENTATION AND USER ASSISTANCE

A major reason for the selection of the specified codes was the availability of user guidance. Key sources of information and help are summarized below.

HSPF was created and is supported by the EPA and is currently available on magnetic tape, as a source code for mainframe computers from:

> The Water Quality Modeling Center
> U.S. EPA, Environmental Research Lab
> Athens, GA 30613
>
> Contact: Mr. Thomas O. Barnwell

The model is currently maintained, under a contract with EPA, by:

> Anderson-Nichols & Co., Inc.
> 2666 East Bayshore Road
> Palo Alto, CA 94303
>
> Contact: Mr. Jack Kittle or Mr. Brian Bicknell

A current release of the user's manual for HSPF can be obtained from the Water Quality Modeling Center. Key references for HSPF design, structure and application include Johanson et al., (1981) and Donigian et al., (1984).

FEMWATER/FEMWASTE were developed at Oak Ridge National Laboratory and are currently maintained by ORNL staff. The source code and documentation are available from:

> Oak Ridge National Laboratory
> Environmental Science Division
> P.O. Box X
> Oak Ridge, TN 37830
>
> Contact: Dr. George T. Yeh

Key references on the design, structure, implementation and use of these codes include: Reeves et al., 1975; Duguid et al., 1976; Yeh and Ward, 1980 and 1981; and Yeh 1982a and 1982b. The last two sources are self-contained training courses.

FE3DGW/CFEST were developed by Battelle, Pacific Northwest Laboratory (PNL) and are currently maintained by PNL staff. The source code and documentation can be obtained from:

> Battelle, Pacific Northwest Laboratory
> P.O. Box 999
> Richland, WA 99352
>
> Contact: Mr. Charles R. Cole

Key references on the design, implementation and use of these codes include: Gupta et al., (1984), Gupta et al., (1982) and Cole et al., (1984).

## A.5 PARAMETER ADJUSTMENTS REQUIRED FOR EACH REMEDIAL MEASURE

This section presents the *specific* parameters and input boundary conditions that must be adjusted in HSPF, FEMWATER/FEMWASTE and CFEST to represent each remedial measure discussed in Section 6. The recommended adjustments are presented in a series of tables for each measure and for each code needed for that measure. The following is a list of the tables.

| Table A.2 | Capping, Grading and Revegetation Parameter Adjustments for HSPF |
|---|---|
| Table A.3 | Capping, Grading and Revegetation Parameter Adjustments for FEMWATER/FEMWASTE |
| Table A.4 | Surface Water Diversion and Collection Parameter Adjustments for HSPF |
| Table A.5 | Ground-Water Pumping and Interceptor Trench Parameter Adjustments for CFEST |
| Table A.6 | Impermeable Barrier Parameter Adjustments for CFEST |
| Table A.7 | Subsurface Drains and Solution Mining Parameter Adjustments for FEMWATER/FEMWASTE |
| Table A.8 | Subsurface Drains and Solution Mining Parameter Adjustments for CFEST |
| Table A.9 | Excavation Parameter Adjustments for FEMWATER/FEMWASTE |

Table A.10  Excavation Parameter Adjustments for CFEST

Table A.11  Hydraulic Dredging and Seepage Basin Parameter Adjustments for FEMWATER/FEMWASTE

Table A.12  Bioreclamation and Chemical Injection Parameter Adjustments for CFEST

Table A.13  Permeable Treatment Bed Parameter Adjustments for CFEST

TABLE A.2  CAPPING, GRADING AND REVEGETATION PARAMETER ADJUSTMENTS FOR HSPF

| Parameter | Purpose | Range/Units | Reference |
|---|---|---|---|
| NSUR | Surface roughness (Manning's n) | 0.25-0.4 | Donigian et al., 1983 |
| LSUR | Slope length | (S) L | --- |
| SLSUR | Slope | 5-18% | JRB Associates, 1982 |
| KRER | Coefficient for soil detachment | 0.08-0.28 | Donigian et al., 1983; Johanson et al., 1981 |
| KGER | Coefficient of soil scour | 0.0-1.0 | Donigian et al., 1983 |
| INFILT | Infiltration capacity | 0.05-1.0 in/hr | Donigian and Davis, 1978 |
| COVER | Canopy development | 0-1.0 | Donigian et al., 1983 |
| LZETP | Lower zone evapotranspiration parameter | 0.2-0.9 | Donigian et al., 1983 |
| CEPSC | Maximum interception | 0.06-0.25 in. | Donigian et al., 1983 |
| AFFIX | Soil compaction factor | 0.1-.001 | Donigian et al., 1983 |

(S) Site-specific
L  Length

TABLE A.3  CAPPING, GRADING AND REVEGETATION PARAMETER ADJUSTMENTS FOR FEMWATER/FEMWASTE

### FEMWATER

| Parameter | Purpose | Range/Units | Reference |
|---|---|---|---|
| NMAT | Number of materials (cap, drainage and filter layers) | (S) | --- |
| PROP(3,I) | Porosity of each material 'I' | (S) % | Section 6.3.2.2.2 |
| HPROP(J,K) | Pressure head of Jth point for material 'K' | (S) L | Section 6.3.2.1.1 |
| THPROP(J,K) | Moisture-content of Jth point for material 'K' | (S) $L^3/L^3$ | Section 6.3.2.1.1 |
| AKPROP(J,K) | Relative conductivity of Jth point for material 'K' | (S) | Section 6.3.2.1.1 |
| PROP(4,I) | xx-component of saturated hydraulic conductivity for material 'I' | (S) L/T | Section 6.3.2.1.2 |
| PROP(5,I) | zz-component of saturated hydraulic conductivity for material 'K' | $Kz/Kx=0.1$ (initial estimate) | Freeze and Cherry, 1979 |

### FEMWASTE

| Parameter | Purpose | Range/Units | Reference |
|---|---|---|---|
| PROP(3,I) | Longitudinal dispersivity for material 'I' | (S) L | Section 6.3.2.2.1 |
| PROP(4,I) | Lateral dispersivity for material 'I' | (S) L | Section 6.3.2.2.1 |
| PROP(6,I) | Porosity of material 'I' | (S) % | Section 6.3.2.2.2 |
| PROP(9,I) | Tortuosity of material 'I' | 0.0-5.0 | Yeh, 1982 |

TABLE A.4  SURFACE WATER DIVERSION AND COLLECTION PARAMETER ADJUSTMENTS FOR HSPF

Pervious Land Segments

| Parameter | Purpose | Range/Units | Reference |
|---|---|---|---|
| NSUR | Surface roughness (Manning's n) | 0.15-0.4 | Donigian et al., 1983 |
| LSUR | Slope length | (S) L | --- |
| SLSUR | Slope | 12-18% | JRB Associates, 1982 |

Channel Segments

| Parameter | Purpose | Range/Units | Reference |
|---|---|---|---|
| NSUR | Channel roughness (Manning's n) | - | Section 6.3.1.1 |
| SLSUR | Channel Slope | 6-12% | JRB Associates, 1982 |

(S) Site-specific
L  Length

TABLE A.5  GROUND WATER PUMPING AND IMPERMEABLE BARRIER PARAMETER ADJUSTMENTS FOR CFEST

| Parameter | Purpose | Range/Units | Reference |
|---|---|---|---|
| NODB | Held head node number | (S) | --- |
| BIV | Value of held head | (S) L | --- |
| NODALQ | Time-constant nodal flux | 2 | --- |
| NQNDOE | Node number having nodal flux | (S) | --- |
| BIVF | Integrated flow volume | (S) $L^3/T$ | --- |
| BIVFC | Concentration of injection fluid | (S) M/T | --- |

(S) Site-specific
L  Length
M  Mass

TABLE A.6 IMPERMEABLE BARRIER PARAMETER ADJUSTMENTS FOR CFEST

| Parameter | Purpose | Range/Units | Reference |
|---|---|---|---|
| MAT | Number of materials (barrier, surrounding media) | (S) | -- |
| XK | Hydraulic conductivity (for 'x' direction) | (S) L/T | Section 6.3.2.1.2 |
| YK | Hydraulic conductivity (for 'y' direction) | (S) L/T | Section 6.3.2.1.2 |
| ZK | Hydraulic conductivity (for 'z' direction) | (S) L/T | Section 6.3.2.1.2 |
| THETAO | Porosity at reference pressure head | (S) % | Section 6.3.2.2.2 |
| HTHETA | Pressure head at which THETAO is defined | (S) L | --- |
| ALPHAL(I) | Longitudinal dispersivity for each material 'I' | (S) L | Section 6.3.2.2.1 |
| ALPHAT(I) | Lateral dispersivity for each material 'I' | (S) L | Section 6.3.2.2.1 |

(S) Site-specific
L  Length

TABLE A.7  SUBSURFACE DRAINS AND SOLUTION MINING PARAMETER
          ADJUSTMENTS FOR FEMWATER/FEMWASTE

| FEMWATER Parameter | Purpose | Range/Units | Reference |
|---|---|---|---|
| NMAT | Number of materials (waste, gravel and surrounding area) | (S) | |
| HPROP(J,K) | Pressure head of Jth data point for material 'K' | (S) L | Section 6.3.2.1.1 |
| THPROP(J,K) | Moisture-content of Jth data point for material 'K' | (S) $L^3/L^3$ | Section 6.3.2.1.1 |
| AKPROP(J,K) | Relative conductivity of Jth data point for material 'K' | (S) | Section 6.3.2.1.1 |
| PROP(3,I) | Porosity of medium 'I' | (S) % | Section 6.3.2.1.2 |
| PROP(4,I) | xx-component of saturated hydraulic conductivity for material 'I' | (S) L/T | Section 6.3.2.1.2 |
| PROP(5,I) | zz-component of saturated hydraulic conductivity for material 'I' | (S) L/T | Section 6.3.2.1.2 |
| THDBF(J,I) | Time of Jth data point on Ith held head profile | (S) T | --- |
| HDBF(J,I) | Total head of Jth data point in Ith profile | (S) L | --- |
| NPDB(I) | Global node number of Ith node | (S) | --- |

(continued)

TABLE A.7 (continued)

| FEMWASTE Parameter | Purpose | Range/Units | Reference |
|---|---|---|---|
| NMAT | Number of materials (waste, gravel and surrounding area) | (S) | --- |
| PROP(1,I) | Distribution coefficient for materials 'I' | (W) $L^3/M$ | Section 6.3.2.2.4 |
| TSOSF(J,I) | Time of Jth data point on Ith Cauchy flux profile | (S) T | --- |
| SOSF(J,I) | Source/sink value of Jth data point in Ith profile | (S) $L^3/T/L$ | --- |
| TCRSF(J,I) | Time of Jth data point on Ith incoming-concentration vs. time-profile | (S) T | --- |
| CRSF (J,I) | Concentration | (S) M/L | --- |

(S) Site-specific
(W) Waste-specific
L  Length
T  Time
M  Mass

TABLE A.8  SUBSURFACE DRAINS AND SOLUTION MINING PARAMETER ADJUSTMENTS FOR CFEST

| Parameter | Purpose | Range/Units | Reference |
|---|---|---|---|
| MAT | Number of materials (waste, surrounding media) | (S) | --- |
| XK,YK,ZK | Hydraulic conductivity for 'x', 'y' and 'z' directions | (S) L/T | Section 6.3.1.2.2 |
| THETAO | Porosity at reference pressure head | (S) % | Section 6.3.2.2.2 |
| HTHETA | Pressure head at which THETAO is defined | (S) L | --- |
| NODB | Held head node number | (S) | --- |
| BIV | Value of held head | (S) L | --- |
| NODALQ | Time-constant nodal flux | 2 | --- |
| NQNDOE | Node number having nodal flux | (S) | --- |
| BIVF | Integrated flow volume | (S) $L^3/T$ | --- |
| BIVCF | Concentration of injected fluid | (S) $M/L^3$ | --- |

(S) Site-specific
L   Length
T   Time
M   Mass

TABLE A.9   EXCAVATION PARAMETER ADJUSTMENTS FOR FEMWATER/
            FEMWASTE

FEMWATER

| Parameter | Purpose | Range/Units | Reference |
|---|---|---|---|
| NMAT | Number of materials (waste and surrounding media) | (S) | --- |
| PROP (3,I) | Porosity of material 'I' | (S) % | Section 6.3.2.2.2 |
| HPROP(J,K) | Pressure head of Jth data point for material 'K' | (S) L | Section 6.3.2.1.1 |
| THPROP(J,K) | Moisture-content of Jth data point material 'K' | (S) L/L | Section 6.3.2.1.1 |
| AKPROP(J,K) | Relative conductivity of Jth data point for material 'K' | (S) | Section 6.3.2.1.1 |
| PROP(4,I) | xx-component of saturated hydraulic conductivity for material 'I' | (S) L/T | Section 6.3.2.1.2 |
| PROP (5,I) | zz-component of saturated hydraulic conductivity for material 'I' | (S) L/T | Section 6.3.2.1.2 |

(continued)

TABLE A.9 (continued)

| FEMWASTE Parameter | Purpose | Range/Units | Reference |
|---|---|---|---|
| NMAT | Number of materials materials (waste and surrounding media) | (S) | --- |
| PROP(1,I) | Distribution coefficient for material 'I' | (W) L/M | Section 6.3.2.2.4 |
| PROP(2,I) | Bulk density | (S) $M/L^3$ | Section 6.3.2.2.3 |
| PROP(3,I) | Longitudinal dispersivity for material 'I' | (S) L | Section 6.3.2.2.1 |
| PROP(4,I) | Lateral dispersivity for material 'I' | (S) L | Section 6.3.2.2.1 |
| PROP(6,I) | Porosity of material 'I' | (S) % | Section 6.3.2.2.2 |
| PROP(9,I) | Tortuosity of material 'I' | 0.0-0.5 | Yeh, 1982 |
| TCDBF(J,I) | Time of Jth data point in Ith held concentration profile | (S) T | --- |
| CDBF (J,I) | Held concentration of Jth data point in Ith profile | (S) $M/L^3$ | --- |

(S) Site-specific
(W) Waste-specific
 L  Length
 T  Time
 M  Mass

TABLE A.10  EXCAVATION PARAMETER ADJUSTMENTS FOR CFEST

| Parameter | Purpose | Range/Units | Reference |
|---|---|---|---|
| MAT | Number of materials (waste and surrounding media) | (S) | --- |
| XK | Hydraulic conductivity in 'x' direction | (S) L/T | Section 6.3.2.1.2 |
| YK | Hydraulic conductivity in 'y' direction | (S) L/T | Section 6.3.2.1.2 |
| THETAO | Porosity at reference pressure head | (S) % | Section 6.3.2.2.2 |
| HTHETA | Pressure head at which THETAO is defined | (S) L | --- |
| ALPHAL(I) | Longitudinal dispersivity for material 'I' | (S) L | Section 6.3.2.2.1 |
| ALPHAT(I) | Lateral dispersivity for material 'I' | (S) L | Section 6.3.2.2.1 |
| NODBC | Held concentration node number | (S) | --- |
| BIVC | Value of held concentration | (S) M/L | --- |

(S) Site-specific
L  Length
T  Time
M  Mass

TABLE A.11  HYDRAULIC DREDGING AND SEEPAGE BASIN PARAMETER
           ADJUSTMENTS FOR FEMWATER/FEMWASTE

**FEMWATER**

| Parameter | Purpose | Range/Units | Reference |
|---|---|---|---|
| THDBF(J,I) | Time of Jth data point in Ith held head profile | (S) T | --- |
| HDBF(J,I) | Total head of Jth data point in Ith profile | (S) L | --- |

**FEMWASTE**

| Parameter | Purpose | Range/Units | Reference |
|---|---|---|---|
| TCDBF(J,I) | Time of Jth data point in Ith held concentration profile | (S) T | --- |
| CDBF(J,I) | Held concentration of Jth data point in Ith profile | (S) M/L | --- |

(S) Site-specific
L  Length
T  Time
M  Mass

TABLE A.12  BIORECLAMATION AND CHEMICAL INJECTION PARAMETER ADJUSTMENTS FOR CFEST

| Parameter | Purpose | Range/Units | Reference |
|---|---|---|---|
| NODB | Held head node number | (S) | --- |
| BIV | Value of held head | (S) L | --- |
| NODBC | Held concentration node number | (S) | --- |
| BIVC | Value of held concentration | (S) $M/L^3$ | --- |
| DECAY | Degradation rate | (W) /T | Section 6.3.2.2.5 |

(S) Site-specific
L  Length
M  Mass

TABLE A.13  PERMEABLE TREATMENT BED PARAMETER ADJUSTMENTS FOR CFEST

| Parameter | Purpose | Range/Units | Reference |
|-----------|---------|-------------|-----------|
| RETARD | Retardation factor | (W) $L^3/M$ | Section 6.3.2.2.4 |

# VOLUME 4

# Analytical and Numerical Models for the Evaluation of Remedial Actions in Surface Water

# 1. Introduction

1.1 BACKGROUND

Releases of hazardous substances into rivers, lakes, and estuaries have been a major concern of the Environmental Protection Agency (EPA) for many years. EPA, the U.S. Coast Guard, and state and local agencies have responded to numerous release episodes which derived from a variety of sources. Agency actions involved problem identification and quantification, assessment of hazards to health and the environment, selection and implementation of appropiate responses, and follow-up monitoring of contaminant levels in the water body. Analytical and numerical predictive tools are available and have been used to varying degrees to (1) identify current chemical locations and concentrations, (2) predict future movement of chemical plumes, and (3) evaluate the responses of the receiving water body and the chemical plume to alternative actions.

Models may be used in the selection and design of removal and other long-term remedial actions. Emphasis is placed on the representation of complex processes in the water bodies and the physical and chemical effects of specific actions. Both analytic and numerical models are considered; however, emphasis is placed on numerical models which are capable of representing a broad variety of complex conditions. Guidance on model selection is intended to assist EPA and other agencies in performing in-house studies and in working with and evaluating the results of studies by other organizations.

This report constitutes one volume of a four volume set of reports designed for the selection and use of models for remedial action assessment at hazardous waste sites. Volume 1: Selection of Models for Remedial Action Assessment provides a model selection methodology for ground-water and surface water contamination problems. This report (Volume 4) is designed to complement Volume 1 by providing guidance on the evaluation of available surface water models for remedial action assessment. Volumes 2 and 3 of the set provide similar guidance for the evaluation of simplified and numerical models for subsurface problems, respectively.

## 1.2 PURPOSE OF REPORT

The primary goal of this report was the development and documentation of a model evaluation and application procedure. This procedure enables the user to identify models which are most appropriate for his or her site-specific needs and apply specific models to specific applications. Development of the model evaluation procedure involved the following steps.

1. Identify in-stream processes which may be impacted by discharges and/or remedial actions.

2. Identify potentially viable remedial actions and relate such actions to specific discharge scenarios and water body types.

3. Relate remedial actions to in-stream processes, including a determination of whether the action will enhance or retard each process.

4. Relate specific hazardous chemicals to in-stream processes, including the importance of each process in determining chemical migration and fate.

5. Identify available analytic and numeric models for chemical transport and fate and evaluate their potential applicability to each scenario (i.e., water body, discharge, and chemical type).

6. Evaluate representative models using matrices relating key model capabilities to in-stream processes, water body characteristics, and remedial action modeling requirements.

7. Identify representative types of models which are suitable for various scenarios.

8. Develop modeling requirements in the form of model type, required dimensionality and grid configuration, and parameter adjustment, for each remedial action group.

1.3 REPORT CONTENT

This report presents the model evaluation procedure and supporting information needed to (1) identify potential remedial actions for a given discharge scenario, (2) identify key processes which should be simulated, (3) evaluate alternative models, and (4) evaluate specific models for application. Sections 2 and 3 summarize migration and fate processes and remedial actions, respectively. These summaries are basic and assume little prior reader familiarity with chemical transport and fate. The experienced reader may wish to use these sections only for reference while reading the remainder of the report.

Section 4 is a summary of eight case histories which show typical discharge scenarios and remedial responses. Where appropriate, use of models in each response process is described. These cases form a background for the model selection process. Analytic and simplified assessment techniques are described in Section 5, while numerical models are discussed in Section 6. Descriptions of each level of technique summarize capabilities, data needs, and general computation method. Section 7 provides modeling requirements for surface water remedial actions. These requirements include specific adjustments required for each remedial measure, as well as parameter estimation guidance. References, including those for models mentioned in the report, follow Section 7.

# 2. Migration and Fate

2.1 OVERVIEW

The migration and fate of chemicals in surface waterbodies results from both physical and chemical processes. Physical processes cause movement of chemicals, while chemical processes cause degradation and transformation of pollutants. Chemical processes affect migration when changes of state or physical properties occur (e.g., precipitation of a chemical or sorption onto sediments). Mills, et al., (1982) provide survey level discussions of all important processes, while Callahan, et al., (1979) discuss chemical processes affecting the fate of priority pollutants. Table 2.1 lists important instream migration and fate processes in the order they are discussed in this chapter. The following subsections provide brief descriptions of each of the key processes and are not intended to be comprehensive. Rather, they offer a basis for later discussions of remedial actions and model applications.

2.2 PHYSICAL PROCESSES

2.2.1 Overview

Physical processes may be lumped into 3 groups.

1. Advection: Transport of the pollutant at the same velocity as surrounding water molecules, in vertical or horizontal directions.

2. Dispersion: Spreading of the pollutant plume in the water column as a result of molecular diffusion, turbulent diffusion and shear-flow dispersion. Molecular diffusion represents the scattering of molecules from random motions, and is dependant on the viscosity of the fluid and the size of the particles. Turbulent diffusion operates on a larger scale. It

TABLE 2.1  IMPORTANT PROCESSES:  PHYSICAL AND CHEMICAL

I.  Physical

- Advection of flow
    Vertical
    Horizontal

- Dispersion
    Longitudinal
    Transverse and vertical

- Sedimentation
    Advection of sediment
    Erosion:  scour of native material and resuspencsion of contaminated sediments
    Deposition:  settling

II.  Chemical

- Hydrolysis
- Oxidation
- Photolysis
- Bio-degradation
- Bio-accumulation
- Precipitation/dissolution
- Volatilization
- Adsorption

represents mixing and spreading of particle clouds from un-steady flow conditions and random velocity fluctuations. Shear-flow dispersion is caused by spatially-averaged gross velocity differences in the flow which create shearing and spreading movement during advection. It applies in 2-D flow when a spatial average is taken. These three mechanisms will often be lumped into a term called "effective" dispersion for each dimension (x, y, z) (Orlob, 1971).

3. Sedimentation: This may be considered a form of advective transport for particulate matter. Pollutants sorbed onto suspended or bed sediments differ in their transport rates from that of the water because of the varying densities and radii of the contaminated sediments. Sediments may be transported, deposited, and resuspended in the water column, depending on the hydrodynamics of the system.

Surface water bodies differ substantially in the way physical processes operate due to the overall size, geometry and boundary conditions of each water body. The full range of surface water bodies can be classified into one of three types: impoundments, rivers and estuaries. The energy needed to drive these processes in water bodies may result from the following sources:

o Density differences in water due to temperature, distribution of dissolved solids (including salinity), and sediment concentration (rare cases)

o Energy gradient due to wind shear hydraulic forces, gravity, or boat/ship traffic

o Coriolis force which is due to the earth's rotation and imposes lateral forces on a flow (Important only when large waterbodies are analyzed)

o Water momentum at boundaries including tidal and fresh water inflow and outflow

o Mechanical energy transfer due to wind shear on the water surface and waves induced by wind

Figure 2.1 is a diagram of the important physical processes in lakes, rivers, and estuaries. The range of important processes for each water body are represented by the horizontal lines along the bottom of the figure.

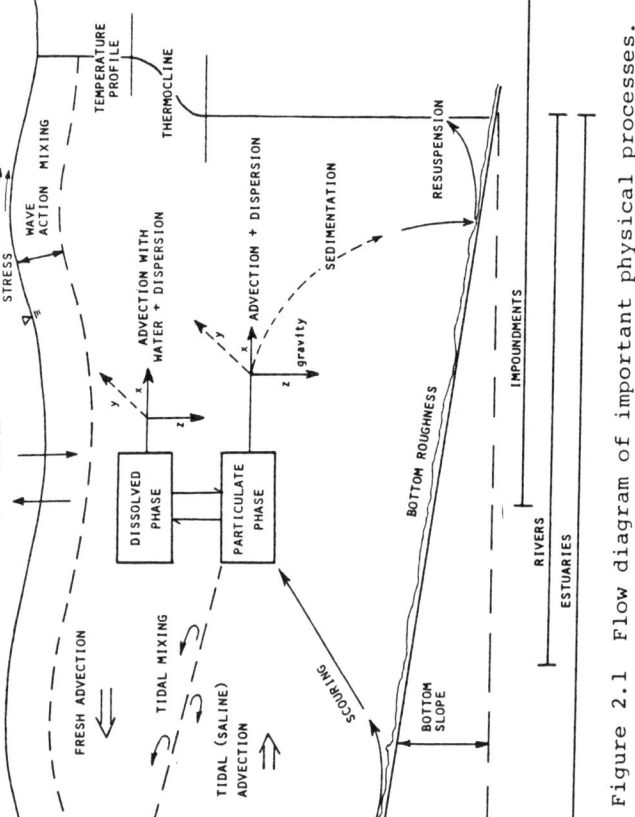

Figure 2.1 Flow diagram of important physical processes.

## 2.2.2 Rivers/Streams

Rivers are characterized by uni-directional flow which is often well-mixed. When a discharge into a river occurs, three stages of mixing may follow (Neely et al., 1976): (1) initial buoyancy and momentum of the spilled material, (2) lateral dispersion across the channel width, and (3) longitudinal dispersion downstream.

> Dominant transport mechanism: Longitudinal advection and longitudinal ("effective") dispersion. Assumes river is relatively shallow, and well-mixed across cross-section and depth.
>
> Significant parameters:
>
> o  Longitudinal dispersion coefficient
> o  Mean velocity
> o  Cross-sectional area
> o  Depth
> o  Bottom roughness and slope
> o  Sediment size
> o  Vertical and lateral locations of inlets/outlets
>
> Environmental conditions of concern:
>
> o  Precipitation and surface runoff of watershed
> o  Evaporation
> o  Scouring of the channel bed
> o  Deposition of sediments
> o  Water temperature (affects sediment transport)

## 2.2.3 Impoundments

Lakes, impoundments, and reservoirs are characterized by relatively low velocities (except at inflow/outflow areas), high retention time, and large surface areas. All of these properties enhance heat transfer with the atmosphere. Of primary concern are impoundments in the North American temperate region: lakes that are monomictic or dimictic (one or two turnovers of lake water per year, respectively). This annual cycle of vertical mixing followed by stratification is caused by wind stress on the surface, density differences in the water caused by solar insolation, and changes in air temperature (Fischer et al., 1979).

> Dominant transport mechanism: Vertical and longitudinal advection and effective dispersion in the "x" and "z" directions. Representation of transport often requires a

2-D (vertical) simulation. In large, shallow lakes, a 2-D x-y simulation may be more appropriate.

Significant parameters (in addition to those for rivers):

o   Wind stress
o   Boundary conditions: Shoreline shape, mixed depth, boundary roughness
o   Vertical and horizontal dispersion coefficients
o   Inflow/outflow rates
o   Rate of heat transfer at surface
o   Detention time

Environmental conditions of concern (in addition to those for rivers):

o   Air temperature
o   Specific humidity
o   Solar insolation
o   Dominant wind direction and speed

## 2.2.4  Estuaries

Estuaries may be the most complex waterbodies in hydrodynamic terms. Primary forces affecting transport are tidal variations in water surface elevation, wind stress, fresh water inflow, and internal density differences. Some estuaries with large surface areas may also be influenced by Coriolis forces.

Dominant transport mechanism: Longitudinal and lateral advection and dispersion. Although a 3-D representation is desirable in many stratified estuaries, most numerical simulations use a 2-D vertical/longitudinal, or lateral/longitudinal model, which is tidally varying. However, pseudo 2-D (network) simulations have been applied to shallow, well-mixed systems.

Significant parameters (in addition to those for impoundments):

o   Tidal exchange at seaward boundary
o   Tidal circulation within estuary
o   Tidal height variation
o   Tidal period
o   Freshwater inflow (inland boundary condition)

Environmental conditions of concern (in addition to those for impoundments):

o   Types and concentrations of suspended material

## 2.3 CHEMICAL PROCESSES

The instream processes listed in Table 2.1 constitute the various means of degradation and transformation of a pollutant in the water body. Figure 2.2 is a flow chart of these processes and their interactions. Descriptions of each process follow in standardized form, including significant parameters, environmental conditions of concern, and relation to other processes. Speciation processes such as acid-base equilibria are different for each pollutant, and are considered implicitly in the descriptions of the chemical/biological processes. Table 2.2, a matrix adapted from Callahan et al. (1979), summarizes the relative importance of degradation processes affecting the aquatic fate of priority pollutants. The matrix originally incorporated only organics; heavy metals and inorganics have been added. As the matrix is reviewed, trends between pollutant group and important processes will be noticed. For example, dominant fate processes for pesticides include sorption, volatilization and bio-degradation; for aliphatic hydrocarbons (compounds with carbon atoms formed in open chains, such as chloroform and vinyl chloride), volatilization; and for metals and inorganics, sorption and bio-accummulation.

### 2.3.1 Hydrolysis

Hydrolysis may be defined as any reaction (without the aid of light or micro-organisms) in which a chemical combines with water molecules to form a new compound. Many hydrolysis reactions are pH-dependent. Significant parameters include hydrolysis rate coefficients which are dependent on the chemical structure of the compound, pH, and temperature. Rate constants for particular compounds can be obtained in literature or be determined by standard laboratory tests (Mills et al., 1982). Environmental conditions of concern include pH and water temperature.

Hydrolysis affects other chemical processes by either creating new, more active compounds or replacing active compounds with relative inert ones. Biodegradation, volatilization and bio-accumulation may be affected in this way.

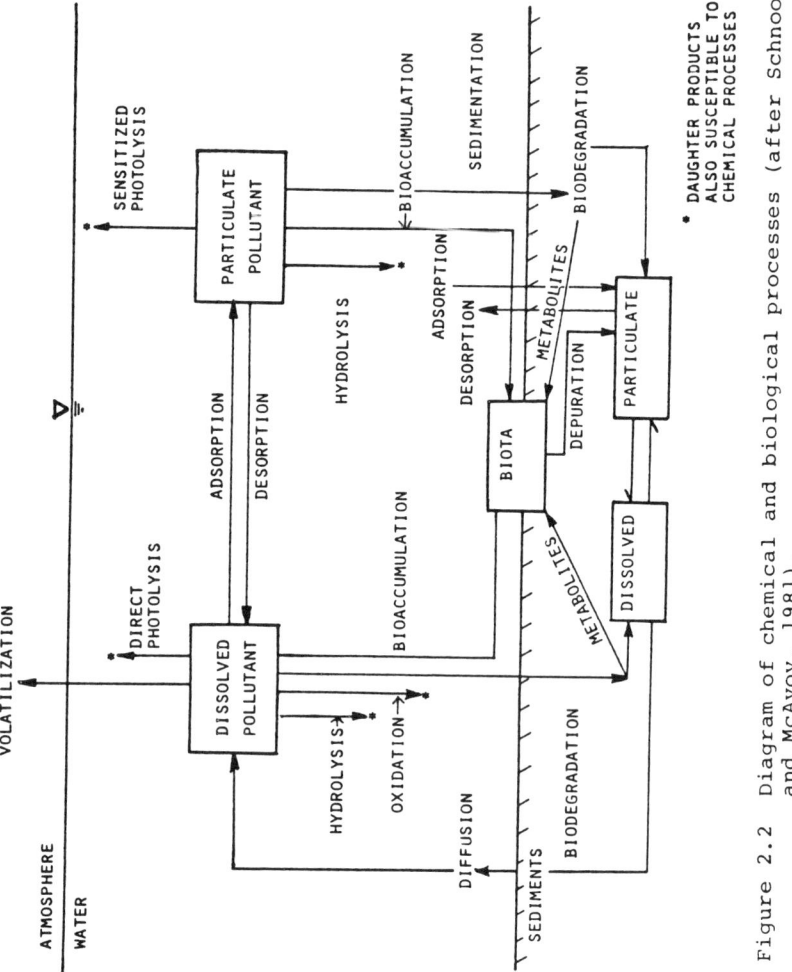

Figure 2.2 Diagram of chemical and biological processes (after Schnoor and McAvoy, 1981).

TABLE 2.2   POLLUTANT VS. PROCESSES MATRIX
(after Callahan et al., 1979)

| Compound | Sorption | Volatilization | Biodegradation | Photolysis-Direct | Hydrolysis | Bioaccumulation |
|---|---|---|---|---|---|---|
| **PESTICIDES** | | | | | | |
| Acrolein | − | + | + | + | − | − |
| Aldrin | + | + | ? | − | − | + |
| Chlordane | + | + | ? | − | − | + |
| DDD | + | + | − | − | − | + |
| DDE | + | + | − | + | − | + |
| DDT | + | + | − | − | + | + |
| Dieldrin | + | + | − | + | − | + |
| Endosulfan and Endosulfan Sulfate | + | + | + | ? | + | − |
| Endrin and Endrin Aldehyde | ? | ? | ? | + | − | + |
| Heptachlor | + | + | − | ? | ↔ | + |
| Heptachlor Epoxide | + | − | ? | ? | − | + |
| Hexachlorocyclohexane (α,β,δ isomers) | + | ? | + | − | − | − |
| -Hexachlorocyclohexane (Lindane) | + | − | + | − | − | − |
| Isophorone | − | − | ? | + | − | − |
| TCDD | + | − | − | ? | − | + |
| Toxaphene | + | + | + | − | − | + |
| **PCBs and RELATED COMPOUNDS** | | | | | | |
| Polychlorinated Biphenyls | + | + | +[a] | ? | − | + |
| 2-Chloronaphthalene | − | ? | + | + | − | − |
| **HALOGENATED ALIPHATIC HYDROCARBONS** | | | | | | |
| Chloromethane (methyl chloride) | − | + | − | − | − | − |
| Dichloromethane (methylene chloride) | − | + | ? | − | − | − |
| Trichloromethane (chloroform) | − | + | ? | − | − | − |
| Tetrachloromethane (carbon tetrachloride) | ? | + | − | − | − | ? |
| Chloroethane (ethyl chloride) | − | + | ? | − | + | − |
| 1,1-Dichloroethane (ethylidene chloride) | − | + | ? | − | − | − |
| 1,2-Dichloroethane (ethylene dichloride) | − | + | ? | − | − | − |
| 1,1,1-Trichloroethane (methyl chloroform) | − | + | − | − | − | − |
| 1,1,2-Trichloroethane | ? | + | − | − | − | ? |
| 1,1,2,2-Tetrachloroethane | ? | + | − | − | − | ? |

Key to Symbols:
↔ Predominant fate determining process   − Not likely to be an important process
\+ Could be an important fate process    ? Importance of process uncertain or not known

(continued)

TABLE 2.2 (continued)

| Compound | Sorption | Volatilization | Biodegradation | Photolysis-Direct | Hydrolysis | Bioaccumulation |
|---|---|---|---|---|---|---|
| Hexachloroethane | ? | ? | ? | ? | ? | + |
| Chloroethene (vinyl chloride) | + | − | − | − | − | − |
| 1,1-Dichloroethene (vinylidene chloride) | ? | + | ? | − | − | ? |
| 1,2-trans-Dichloroethene | − | + | ? | − | − | − |
| Trichloroethene | − | + | ? | − | − | − |
| Tetrachloroethene (perchloroethylene) | − | + | + | − | − | − |
| 1,2-Dichloropropane | ? | + | − | ? | + | ? |
| 1,3-Dichloropropene | ? | + | − | ? | + | − |
| Hexachlorobutadiene | + | + | ? | − | ? | + |
| Hexachlorocyclopentadiene | + | + | − | + | + | + |
| Bromomethane (methyl bromide) | − | + | − | − | + | − |
| Bromodichloromethane | ? | ? | ? | ? | − | + |
| Dibromochloromethane | ? | + | ? | ? | − | + |
| Tribromomethane (bromoform) | ? | + | ? | ? | − | + |
| Dichlorodifluoromethane | ? | + | − | ? | − | ? |
| Trichlorofluoromethane | ? | + | − | − | − | ? |
| **HALOGENATED ETHERS** | | | | | | |
| Bis(choromethyl) ether | − | − | ? | − | ++ | − |
| Bis(2-chloroethyl) ether | − | + | − | − | − | ? |
| Bis(2-chloroisopropyl) ether | − | + | − | − | − | ? |
| 2-Chloroethyl vinyl ether | − | + | ? | − | + | − |
| 4-Chlorophenyl phenyl ether | + | ? | ? | + | − | + |
| 4-Bromophenyl phenyl ether | + | ? | ? | + | − | + |
| Bis(2-chloroethoxy) methane | − | − | ? | − | + | ? |
| **MONOCYCLIC AROMATICS** | | | | | | |
| Benzene | + | + | − | − | − | − |
| Chlorobenzene | + | + | − | ? | − | + |
| 1,2-Dichlorobenzene (o-dichlorobenzene) | + | + | − | ? | − | + |
| 1,3-Dichlorobenzene (m-dichlorobenzene) | + | + | ? | ? | ? | + |
| 1,4-Dichlorobenzene (p-dichlorobenzene) | + | + | − | ? | − | + |
| 1,2,4-Trichlorobenzene | + | + | − | ? | − | + |
| Hexachlorobenzene | + | − | − | − | − | − |

Key to Symbols:
++ Predominant fate determining process
+ Could be an important fate process
− Not likely to be an important process
? Importance of process uncertain or not known

(continued)

TABLE 2.2 (continued)

| Compound | Sorption | Volatilization | Biodegradation | Photolysis-Direct | Hydrolysis | Bioaccumulation |
|---|---|---|---|---|---|---|
| Ethylbenzene | ? | + | ? | - | - | - |
| Nitrobenzene | + | - | - | + | - | - |
| Toluene | + | + | ? | - | - | - |
| 2,4-Dinitrotoluene | + | - | - | + | - | ? |
| 2,6-Dinitrotoluene | + | - | - | + | ? | ? |
| Phenol | - | + | + | + | - | - |
| 2-Chlorophenol | - | - | ? | + | - | - |
| 2,4-Dichlorophenol | - | - | ++ | - | - | - |
| 2,4,6-Trichlorophenol | ? | - | ? | ? | - | - |
| Pentachlorophenol | + | - | + | + | - | + |
| 2-Nitrophenol | - | - | - | ++[b] | - | - |
| 4-Nitrophenol | + | - | - | ++ | - | - |
| 2,4-Dinitrophenol | + | - | - | ++[b] | - | - |
| 2,4-Dimethyl phenol (2,4-xylenol) | - | - | ? | + | - | - |
| p-chloro-m-cresol | - | - | ? | ++ | - | - |
| 4,6-Dinitro-o-cresol | + | - | - | ++ | ? | ? |

**PHTHALATE ESTERS**
| | | | | | | |
|---|---|---|---|---|---|---|
| Dimethyl phthalate | + | - | + | - | - | + |
| Diethyl phthalate | + | - | + | - | - | + |
| Di-n-butyl phthalate | + | - | + | - | - | + |
| Di-n-octyl phthalate | + | - | + | - | - | + |
| Bis(2-ethylhexyl) phthalate | + | - | + | - | - | + |
| Butyl benzyl phthalate | + | - | + | - | - | + |

**POLYCYCLIC AROMATIC HYDROCARBONS**
| | | | | | | |
|---|---|---|---|---|---|---|
| Acenaphthene[c] | + | - | + | + | - | - |
| Acenaphthylene[c] | + | - | + | + | - | - |
| Fluorene[c] | + | - | + | + | - | - |
| Naphthalene | + | - | + | + | - | - |
| Anthracene | + | + | + | + | - | - |
| Fluoranthene[c] | + | + | + | + | - | - |
| Phenanthrene[c] | + | + | + | + | - | - |
| Benzo(a)anthracene | + | + | + | + | - | - |
| Benzo(b)fluoranthene[c] | + | - | + | + | - | - |
| Benzo(k)fluoranthene[c] | + | - | + | + | - | - |
| Chrysene[c] | + | - | + | + | - | - |

Key to Symbols:
++ Predominant fate determining process
+ Could be an important fate process
- Not likely to be an important process
? Importance of process uncertain or not known

(continued)

TABLE 2.2 (continued)

| Compound | Sorption | Volatilization | Biodegradation | Photolysis-Direct | Hydrolysis | Bioaccumulation |
|---|---|---|---|---|---|---|
| Pyrene[c] | + | - | + | + | - | - |
| Benzo(ghi)perylene[c] | + | - | + | + | - | - |
| Benzo(a)pyrene | + | + | + | + | - | - |
| Dibenzo(a,h)anthracene[c] | + | - | + | + | - | - |
| Indeno(1,2,3-cd)pyrene | + | - | + | + | - | - |
| **NITROSAMINES AND MISC. COMPOUNDS** | | | | | | |
| Dimethylnitrosamine | - | - | - | ++ | - | - |
| Diphenylnitrosamine | + | - | ? | + | - | ? |
| Di-n-porpyl nitrosamine | - | - | - | ++ | - | - |
| Benzidine | + | - | ? | + | - | - |
| 3,3'-Dichlorobenzidine | ++ | - | - | + | - | - |
| 1,2-Diphenylhydrazine (Hydrazobenzene) | + | - | ? | + | - | + |
| Acrilonitrile | - | + | ? | - | - | + |
| **METALS AND INORGANICS** | | | | | | |
| Asbestos | + | - | - | - | - | - |
| Antimony | + | - | - | - | + | + |
| Arsenic | + | + | + | - | + | + |
| Berylumm | + | - | ? | - | + | + |
| Cadmium | + | - | - | - | + | + |
| Copper | + | - | - | - | + | + |
| Chromium | + | - | - | - | + | + |
| Cyanides | - | + | + | + | - | - |
| Lead | + | - | + | + | - | + |
| Mercury | + | + | + | + | - | + |
| Nickel | + | - | - | - | + | - |
| Selenium | + | + | + | - | + | + |
| Silver | + | - | - | - | - | - |
| Thallium | + | - | - | - | + | + |
| Zinc | + | - | - | - | + | + |

Key to Symbols:
++ Predominate fate determining process  - Not likely to be an important process
+ Could be an important fate process   ? Importance of process uncertain or not known

**Notes**

[a] Biodegradation is the only process knoen to transform polychlorinated biphenyls under environmental conditions, and only the lighter compounds are measurably biodegraded. There is experimental evidence that the heavier polychlorinated biphenyls (five chlorine atoms or more per molecule) can be photolyzed by ultraviolet light, but there are no data to indicate that this process is operative in the environment.

[b] Based on information for 4-nitrophenol

[c] Based on information for PAH's as a group. Little or no information for these compounds exists.

## 2.3.2 Oxidation

Two general types of chemical oxidation occur in the aquatic environment:

o   photo-oxidation, in which photolysis, either direct or by interaction with a photosensitizer, serves as the driving force

o   thermal or auto-oxidation, known simply as oxidation (occurs when the pollutant reacts with oxidants in solution)

The term oxidation in this report will refer to all oxidizing processes except photo-oxidation. Significant parameters include the base oxidant rate coefficient for a pollutant and the free radical oxygen concentration. Environmental conditions of concern include water temperature and reaeration rate which affect oxygen concentration. Oxidation can affect other processes in three ways: by producing reducing conditions (inhibits bio-degradation), by altering solubility (affects precipitation), and by lowering reactivity (affects volatilization and photolysis).

## 2.3.3 Photolysis

Photo-chemical transformation may occur directly or indirectly. Direct photolysis involves the absorption of light by the pollutant, placing electrons in an excited state from which reactions can transpire. Indirect photolysis occurs when another chemical absorbs light, and in its excited state, undergoes reaction with the pollutant (Mills et al., 1982). Significant parameters include the molar absorption coefficient (specific to each chemical) and the incident light intensity at a specific wavelength, which is a function of the mixed depth of water and attenuation of light by natural waters. Environmental conditions of concern include vertical mixing of the water column, turbidity caused by suspended sediments, water temperature, and incident light at the water surface. The oxidation of material may result in reducing conditions, inhibiting bio-degradation.

## 2.3.4 Volatilization

Volatilization is actually a physical process in which the dissolved pollutant changes state and is transported from the water to the atmosphere. Current evidence indicates that it is the dominant aquatic fate process for low molecular weight, non-polar compounds that don't easily degrade biologically or chemically (Callahan et al., 1979). Significant parameters include Henry's Law constant for compounds (essentially a partitioning coefficient between the gas and liquid phases), and reaeration rate, which is a function of wind speed and the mixed depth of water. Environmental processes of concern include water temperature, dissolved oxygen concentration, and vertical mixing. Increased turbulence increases the reaeration rate, enhancing volatilization (Smith et al., 1981).

## 2.3.5 Adsorption

The adsorption process involves the exchange of a pollutant between the dissolved and adsorbed states. Usually this includes chemi-adsorption, or chelation with the sorbent, as well as physical adsorption, in which the sorbate is loosely held by ionic attraction. Consequently, the type and amount of suspended sediments strongly influence the type of adsorption isotherm (graph of sorbed material vs. material dissolved at a specific temperature) that describes the sorption process. A linear isotherm is often assumed at low pollutant concentrations (Karickhoff, 1979). Because contaminated particulates may settle out of the water column, the bed sediment may serve as a repository or sink allowing release and/or resuspension over a long time period. For most organic and non-polar compounds the amount of organic carbon in the sediment determines the extent of sorption (Mulkey et al., 1982).

Significant parameters include partition coefficient at equilibrium (for hydrophobic or low solubility pollutants) or expressed on an organic carbon basis, and dissolved concentration of pollutant. Environmental conditions of concern include pH (particularly important when particulates are clay or organic material), water temperature, and sediment concentrations and organic content. Adsorption rates may be increased by vertical mixing and turbulence which causes suspension of sediments or may be decreased by deposition of sediments. Sorbed chemicals are not generally subject to photolysis or volatilization, but may be more or less available to bio-degradation.

## 2.3.6 Bio-degradation

Microbial breakdown is significant because of the high species diversity and metabolic rates of microbes in the natural environment. Pollutants are most susceptible to breakdown when they sorb with suspended sediments or settle out of the water column onto the bed. This resulting increased surface area can cause an increase in biodegradation (Mills et al., 1982). Significant parameters include pollutant concentration, standing microbial biomass, specific growth rate constant for the bacterial group, and metabolic pathways. Environmental conditions of concern include pH, water temperature, reaeration and resultant dissolved oxygen concentrations, trace nutrient concentrations, and specific toxicity to bacteria.

### 2.3.7 Bio-accumulation or Bio-magnification

Bio-accumulation or magnification is an important process for the partitioning of hydrophobic pollutants. Such pollutants are usually lipid-soluble; hence, uptake via absorption or ingestion results in the accumulation of the pollutant in the fatty tissue of an organism. An octanol-water partition coefficient is used to describe the uptake as octanol resembles body fat (Neely et al., 1974).

Significant parameters include an octanol-water partition coefficient (determined from laboratory test or structure-activity relationship) and solubility of pollutant in water. Environmental conditions of concern include fish and other biomass standing crops, water temperature (can affect rates of uptake and metabolism of organisms), and food chain order.

### 2.3.8 Precipitation/Dissolution

The solubility of a contaminant in water is defined as the maximum amount of that chemical that will dissolve in pure water at a specified temperature (Lyman et al., 1982). Above this amount, two phases may exist: the saturated aqueous solution and the precipitated solid. Most organic pollutants have low solubilities (Lyman et al., 1982). It is probable that their maximum solubility would not be reached in the aquatic environment except where high, localized, concentrations exist (as in a spill). However, fluctuating environmental conditions, such as pH or temperature, may alternately cause a pollutant to dissolve or precipitate and, as a result, will affect the mode of transport and importance of some chemical or biological processes.

Significant parameters include octanol-water coefficient ($K_{ow}$), solubility product ($K_{sp}$), and distribution (partition) coefficient ($K_d$). Environmental conditions of concern include pH, temperature, total dissolved solids, dissolved organic matter, degree of mixing in water column, and pressure (rare cases).

The form of the contaminant (dissolved or solid) will control the transport mechanisms in the aquatic system. Soluble pollutants can be easily distributed, as they move with water molecules. These pollutants usually exhibit low sorption and bio-concentration characteristics (Lyman et al., 1982). Insoluble pollutants may behave similarly to suspended sediments; they may be deposited, resuspended, and partitioned between the sediment and biotic compartments. The solubility will affect other processes: photolysis, hydrolysis and oxidation are enhanced by high solubility, while sorption and bioconcentration are often enhanced by precipitation.

# 3. Remedial Actions and Affected Critical Processes

## 3.1 OVERVIEW

Remedial actions may be classified into four groups: dilution, containment, removal and treatment. Individual remedial actions (such as mechanical dredging) are divided into these groups and described herein, with attention given to the dimensionality required for simulating that action, as well as the affected critical processes. Table 3.1 provides an outline of the remedial actions considered.

The purpose of this section is to: 1) briefly overview the design objectives of each of the measures listed in Table 3.1 and 2), identify which water bodies and processes are affected by these measures and how they are affected. This type of information is needed to support the development of guidance on the use of models to evaluate remedial action performance. Detailed information regarding design of these remedial actions, potential applications and their effect on surface water bodies can be found in JRB (1982), Raj and O'Farrel (1977), Thibodeaux (1979), and other sources.

Table 3.2 is a matrix of environmental processes vs. remedial actions. Environmental processes are grouped as either chemical/biological or physical processes, similar to the format for descriptions of processes presented earlier. Remedial actions are grouped in a fashion similar to their descriptions earlier. This matrix will allow the reader to identify specific remedial actions with affected processes. This information should be kept in mind while the following matrices are reviewed.

As an example of matrix interpretation, consider the following example. Mechanical dredging is a common form of waste source removal for contaminated sediments in a shallow, low flow waterbody. The use of this action may enhance the following in-stream processes, as denoted by a "+" on the matrix: photolysis, volatilization, sedimentation, and dispersion. These processes, then, are more important in assessing

TABLE 3.1   OUTLINE OF REMEDIAL ACTIONS

I.   Dilution

II.  Containment

   o   Booms
   o   Silt Curtains
   o   Cofferdams
   o   Barriers/diversions
   o   Capping

III. Removal

   o   Skimming
   o   Hydraulic dredging
   o   Mechanical dredging
   o   Excavation

IV.  Treatment

   o   In-situ
   o   On-site

TABLE 3.2  REMEDIAL ACTION VS. PROCESSES MATRIX

| PROCESSES / ACTIONS | TRANSFORMATION ||||||| PHYSICAL ||||
|---|---|---|---|---|---|---|---|---|---|---|
| | HYDROLYSIS | OXIDATION | PHOTOLYSIS | VOLATILIZ. | BIO-DEG. | BIO-ACC. | ADSORPTION | SEDIMENT | ADVECTION | DISPERSION |
| NO ACTION | 0 | 0 | 0 | 0 | 0 | 0 | 0 | 0 | -,+ | 0 |
| **REMOVAL** | | | | | | | | | | |
| MECHANICAL DREDGING | 0 | 0 | + | + | - | 0 | - | + | 0 | + |
| EXCAVATION | 0 | 0 | 0 | 0 | - | 0 | - | + | + | + |
| HYDRAULIC DREDGING | 0 | 0 | 0 | 0 | - | 0 | - | + | + | + |
| BARRIERS/ DIVERSIONS | 0 | 0 | 0 | 0 | 0 | 0 | 0 | 0 | + | + |
| SKIMMING | 0 | 0 | - | - | 0 | 0 | 0 | 0 | - | - |
| DILUTION | 0 | 0 | 0 | 0 | 0 | 0 | 0 | 0 | + | + |
| **CONTAINMENT** | | | | | | | | | | |
| COFFERDAMS | 0 | 0 | 0 | 0 | 0 | 0 | 0 | + | + | + |
| BOOMS | 0 | 0 | - | + | 0 | 0 | 0 | 0 | - | - |
| SILT CURTAINS | 0 | 0 | 0 | 0 | 0 | 0 | - | - | + | + |
| CAPPING | 0 | 0 | 0 | 0 | - | 0 | - | - | + | + |
| **TREATMENT** | | | | | | | | | | |
| IN-SITU | 0 | 0 | 0 | 0 | 0 | 0 | - | - | 0 | 0 |
| ON-SITE | 0 | 0 | 0 | + | 0 | 0 | + | + | + | + |

LEGEND:
+ = ENHANCES THE PROCESS IN RELATION TO NO ACTION
- = MITIGATES THE PROCESS IN RELATION TO NO ACTION
0 = DOES NOT AFFECT THE PROCESS

transport and fate of a pollutant using this remedial action. Also of importance, the processes of bio-degradation and adsorption may be decreased (denoted by "-"). The rest of the in-stream processes should not be affected (denoted by "0").

Model dimensionality required to adequately represent the effects of these actions is not typically dependent on the action, but relates to the water body shape, size and type. Exceptions to this are noted.

## 3.2 DILUTION

This action can reduce in-stream concentrations by increasing flow and reducing hydraulic retention time. This may be accomplished in water systems with controllable flows, as in rivers with dams upstream. Dilution will not appreciably affect the geometry or dimensionality of the flow but may increase dispersion due to mixing and higher velocities. Affected processes are limited to dispersion and advection, which are both increased.

## 3.3 CONTAINMENT ACTIONS

Containment actions separate chemicals from the rest of the waterbody. Consequently, they often alter the geometry of the body and flow direction. Both advection and dispersion change as a result. These changes can usually be represented by a 2-D (horizontal plane) model. Changes in chemical processes depend on the action, as discussed below.

### 3.3.1 Booms

Booms can be used to intercept or contain light, miscible pollutants (Specific Gravity <1) in a surface slick. This limits their use to a period immediately after the spill before the plume disperses or to small impoundments or dead-end branches in estuaries where surface wave action and wind shear are at a minimum (Raj and O'Farrel, 1977). Skimming may be used in conjunction for removal of pollutants. Figure 3.1 shows possible deployments of booms. Advection and dispersion will typically be decreased in the surface layer due to the blockage of wave action and surface currents by the booms. Because the chemical slick is contained, chemical concentrations will remain high and processes which depend on concentration (e.g., volatilization) may increase.

Conversely, self shading due to slick capacity may decrease photolysis.

### 3.3.2 Silt Curtains

Silt curtains and nets serve a function similar to booms, but may also trap suspended material, such as plumes downstream from a dredging operation. Figure 3.1 shows possible deployments of silt curtains. These actions have much the same effects on processes as booms, except that they can affect the entire water column and contain chemicals and sediments from surface to bottom. Use of silt curtains should be limited to situations with low velocities and minimal wave action to avoid failure.

### 3.3.3 Cofferdams

Cofferdams are single wall barriers usually made out of earth or steel, and constructed for shallow streams or rivers, or for those waterbodies with low flow. The dams divert or contain streamflow so that an area can be dewatered or isolated in preparation for excavation or dredging actions. Two possible configurations for cofferdams are shown in Figure 3.2. These structures typically confine the flow and, especially in rivers, will cause increased velocities and concomitant increases in dispersion, scour, and sediment deposition. If a cofferdam isolates contaminants from the water body all contaminant release and transport processes are minimized.

### 3.3.4 Barriers/Diversions

This group includes all physical structures that impede flow and divert water away from contaminated area by using a separate diversion channel. An example of this is a complete stream flow diversion around a contaminated area described by Zaccor (1981) or as shown in Figure 3.3. Complete diversions are usually required when an entire stream cross-section is heavily contaminated and removal of the contaminants is required. The waterbody boundaries are changed, the flow is entirely removed from the vicinity of the contaminants, and chemical processes are stopped.

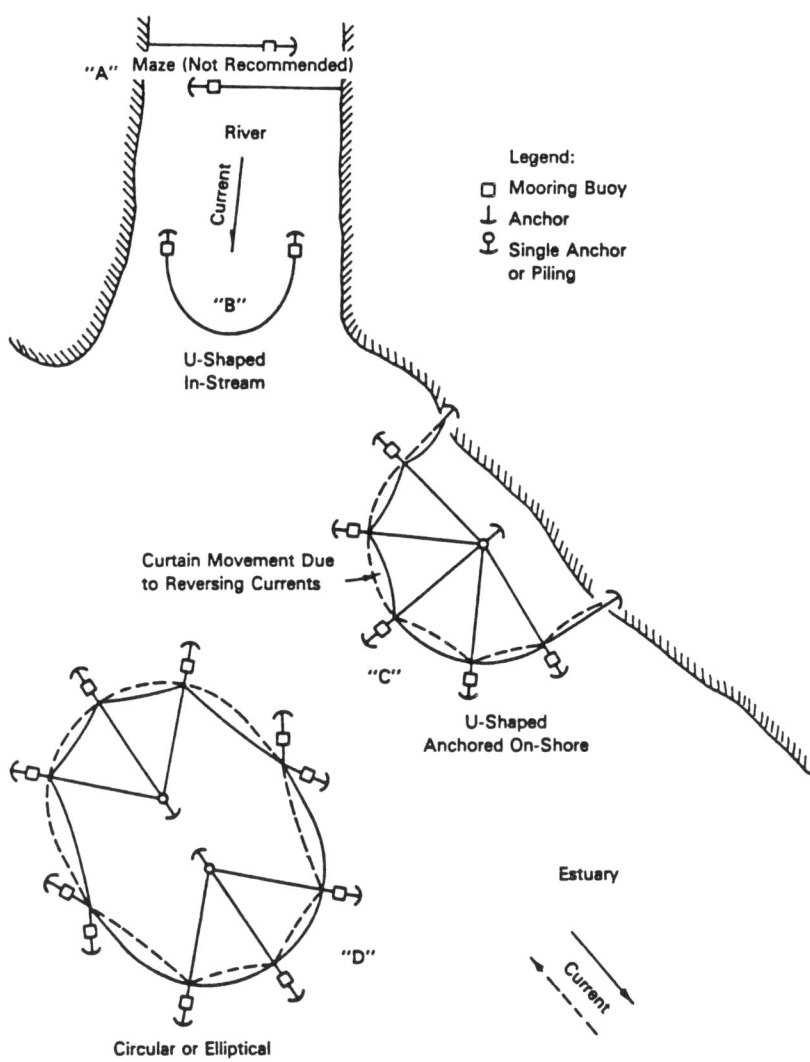

Figure 3.1 Typical boom or silt curtain deployment configurations (from Barnard, 1978).

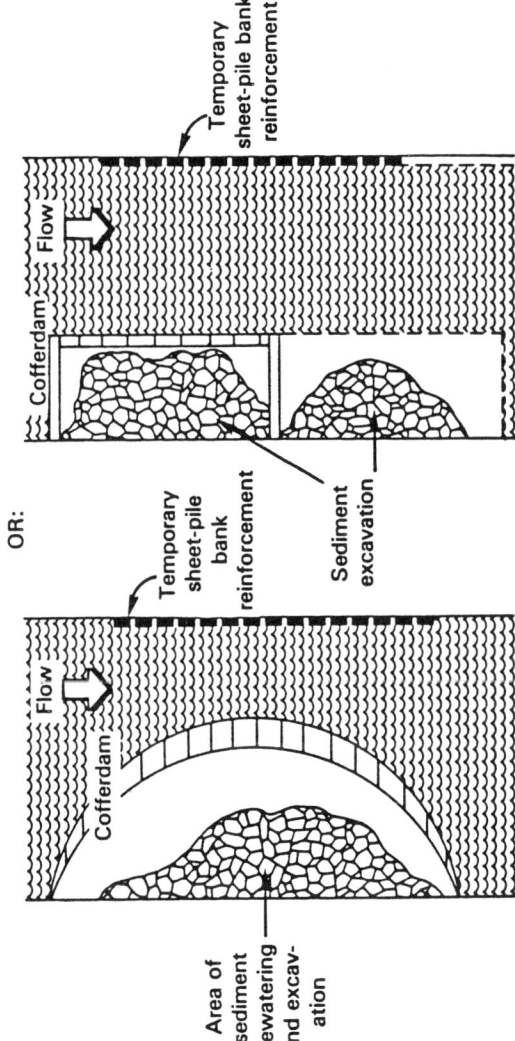

Figure 3.2  Isolation for sediment excavation using single cofferdam (from JRB, 1982).

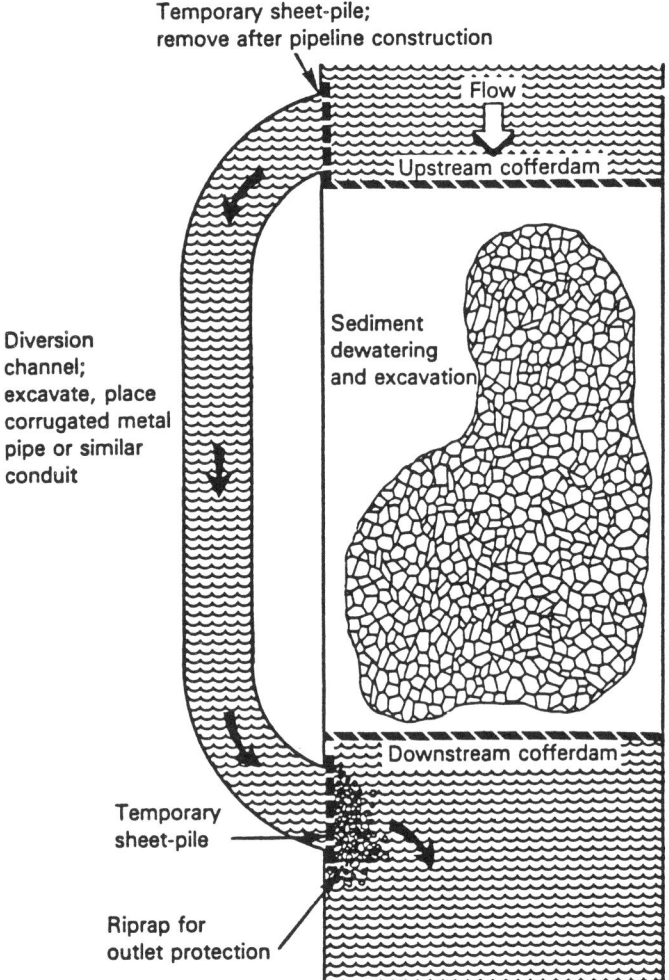

Figure 3.3 Streamflow diversion for sediment excavation using two cofferdams and diversion channel (source: JRB, 1982).

## 3.3.5 Capping

Capping with impervious materials may be applied to localized "hotspots" on the sediment bed, particularly where indentations occur. Problems can occur if stream velocity causes scour or depth is great, making verification of effectiveness difficult. Other problems include locating and treating waste deposits. Once installation is complete, movement of contaminants into the water column by scour, desorption or other processes will effectively cease. No other processes will be affected. During installation, scour and mixing may temporarily increase contaminant mobilization.

## 3.4 REMOVAL MEASURES

Removal measures are designed to eliminate the source of contamination from the water body. All contaminant-related processes will, consequently, be minimized. Four types of methods are available (see Table 3.1) and can be chosen to fit particular water body and contaminant conditions. These measures are often used in conjunction with containment measures to ensure that any chemicals mobilized by the removal process are retained at the site.

### 3.4.1 Skimming

Skimming is used when the pollutant has a specific gravity less than 1 and is contained within an impoundment or by means of surface barriers (booms) (Raj and O'Farrel, 1977). It is not as efficient when there is significant turbulence near the surface as in an estuary or fast moving stream or when strong winds are present. During skimming, increased turbulence and higher local velocities will tend to disperse chemicals unless effectively contained.

### 3.4.2 Hydraulic Dredging

This type of dredging includes the use of centrifugal pumping systems and portable hydraulic pipeline dredges. Centrifugal pumping systems can cut and chop heavy, viscous material (JRB, 1982). It may be applicable to spills of immiscible, high specific gravity material that settles in pools (Thibodeaux, 1979). Both types of hydraulic dredges may be used in impoundments or streams. Advantages over mechanical dredging include: minimal turbidity is created, dewatering of spoils isn't necessary, and it is suitable for removal of material in a wide range of consistencies, from floating liquid to sediment/sludges. However, spoil management actions are more important due to the large volume of material removed and must be included along with the use of diversions or barriers in any dredging plan. During dredging, turbulence and locally high velocities may resuspend and/or disperse chemicals unless effectively confined. These effects are less severe than those caused by mechanical dredging. Disposal of spoils involves large quantities of water, which may contaminate the same or other water bodies unless proper storage or treatment is implemented.

3.4.3 Mechanical Dredging

This measure may be used under conditions of slow, shallow flow. It should be used conjunctively with either streamflow diversion or silt curtains to prevent uncontrolled transport of resuspended contaminated sediments. Applicable waterbodies include streams, small rivers, lake shorelines, and small and then dewatered. However, supernatant from the dredge spoil poses an additional problem. Mechanical dredging will disturb bottom sediments and distribute them over the water column, resulting in increases in all migration and fate processes.

3.4.4 Excavation

This action may be used in conjunction with barriers and diversions, or may be applied to marshes or soil where contaminants are entering the surface water via leaching or runoff. Since excavation implies the removal of dry soil, the dewatering action (containment, diversion) is always considered as a conjunctive measure. Because the excavation site is isolated from the water body, removal of materials causes no changes in processes.

3.5 TREATMENT MEASURES

Remedial actions relating to the treatment of discharged hazardous materials in waterways are minor in importance. Similar actions are much more important at uncontrolled hazardous waste sites. Quite often treatment actions will be used in conjunction with a removal action, such as dredging. In these cases, the removal action has the greatest impact on in-stream processes.

Treatment methods may be physical, chemical, or biological. They may be applied in-situ, or on-site. In-situ treatment applications are rare, and limited to physical or activated carbon filtration systems. On-site applications are more common. Because on-site actions are outside of the waterway, their effect on in-stream processes is rarely felt, especially if the contaminated sediments and/or water are hauled offsite. However, if the material is treated and then released back into the waterbody, some impacts may be felt. Advection and dispersion may be increased locally by the discharge. All chemical migration and fate processes will operate on the discharge plume as they would on any point source of contamination.

# 4. Use of Remedial Actions and Modeling: Case Histories

## 4.1 OVERVIEW

In order to illustrate the need for remedial action assessment tools, case histories of discharge incidents and EPA responses are described below. These represent "typical" or common discharge scenarios that may occur in rivers, lakes, and estuaries. Descriptions of discharge scenario types are provided in Table 4.1. A hypothetical release incident that illustrates some of the release mechanisms is provided in Figure 4.1.

## 4.2 CASE HISTORIES

Eight case histories are briefly described below. Contained within each is an identification of the critical environmental processes, remedial actions considered, and modeling efforts.

### 4.2.1 Hudson River PCB Spill

Approximately 500,000 lbs. of PCBs were discharged into the Hudson River near Troy, New York over a period of time. It was estimated through an Environmental Impact Statement that $40 million would be needed for remedial actions to get the PCB concentration down to 50 ppm. Critical processes were identified as sorption and sedimentation; at high flows, the PCBs would desorb from scouring action on the sediment bed. Remedial actions chosen were: mechanically dredge (using a clamshell dredge) 40 "hotspots" and discharge off-site. Another remedial action of "capping" was considered infeasible due to costs and the fact that the river is a navigable waterway. The models were used to estimate PCB transport. The numerical sediment model HEC-6 (Hydrologic Engineering Center, 1977) was used with the WASP model (Water Quality

TABLE 4.1 TYPES OF DISCHARGE SCENARIOS (after Mills et al., 1982)

DIRECT
- MAY EMANATE FROM BARGE/SHIP DUMPING, OR PIPELINE RUPTURE
- SPECIFIC GRAVITY >1.0, HYDROPHOBIC, OR HAVING HIGH SORPTION; POLLUTANT SETTLES ON BED
  - ADVECTED ALONG BOTTOM
  - RE-ENTRAINED BY RESUSPENSION OF SEDIMENTS
  - DIFFUSION FROM SEDIMENT BED
  - MAY UNDERGO REDUCTION OXIDATION VIA MICROBIAL ACTIVITY IN THE BED
- SPECIFIC GRAVITY ≤1.0, HYDROPHILIC, OR HAVING LITTLE SORPTION; POLLUTANT IS ENTRAINED IN WATER COLUMN
  - VOLATILIZATION AND PHOTOLYSIS MAY BE IMPORTANT
  - OTHER REACTIONS (I.E., HYDROLYSIS) MAY AFFECT SOLUBILITY OR ABILITY TO SORB
  - ADVECTED AND DISPERSED ACCORDING TO BUOYANCY, MOMENTUM (NEAR FIELD), AND DOMINANT MIXING PROCESSES

INDIRECT
- MAY RESULT FROM TRUCK/RAIL OR WASTE SITE ACCIDENT, OR FROM STORM EVENT. SIMILAR BEHAVIOR AS "DIRECT" DISCHARGES
- SURFACE RUNOFF FROM SPILL ON LAND
  - TRANSPORT VIA FIRST STORM EVENT
- CONTAMINATED TRIBUTARY INFLOW
  - SMALL ENOUGH TO BE CONSIDERED A POINT SOURCE, OR,
  - OUTSIDE SYSTEM BOUNDARIES
- GROUND WATER RECHARGING SURFACE WATER, OR DIRECT LEACHING
  - RECHARGE DEPENDANT ON WATER TABLE LEVEL AND STREAM FLOW
  - VIEWED AS CONTINUOUS INPUT
- WET/DRY DEPOSITION FROM AIR TO SURFACE WATER (I.E., ACID RAIN ON LARGE LAKES)

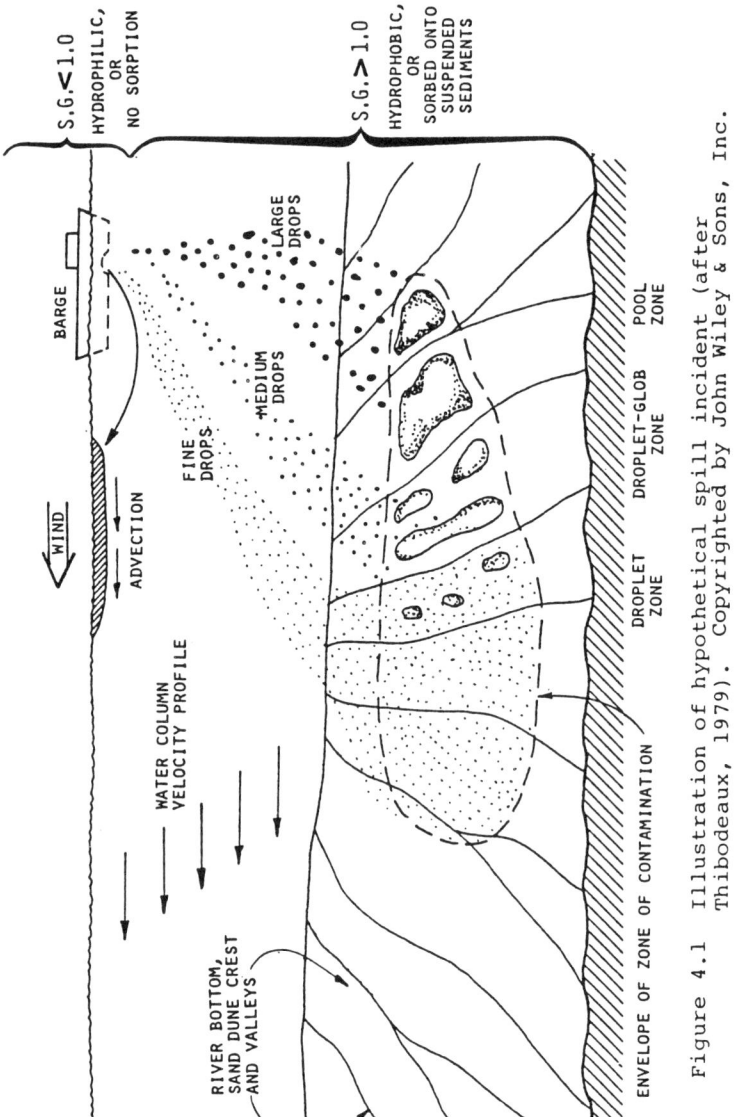

Figure 4.1 Illustration of hypothetical spill incident (after Thibodeaux, 1979). Copyrighted by John Wiley & Sons, Inc.

Analysis Simulation Program by DiToro et al.,1982) to determine PCB distribution in the biotic and abiotic compartments.

### 4.2.2 Waukegan Harbor PCB Spill

A corporation was discovered in 1975 to be discharging wastes containing PCBs into Waukegan Harbor, on Lake Michigan. Total amount of PCBs dumped was estimated to be 1.3 to 1.7 million pounds. Levels of contamination exceeded the F.D.A. fish tissue and sediment criteria level of 50 ppm (EPA, 1982).

Simulations were performed using the WASP program by R. V. Thomann at Hydroqual, Inc. The objective of the study was to quantify loading into Lake Michigan from Waukegan Harbor and the drainage ditch where most of the waste had been dumped. Again the critical process was identified as sorption to bed sediments in the harbor, with sediment and advective transport from natural flushing and dredging operations affecting desorption from the bed. Bio-accumulation was important, also, in light of fish tissue concentrations, but was simulated as a source/sink; depuration (excretion and death) and uptake ratios were simplified. Remedial actions chosen were to mechanically dredge the harbor with turbidity control (barriers) and to excavate the ditch.

### 4.2.3 Iron Mountain Mine Site

Iron Mountain Mine is defunct, and drains to the Sacramento River via a tributary creek near Redding, California. Tailing ponds, portals, and a pit on top of the mountain contribute a variety of heavy metals such as zinc, copper, and cadmium in point-source and non-point source pollution. Problems occur in the spring when snow-melt and rains lead to a large contaminated runoff flow. Two treatment plants can remove 75% of the copper from controlled flows (little runoff) and as much zinc and cadmium required with the control technology; the problem then is exacerbated when high runoffs can by-pass the plants. The only remedial action taken to date is the construction of a dam on Cheswick Lake, leading into the Sacramento River, to control or dilute flows downstream. Critical processes are identified as advective transport (dilution of acid drainage), and hydrolysis (metal mobility). No modeling efforts have been done, although funding may be provided via a feasibility study for clean-up through Superfund.

### 4.2.4 Kepone Contamination in the James River

In the 1970's high concentrations of the pesticide Kepone were discovered in fish tissue and sediments in the tidally-influenced James River, VA. The material had been discharged over a period of time into Bailey Creek (river mile 120). Kepone is hydrophobic (low solubility/high sorption) and is predominately transported by contaminated sediments. Distribution is also complicated by the facts that Kepone may bio-accumulate and that the James River is essentially an estuarial system with complex hydrodynamics. The most critical process identified was sorption onto specific sediments. Estuary systems have a number of sediment types, each with specific sorption capacities. Modeling efforts were conducted by Onishi of Battelle using FETRA (Finite Element Transport Model, (Onishi et al., 1979), and by O'Conner of Hydroqual (O'Connor and Farley, 1981). The FETRA model was used to simulate the transport of Kepone and sediments utilizing simulated velocities and flow depth from the EXPLORE-I Model. Organic sediments are considered to be important carriers of organic pollutants. Hence, the FETRA code simulated dissolved Kepone and particulate Kepone with their sorption and transport mechanisms for noncohesive (sand), cohesive (clay and silt), and organic sediment separately. O'Conner also simulated transport mechanisms but examined bio-accumulation also. No simulation of remedial actions was conducted. Drinking water and fish harvesting bans were temporarily enacted as remedial measures.

### 4.2.5 Formalin Spill on the Russian River

A one-time, finite duration spill of formalin occurred on the Russian River in Cloverdale, CA, in May 1982. The pollutant entered the river via surface flow and leaching into a tributary. Critical processes were identified as volatilization and sorption. Because of the number of drinking water intakes along the river, a drinking water ban was enacted in conjunction with controlling the flow by closing the upstream dams, allowing the discharge of formalin over time into the Pacific Ocean. However, afterwards the in-stream concentrations were found still to be high, so the Army Corps of Engineers decided to use dilution as a remedial action, opening the upstream dams. No modeling efforts were initiated.

### 4.2.6 Triana DDT Site

This site has been releasing DDT over a long period into a stream which leads into wetlands in Wheeler Reservoir near Decatur, Alabama. The most critical process identified was bio-accumulation as fish tissue concentrations were high (50 ppm) and the fish were a staple food item to the indigenous population. The Army Corps of Engineers evaluated remedial actions including dredging, coating of the sediments with an impervious layer (capping) and the creation of a sediment dam with channel diversion around the wetlands area. Modeling the systems with EXAMS (Exposure Analysis Modeling System, Burns et al., 1982) was suggested by TVA but has not been done to this date.

### 4.2.7 Marathon Oil Spill

In July 1982, between 150,000 - 250,000 gallons of heavy crude oil spilled into an irrigation ditch that leads to the Shoshone River in northern Wyoming. The spill occurred as a result of a pipeline accident with the Marathon Oil Company. The critical process was considered to be advective transport. Deflection booms and siphon trucks were utilized as part of a clean-up program. Response was more of an emergency nature than a long-term remedial one.

### 4.2.8 Chlorine Barge Spill

A barge containing chlorine gas ruptured in San Francisco Bay in 1981. Hazard from toxic fumes was considered imminent, so the emergency response team at EPA took charge. Data from CHRIS (Chemical Hazard Research Information System) were utilized, as well as a gas dispersion model. No attempt was made to examine in-stream processes as the immediate need was to assess the toxic cloud formation.

### 4.3 SUMMARY

Clean-up programs have traditionally been used in response to emergency conditions where limited time and data require rapid, simple screening techniques. However, most of the earlier spill incidents of note were petroleum products with known or simple chemical properties. Modeling efforts concerned the simulation of circulation processes in open waters. The influx of more complex and toxic materials that degrade slowly, however, now presents additional problems over a longer time period.

Critical processes identified in the case histories most often were sorption, sediment migration (transport, scouring and deposition), advective transport, volatilization and some degradation processes such as bio-accumulation. Indeed, for the 103 organic priority pollutants, sorption processes are important for 60, and volatilization is important for 52 (Mills et al., 1982). Many of these pollutants are hydrophobic and thus sorb readily and can be transported with sediment. Advection and dispersion are also quite important, as evidenced by the James River and Russian River cases, and are specific for each waterbody.

Modeling efforts are not commonplace in remedial action programs, as seen in the case histories. Simulations were applied where it was apparent that long-term hazards could arise from fluctuating environmental conditions and the slow degradation of the pollutant (i.e. Kepone and PCB). In the future, models may provide guidance for implementation of remedial actions, including design considerations, such as placement and size of barriers. Simplified assessment techniques and analytical models may also be used for screening purposes and to characterize the site conditions.

# 5. Use of Analytical and Simplified Assessment Techniques for Remedial Action Screening and Assessment

## 5.1 OVERVIEW

Simplified assessment techniques and analytical models play an important part in the screening of hazards and assessment of exposure from contaminant discharges. These simpler models are presented to allow the user a choice between levels of model complexity, depending upon the nature of the problem. Their relatively simple nature allows application with little data and resources. They can, consequently, be used on a site for initial evaluation of site conditions and testing of hypotheses regarding gross contaminant transport processes. However, they can be considerably less accurate than numerical models and are not able to adequately simulate complex environmental conditions or the detailed effects of remedial actions. Despite their simplicity they may require substantial user experience and judgment to estimate appropriate parameter values and to apply the procedures effectively.

Simplified techniques and analytical models are similar in that they use analytical solutions for the flow and transport equations. Such solutions require that numerous assumptions be made, including steady-state conditions, homogeneous physical and chemical properties and simple flow geometries. The simplified techniques usually produce one value, because they are essentially comprised of one equation. These techniques are most useful for predicting steady-state contaminant concentrations under fixed environmental conditions. While they may use the same equations, analytical models can calculate concentrations over extended time periods with variations in parameters such as flow rate. A computer program is used to solve the analytic equation(s) repeatedly as time steps are taken. This allows the use of analytic models for time-dependent problems and for sensitivity analyses where the effects of parameter uncertainty are evaluated.

5.2  USES OF SIMPLIFIED ASSESSMENT TECHNIQUES

Simple methods are useful for screening and preliminary exposure assessments where the primary goal is to determine compliance with instream water quality standards. They can also be used to better define objectives, estimate the level of study required to attain the objectives, and to determine the nature of analysis required (i.e., numerical, analytical or physical modeling). Specific uses include: mixing zone water quality criteria development and determination of peak concentration, travel time, and concentration as a function of distance. These uses are referred to in the simplified assessment techniques vs. use and data matrix (Table 5.1).

The mixing zone or near-field is that area where initial dilution of the contaminant takes place. The degree of dilution and mixing determines the initial concentration (Co). Because of the limited dispersion that occurs near the discharge site, concentrations tend to be high and chronic toxicity to biota is often a problem if the discharge is continuous. For this reason, mixing zone criteria are established. Simple techniques can be used, based upon the buoyancy and momentum of the discharge, water depth, and stream velocity, to determine whether and how compliance can be attained. Determination of peak concentration is important when a worst case scenario is assumed. The contaminant is considered to be conservative (no degradation and minimal mixing is assumed), so that a maximum concentration is predicted.

Determination of travel time in regions of the waterbody away from the discharge is probably the most common use of simplified techniques. Velocity and distance are used to determine the time it takes for a slug input to reach a given point downstream. This point could be a drinking water intake, or other area where health effects may be felt. Flow is assumed to be steady and non-dispersive (plug flow). Dispersion is considered for time of travel of a slug input. Degradation of the contaminant is represented as a function of time. Far field techniques are designed for this use and are particularly applicable to rivers where advection dominates.

Variations in far field concentrations with distance from the source can be readily determined through solution of the analytic equation(s) at different locations. Such profiles provide a one-, two-, or three-dimensional picture of the effluent plume for either continuous or short duration discharges.

Simplified Assessment Techniques for Remedial Action 383

**TABLE 5.1 SIMPLIFIED ASSESSMENT TECHNIQUES VS. USE AND REQUIRED DATA**

| Simplified Assessment Techniques | Type of Release | | Waterbody | | | Capabilities and Required Data | | | | | | | | Use | | | |
|---|---|---|---|---|---|---|---|---|---|---|---|---|---|---|---|---|---|
| | Instantaneous | Continuous | Estuary | Lake | River | Velocity | Longitudinal dispersion | Lateral dispersion | Channel geometry | Buoyancy momentum | Transformation | Sorption | Bed-water interface | Mixing zone WQ criteria | Peak concentration | Time of travel | Concentration profile |
| **Mixing Zone Analysis** | | | | | | | | | | | | | | | | | |
| Degree of initial mixing | ●●● | | ● | | ● | ●● | | ●●● | ●●● | | | | | ●● | | | |
| Initial dilution | | | | | | | | | | | | | | | | | |
| Mixing across width | | | | | | | | | | | | | | | | | |
| **Far Field Approaches** | | | | | | | | | | | | | | | | | |
| Fraction of fresh water | ●●● | ●● | ●● | ●●● | ●●● | ●●● | ●●● | ●●● | | ●●● | | | | ● | ●● | ●● | |
| Modified tidal prism | | | | | | | | | | | | | | | | | |
| Point source | ● | | ●● | ●●● | ● | | | | | | ● | ●● | | ● | | ● | |
| Non-point source | | | | | | | | | | | | | | | | | |
| Discharge of pollutants with finite solubility | | | | | | | | | | ● | ● | | | | | | |
| **Transformation Equations** | | | | | | | | | | | | | | | | | |
| **Sediment-Water Interactions** | | | | | | | | | | | | | | | | | |
| Vertical distribution of sorbate | | | | | | | | | | | | | | | | | |
| Desorption from sediment bed | | | | | | | | | | | | | | | | | |
| Transport of high density/sorbed pollutants | | | | | | ● | | | | ● | ●●● | | | ● | | ● | |

5.3 CLASSIFICATION OF SIMPLIFIED ASSESSMENT TECHNIQUES

These techniques include computations that require few parameters and may be performed on a hand calculator. They may be used for site screening purposes to provide an initial assessment of the extent of the hazard and to determine what, if any, subsequent analyses should be employed. Table 5.2 is a list of these methods with references and uses, and includes the general groups of mixing zone and far field approaches, transformation equations, and sediment-water interactions. These groups are described below.

### 5.3.1 Near-Field Analyses

Several techniques can be used to determine the discharge concentration after initial mixing: degree of initial mixing, initial dilution, and mixing across width. Critical parameters are usually stream velocity and the buoyancy, momentum, and flow rate of the discharge. This group uses initial dilution processes to determine the maximum concentration after near field mixing. The calulated concentration allows determination of mixing zone water quality criteria, and is used as an initial concentration (Co) in far field analyses.

The degree of initial mixing analysis can be used on rivers to determine the distance downstream below a point source where complete mixing occurs, or to define the boundaries of the mixing zone. Pollutant loading is assumed to be instantaneous. River width is a sensitive parameter in the analysis. The simple equation computes downstream distance as a function of lateral dispersion, river width, and stream velocity.

Initial (near-field) dilution analysis is designed for estuaries or coastal waters where the pollutant is discharged through submerged diffusers (Frick, 1981). The dominant mixing process is different from that of a river, where width and velocity govern mixing. Mixing occurs as the buoyant effluent plume rises from the diffuser and entrains the ambient fluid (Mills et al., 1982). Critical to the calculation is the degree of density stratification, port spacing, effluent velocity to current velocity ratio, and depth. Initial dilution values as a function of depth and Froude number have been developed by Frick (1981) using a plume model under various physical conditions.

The mixing across width analysis is designed to determine the mixing zone size for lakes and wide rivers with irregular

TABLE 5.2  SIMPLIFIED ASSESSMENT TECHNIQUES
          FOR SURFACE WATER

Technique | Reference
---|---
I. Near-Field Analysis | Codell et al., 1982
 | Mills et al., 1982
   o Degree of initial mixing | Fischer et al., 1979
   o Initial dilution |
   o Mixing across width (lateral dispersion) |
II. Far-Field Approaches | Mills et al., 1982
 | Fischer et al., 1979
   o Estuaries | Tracor, 1971
      Fraction of freshwater |
      Modified total prism |
   o Rivers/Lakes | Mills et al., 1982
      Point source-continuous | Codell et al., 1982
 | Neely et al., 1976
      Non point source-continuous | Raj and O'Farrell, 1977
 | Krenkel and Novotny, 1980
      Spills of pollutants | Thomann, 1972
 | Csanady, 1973
III. Transformation Equations | Mills et al., 1982
IV. Sediment-Water Interactions | Mills et al., 1982
   o Vertical distribution of sorbate |
   o Desorption from sediment bed |
   o Transport of high density/sorbed pollutants |

geometries, especially where it's not apparent that the far shore affects mixing. This method is similar to the degree of initial mixing approach, except that the discharge velocity and geometry control near field pollutant dispersion, because of the relatively low ambient velocities present. Critical parameters also include depth and width of the waterbody.

### 5.3.2  Far-Field Analyses

Far field approches are used to determine downstream transport of pollutants, including time of travel of a pulse input, peak concentrations, and concentration profiles or extent of plumes. Most often, results from mixing zone analysis (such as Co) are used as input because far field methods do not consider such parameters as buoyancy or momentum of the discharge. Geometry is usually simplified, and complete mixing across a stream width is assumed. An analytical solution is derived from the one-dimensional transport or mass balance equations using steady flow parameters (velocity, depth, and cross-sectional area).

The *fraction of freshwater method* is a simple calculation for pollutant transport in estuaries. Transport is determined using the flushing time, which is the time of travel required to move a pollutant to the mouth of the estuary. The calculation assumes that the salinity is uniform throughout the estuary and that net seaward flow of saline water is proportional to the river discharge for that tidal cycle. Mixing is assumed to be instantaneous within each estuary segment. Plume movement is calculated based on net seaward velocity during a tidal cycles.

The *modified tidal prism approach* is used to calculate flushing time in estuaries also. Flushing time is calculated by dividing the estuary into segments with lengths determined by the maximum flow path of water during a tidal cycle (Mills et al., 1982). The tidal prism is compared to the total volume for each segment, as a measure of flushing potential. Salinity distribution is not required. A disadvantage is that in order to predict the flushing time of a pollutant midway in the estuary, the method has to be applied to the whole system. Parameter requirements include the river discharge over each tidal cycle and segment dimensions.

The *point source analysis* is applicable to both a continuous source effluent and a finite duration release of a pollutant. Uses include prediction of: steady-state and transient concentrations as a function of distance, advection rate past a specified location, and transformation to other species over a specified reach. Plug flow (no dispersion) is sometimes assumed. Concentrations are calculated by the transformation of a given initial concentration over time. This transformation rate is represented by an exponential term containing transformation coefficients and a distance/velocity ratio (which denotes the time of travel). Thus the amount of data required is not extensive. Transformation of the dissolved fraction can be calculated provided that the partition coefficient for the pollutant is known.

The nonpoint source analysis is designed to calculate steady-state or transient concentration profiles and time of travel. The far field analysis for downstream transport is similar to that of the point source assessment; however, the initial concentration is calculated by estimating loading into a specified volume of water from an adjacent land segment. Mixing is assumed to be complete and instantaneous for each event. Besides the data mentioned for the point source method, river and runoff flows as well as segment length are needed. Runoff flow may be estimated using SCS runoff-infiltration curves. If the pollutant is not highly soluble, a partition coefficient is needed and runoff of contaminated sediments must also be estimated. The user is referred to Donigian (1981), O'Connor (1967) or Mills et al., (1982) for more information.

A number of specific methods for one-time discharges of highly soluble contaminants are available for determining time of travel, concentration profile, and peak concentration. The analyses are designed for calculating initial concentration and downstream transport. Because the contaminant is released as a "slug" input and not a continuous release, a different solution technique from the continuous point source analysis is required. The dissolved phase concentration is calculated by using an expression containing the dissolved mass fraction, cross-sectional area (assumed to be constant) and time. This expression replaces initial concentration as used in continuous effluent analyses. The transformation exponential expression is also more complex, utilizing a simplified form of the advection/dispersion equation (containing a steady velocity, distance, and longitudinal dispersion coefficient), transformation coefficients, and elapsed time. An instantaneous mixing analysis can be performed first, in order to find the volume of water needed to dilute the pollutant to its solubility limit. Assuming concentrations near the solubility limit are rapidly attained, the far field analysis can be performed. The user is referred to Mills et al. (1982) for further detail.

### 5.3.3 Transformation Equations

Transformation equations primarily serve as screening tools based on chemical characteristics. They may also be used in lakes where advection is not a dominant means of transport and fate. Point source and nonpoint source loading data, as well as other hydrological data, must be compiled if the application is for a specific waterbody.

These equations describe the fate of pollutants over time rather than over space. For simplicity, these removal processes are based on equilibrium rates, and are first-order reactions (e.g., dependent only on the concentration of the pollutant and a fixed coefficient). Because waterbody parameters such as advection and mixing are not part of the analyses, the equations are usually not suitable for assessing water quality criteria. However, once travel time is known, transformation equations can be used to obtain concentration. In addition, they can be used as screening tools for persistence of pollutants. Lyman et al., (1982), Callahan et al. (1979), and Mills et al. (1982) can provide additional details.

### 5.3.4 Sediment-Water Interactions

Hydrophobic (low solubility) pollutants are subjected to different transport and fate mechanisms than are hydrophilic (highly soluble) ones. They may be more dense and/or sorb strongly to sediments. A dissolved phase may exist and can present an environmental hazard, although it is usually small compared to the sorbed phase. A series of specific analyses may be performed to determine peak concentrations, concentration profiles, and plume extent of these pollutants.

Before downstream transport can be calculated, a vertical distribution of suspended material analysis must be determined. It is particularly useful when the pollutant's partition coefficient is high (the dissolved phase is small or neglible). Required data and parameters include: settling velocity of the sediments or particulate phase, pollutant density, hydraulic radius of the reach, slope, and shear velocity (related to flow velocity and bottom roughness).

The desorption from sediment bed analysis is used to calculate contaminant concentrations in the water column. A high percentage of the dense or sorbed pollutant can be deposited on the sediment bed. If the pollutant is not very susceptible to degradation, it may slowly desorb back into the water column over a long time period. This desorption process can be calculated based on an initial concentration in the sediment bed, dissolved concentration in the water column, and desorption rate coefficients. The water column concentration is derived using the stream velocity, mass of the pollutant per unit area of bed, equivalent depth of water in the sediment, and a partition coefficient (Mills et al., 1982).

The analysis for a spill of low solubility/high density pollutants provides a means to calculate the water column concentration (dissolved and particulate) that is subject to downstream advection. Primary to the calculation is the diffusion coefficient and thickness of the diffusive sublayer over the bed. Depth and stream velocity will affect this thickness. Before this analysis is used, however, the dimensions of the contaminated zone must be known or calculated (using a mixing zone analysis), as well as the solubility limit of the pollutant in water. Refer to Raj and O'Farrel (1977) or Mills et al. (1982) for more information.

## 5.4 ANALYTICAL MODELS

### 5.4.1 Overview

Analytical models are presented as an intermediate technique in terms of level of complexity, between simplified assessment techniques and numerical models. The difference between simplified assessment techniques and analytical models is often small: analytical models often employ the same equations as simplified techniques but require computers because of the number of calculations to be solved, as for a multi-reach stream. They can be applied to more complex problems where some variation in properties occur.

These models require steady-state flow conditions and uniform geometry. They have limited applicability to remedial actions, given the unsteady flow regimes, non-uniform geometry, and complex sediment-water interactions that characterize environmental conditions when remedial actions are implemented. They are used for time of travel, peak concentration, and concentration profile determinations. Within this group of models, differences can include: complexity of geometry allowed, mode of pollutant loading (instantaneous or continuous), degree of mixing and dispersion (if any), ability to calculate transfer of mass between the sediment bed and the water column, method of estimating sediment transport (user input suspended sediment concentrations, or concentrations calculated for each reach separately), lumped or specific first order decay reactions, and the range of default values available for model parameters.

Table 5.3 is a matrix comparing selected analytical models with respect to model capabilities and required data. The model group is not meant to be comprehensive; rather, it represents a cross-section of available analytical models and is designed for comparison purposes only. Descriptions of models follow.

### 5.4.2 Selected Analytical Models

STTUBE and TUBE (Codell et al., 1982) are steady-state, conservative river models which are used in conjunction. Both use simple geometries (representing the river as a rectangular channel) and constant coefficients to analytically solve a standard dispersion equation. TUBE generates dispersion coefficients and the velocity field for STTUBE, which then simulates dilution and travel times. Computations are performed for stream-tube coordinates, in which the cross-sectional areas are mapped onto a new river discharge based coordinate system, thus simplifying the mathematical representation. STTUBE simulates a steady release of pollutants, and is restricted to portions of the river removed from the influences of discharge (far field). These models do not simulate sorption or transformation processes.

RIVLAK (Codell et al., 1982) computes concentrations in a river or near shore region of a large lake from a non-steady source. RIVLAK requires uniform geometry, steady flow, pulse input, constant dispersion coefficients, and release of the contaminant from a vertical line source. This type of analysis provided is more applicable to near field concentrations, and is useful for mixing zone criteria as well as for determining peak concentration. STTUBE, TUBE, and RIVLAK program listings, as well as user manuals, are provided in Codell et al. (1982).

HACS, or Hazard Assessment Computer System (Raj and O'Farrel, 1981) contains eight analytical models designed for water quality assessment as well as explosion and flammability hazards and toxic cloud formation assessment. Four of the models are suitable for distribution of water borne pollutants. These models all: assume instantaneous release of pollutants; are unable to simulate dispersion, degradation, or sorption processes; and provide peak concentration results. The primary differences between the models are based on chemical characteristics such as density and solubility. The four water quality assessment models are subsequently described.

TABLE 5.3 SELECTED ANALYTICAL MODELS VS. MODEL CAPABILITIES AND REQUIRED DATA/FACTORS

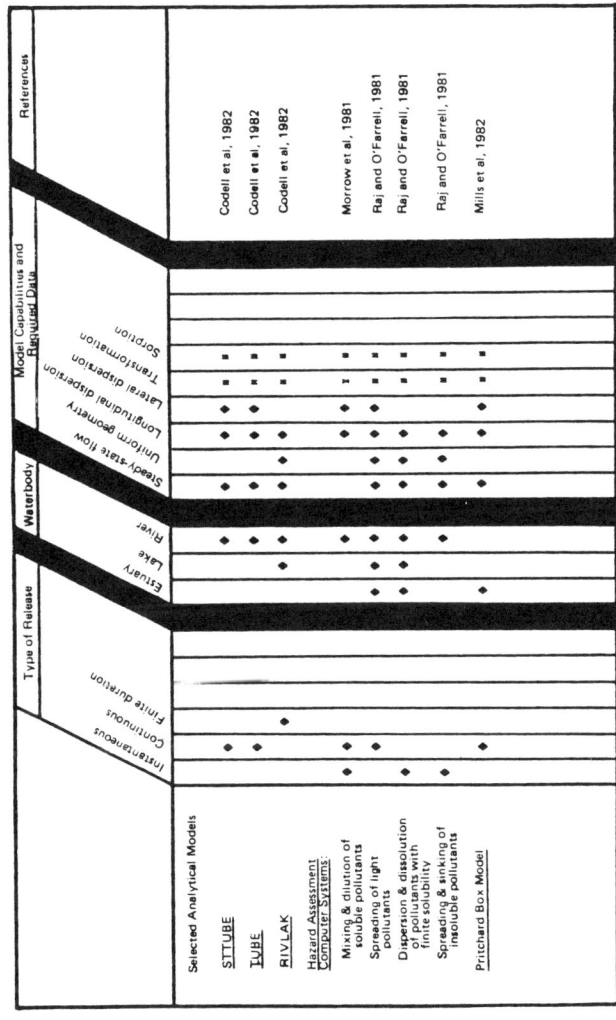

The mixing and dilution of the soluble pollutants model (Morrow et al., 1981) simulates instantaneous and continuous releases of hazardous chemicals into navigable non-tidal rivers. Very near field, near field, and far field computations are performed. The near field analyses are based on buoyancy, momentum of discharge, and turbulence; far field analyses predict steady-state concentrations as a function of distance downstream. Volatilization is the only transformation process simulated.

The spreading of light pollutants model (Raj and O'Farrel, 1977) examines the dispersion of low density (specific gravity less than one), low solubility pollutants on the surface of a waterbody. The pollutant is mixed based on eddy diffusivity so river turbulence parameters are required for simulation.

The dissolution and dispersion of pollutants with finite solubility model (Raj and O'Farrel, 1977) was designed to simulate pollutants that are soluble in low concentrations. The dissolution rate is predicted using solubility and dilution parameters, then dispersion is predicted for uni-directional flow.

The spreading and sinking of insoluble pollutants model (Raj and O'Farrel, 1981) is based on two stages: gravity-inertia and hydrodynamics. The model predicts the shape of the pool and duration of pool spreading. Its use is limited to turbulent rivers. No slope effects, complex geometry, or long-term bed/water interactions are included. HACS is operational on the Cybernet System of Control Data Corporation. Authorization and access procedures for the system are provided by the National Response Center of the U.S. Coast Guard in Washington, DC.

The Pritchard Box Model (Pritchard, 1969) is a steady-state, conservative, 2D (x-z) estuary model. It is designed for stratified estuaries and is sensitive to longitudinal salinity profile inputs. If the estuary is uniform and has little variation in salinity along its axis, it may be divided into two segments, whereupon a hand calculator can suffice for performing the analyses. The model should be implemented on a computer if more than five segments are defined. The model accepts only continuous pollutant release, preferably from the head of the estuary.

# 6. Use of Numerical Models for Remedial Action Assessment

6.1 OVERVIEW

Numerical models provide the investigator with the ability to represent chemical transport in complex water bodies where multi-dimensional flow, stratification, tidal variations and/or complex boundary conditions are important. Although such models involve substantially greater resources, their use may be justified where the effects of candidate remedial actions cannot be adequately represented by simplified methods. This section introduces a number of potentially useful models, discusses their capabilities, and provides a framework for their application.

Numerical models, in contrast to conceptual models (physically-based equations representing key processes) and analytic models (simplified process equations solved exactly using direct mathematical manipulation), approximate the process equations using finite difference or finite element techniques and separate the site into discrete segments. In this way, the full process equations can be solved with a minimum of restrictive assumptions. The solution, however, will not be exact. Consequently, a trade-off must be made between 1) ease of solution, computational accuracy, simplicity and limited applicability for analytical models and 2) greater resolution, more general applicability, increased complexity and increased solution costs for numerical models. Key attributes of numerical models can be summarized as follows.

1. Few simplifying assumptions are required, although the simplicity and computational efficiency of the solution algorithm depend, in part, on assumptions made.

2. Values of key quantities (e.g., velocity and chemical concentration) are computed at discrete space and time intervals selected by the user. These intervals (i.e., model resolution) can be adjusted to achieve

the accuracy and specificity required by the site and problem being addressed.

3. Numerical solutions to the governing equations are approximate and subject to computational errors due to truncation, roundoff and numerical dispersion. Choice of solution scheme can have a substantial effect on these errors.

4. Resources required to implement numerical models depend on the dimensionality, resolution, number of independent variables being predicted, and solution scheme. Required resources include: user expertise in developing and applying such models, field data, data on chemical sources, sinks and reaction rates, personnel time, and (typically) substantial computer facilities. It is reasonable to expect that needed resources will be two to ten times those required for analytic model applications.

5. Multiple independent variables (e.g., velocity, temperature, chemical concentration, etc.), can be simulated simultaneously, including interactions between these variables.

A number of authors provide overviews of numerical models and their use in problems related to surface water bodies. Donigian (1981) reviews runoff and instream contaminant transport and fate models, Onishi, et al., (1981) review sediment transport and water quality models, and Orlob (1971) discusses estuary models. Other current model reviews include Basta and Bower (1982) and EPA (1983). Additional information on surface water models can be obtained from the Center for Water Quality Modeling, EPA Environmental Research Laboratory, Athens, GA.

## 6.2 CAPABILITIES OF AVAILABLE CODES

The development of numerical models for surface water hydrodynamics and chemical transport has been ongoing since the early 1960's. Consequently, a large number of codes providing various degrees of sophistication are available. Some 35 codes were screened for possible use in remedial action assessment. Eleven codes were selected for further evaluation and inclusion in this discussion based on recent applications to toxic pollutant transport and fate studies, or ability to represent complex flow and mixing processes. These models serve as examples of codes which are potentially useful in remedial action assessment and a starting point for evaluations of suitable codes. Other codes (both existing and under development) may be of similar use.

Numerical codes can be differentiated by several aspects of their capabilities: type of water body that can be simulated, spatial domain (dimensionality), temporal domain (steady state versus dynamic time frame), and ability to represent chemical fate. A code is typically written for a certain type of water body (river, lake, or estuary); this target water body often defines the dimensionality and time frame of the code. In many cases a code written for estuaries can also be used for lakes or rivers or a code written for lakes can be used for rivers because all of the required elements of the less complex water body may be contained in the code. A disadvantage of using a complex code on a simple water body is the need to input parameters and data which may be extraneous to the problem and the added computer costs associated with a more sophisticated model.

Unlike ground-water models, which tend to use separate codes for flow modeling and chemical transport modeling, surface water models typically solve both flow and transport equations at the same time. There are two primary reasons for this: 1) there is usually limited interest in water movement without transport of heat, salinity, or chemicals and 2) the movement of heat, salinity and some chemicals affects hydrodynamics and so cannot be separated from the computation of flows. Most of the models discussed here are combined flow and transport models.

The majority of surface water codes provide dynamic (time varying) simulation of flow and transport. Dynamic simulations allow variations in chemical loadings due to changes in meteorology and discharge rates and in water flow rates due to the effects of tides, reservoir operation and streamflow.

Certain estuary models use tidally-averaged flow conditions to eliminate the effect of tides and reduce model complexity and run costs. Such an approach can produce meaningful results when the effects of flow reversals, movement of salinity gradients, and tidally-induced mixing can be ignored or approximated by steady-state parameters. Similarly, conditions in rivers and lakes which are steady over time (i.e., no significant variations in flows, temperature, or chemical inputs) can be simulated by steady-state models. Such models provide results similar to those obtained from analytic models.

Table 6.1 is a matrix of selected codes vs. environmental processes and waterbody conditions. Models chosen for detailed evaluation include: DEM: Dynamic Estuary Model (Ambrose and Roesch, 1982); FETRA: Finite Element Transport

396  Modeling Remedial Actions at Uncontrolled Hazardous Waste Sites

TABLE 6.1  PROCESSES VS. MODELS MATRIX

| PROCESSES / SELECTED NUMERICAL MODELS | WATERBODY | | SPATIAL DOMAIN | | | DISPERSION | | TEMPORAL DOMAIN | | SEDIMENTATION | | | TRANSFORMATION PROCESSES | | | | | | |
|---|---|---|---|---|---|---|---|---|---|---|---|---|---|---|---|---|---|---|---|
| | ESTUARY | LAKE / RIVER | 1-D | 2-D | 3-D | TURBULENCE | EMPIRICAL MIXING | STEADY STATE, TIDALLY AVG., QUASI-DYNAMIC | DYNAMIC | SEDIMENT TRANSPORT | SUS. SED. SORPTION DIRECT BED EXCHANGE | "ARMORING" | PHOTOLYSIS | OXIDATION | HYDROLYSIS | VOLITILIZATION | BIO-DEGRADATION | BIO-ACCUMULATION | LUMPED DECAY OR SINGLE POLLUTANT DAUGHTER PRODUCTS |
| TOXIWASP | • | • | C | | | • | • | | • | • | • | | • | • | • | • | • | • | • |
| HSPF | • | • | B | | | • | | | • | • | • | | • | • | • | • | • | • | • |
| TODAM | | • | • | | | | | | • | • | • | | | | | | | | |
| EXAMS | • | • | C | | | | •¹ | • | | | | | • | • | • | • | • | • | • |
| SERATRA | | • | | V | | | | | • | • | • | | • | • | • | • | • | • | • |
| DEM | • | • | B | | | • | | | • | | | | | | | | | | |
| FLESCOT | | • | | | L | • | • | | • | • | • | | | | | | | | |
| FETRA | • | • | | | | • | • | | • | • | • | | | | | | | | • |
| LEENDERTSE 2D | • | • | | V | | • | • | | • | | | | | | | | | | |
| LEENDERTSE 3D | • | • | | | V | • | • | | • | | | | | | | | | | |
| LARM | | • | | | | • | | | • | | | | | | | | | | |

LEGEND
S = SINGLE ITEM
B = BRANCHING OR NETWORK
L = LATERALLY INTEGRATED
V = VERTICALLY INTEGRATED
C = COMPARTMENTS

FOOTNOTES:
1. FETRA DOES EMPLOY FIRST-ORDER DECAY FOR A POLLUTANT, HOWEVER, THIS LUMPED PARAMETER MAY BE TOO SIMPLE FOR POLLUTANTS WITH MULTIPLE DEGRADATION PROCESSES.

Model (Onishi et al., 1979); TODAM: Transient One Dimensional Degradation and Migration Model (Onishi et al., 1982); SERATRA: Sediment-Radionuclide Transport Model (Onishi and Wise, 1979); FLESCOT: Flow Energy Salinity Sediment Transport Model (Onishi and Trent, 1982); HSPF Hydrologic Simulation Program - Fortran (Johanson et al., 1981); TOXIWASP: Water Quality Analysis Program (Ambrose et al., 1983 and Ditoro et al., 1982); EXAMS: Exposure Analysis Modeling System (Burns et al., 1982); Leenderste two-dimensional and three-dimensional circulation models (Liu and Leenderste, 1972); and LARM: Laterally-Averaged Reservoir Model (Edinger and Buchak, 1982).

The codes represented in Table 6.1 are divided into three groups based on the types of processes represented. The first group (TOXIWASP, HSPF, etc.) model water flow, chemical advection, sedimentation processes, and chemical transformation. The second group (DEM, FLESCOT, and FETRA) represent all other processes. The third group provides only hydrodynamic modeling, with some capability to advect and degrade single pollutants. These three groups also differ in the sophistication of their hydrodynamic computations: the first group uses compartmental or simple branching 1-D models (except for SERATRA), the second group provides a wide range of hydrodynamic solution techniques, and the third group provides relatively sophisticated two-dimensional and three-dimensional hydrodynamic codes. The model user must, in most cases, make a trade-off between detailed representation of chemical transport and transformation and representation of complex flows.

The parameters on the top axis of Table 6.1 may require further explanation. Spatial domain refers to the number of dimensions (1, 2, or 3) that the model may simulate. The two-dimensional models are further described as either lateral ("y" direction) or vertical ("z" direction) along with the normal longitudinal ("x") direction. Dispersion may be simulated by turbulence calculated within the program (generated by velocity differences or shear within the flow field); or it may be simulated empirically via user input dispersion coefficients. Temporal domain refers to the model's ability to simulate steady, continuous events or unsteady, pulse events. Steady-state refers to continuous waste input and flows over the duration of the simulated time period. Tidally averaged is also steady state but refers to simulating steady estuary hydrodynamics for each tidal period. Quasi-dynamic refers to the model's ability to simulate some variables in a steady-state and others dynamically in the same simulation. A dynamic simulation means that flows and waste loading may vary for each time step within a simulation.

Sedimentation refers to the whole range of sediment-water interactions (sediment transport, deposition, and erosion) that may occur. Sediment transport and suspended sediment sorption were described in Section 2. Direct bed exchange encompasses diffusion, scouring, deposition, and resuspension of contaminated material between the sediment bed and the water column. Armoring refers to the sorting of bed sediments during flows such that the bed surface is more resistant to scour than the underlying material. This situation may affect contaminant concentrations in the water and in the bed through modification of exchange rates.

Transformation processes have been described in Section 2. The lumped decay refers to a simple (usually first-order) reaction that accounts for the pollutants' aquatic fate. For some complex pollutants however, this degradation model formulation is an over-simplification and may not provide an accurate picture. "Daughter Products" refers to the model's ability to track the pollutant after it has degraded to another form. This "new" pollutant may be susceptible to the same physical, biological, and chemical processes as its parent. An example of this process is the degradation of DDT. The metabolites (or products) of chemical/biological degradation are DDD or DDE. Both of these compounds are more toxic than DDT, and warrant examination of transport and fate.

Table 6.2 is a matrix of the type of simulations needed for remedial actions and specific waterbodies. The waterbodies are grouped as estuary, lake, or river, with subgrouping within each according to system geometry and degree of mixing. Numbers and letters in the matrix denote the type of simulation needed for that remedial action in the specific waterbody. For example, "2L" denotes that a two-dimensional (lateral-longitudinal) simulation is required for that remedial action/waterbody scenario. A "0" indicates that the remedial action is not suited for use under the specific waterbody conditions. The simpler remedial actions such as dilution and the use of barriers or diversions, often may be simulated by adjusting the boundary conditions and system geometry. Most of the remedial actions require a two-dimensional (longitudinal-lateral) simulation. However, as the mixing becomes more turbulent or complex (as in estuaries and large lakes), a two-dimensional (longitudinal-vertical) simulation with coefficients for the horizontal or lateral (third) dimension, or a full three-dimensional simulation may be required.

The remedial actions vs. models matrix (Table 6.3) is a culmination of the previous two matrices. The critical processes of transport and fate of each remedial action are matched against model capabilities. As the matrix is reviewed the reader should refer to the previous matrices and the

TABLE 6.2  REMEDIAL ACTIONS VS. WATERBODY MATRIX

| REMEDIAL ACTIONS | ESTUARIES | | | | LAKES | | | RIVERS | | |
|---|---|---|---|---|---|---|---|---|---|---|
| | NARROW, WELL-MIXED, SHALLOW | NARROW, STRATIFIED | BAY, WELL-MIXED, SHALLOW | BAY, STRATIFIED | MONO- OR DIMICTIC | WELL-MIXED | RESERVOIRS | SINGLE STEM, WIDE, UNEVEN | SINGLE STEM, NARROW | BRANCHING/ DELTA |
| NO ACTION | | | | | | | | | | |
| **REMOVAL** | | | | | | | | | | |
| MECHANICAL DREDGING | 1 | 2V | 0 | 0 | 3 | 2L | 2P | 0 | 1 | 2L |
| EXCAVATION | 1 | 2V | 0 | 0 | 2L | 2L | 2L | 0 | 1B | 2L |
| HYDRAULIC DREDGING | 1 | 2V | 2L | 3 | 3 | 2L | 2L | 2L | 1B | 2L |
| BARRIERS/ DIVERSIONS | 2L | 3 | 2L | 3 | 3 | 2L | 3 | 2L | 2L | 2L |
| SKIMMING | 0 | 0 | 0 | 2V | 2V | 2V | 2V | 0 | 0 | 0 |
| DILUTION | 1 | 2V | 2L | 3 | 0 | 0 | 2V | 1B | 1 | 1B |
| **CONTAINMENT** | | | | | | | | | | |
| COFFERDAMS | 2L | 3 | 2L | 3 | 3 | 2L | 3 | 2L | 2L | 2L |
| BOOMS | 3 | 3 | 0 | 0 | 3 | | 3 | 0 | 0 | 0 |
| SILT CURTAINS | 2V | 2V | 0 | 0 | 2V | 2V | 2V | 0 | 2V | 0 |
| CAPPING | 0 | 2L | 0 | 0 | 0 | 0 | 2L | 0 | 2L | 0 |
| **TREATMENT** | | | | | | | | | | |
| IN-SITU | 0 | 0 | 0 | 0 | 0 | 0 | 0 | 2L | 2L | 2L |
| ON-SITE | * | DEPENDANT ON REMOVAL ACTION USED IN CONJUNCTION | | | | | | | | * |

LEGEND:
- 1 = 1-DIMENSIONAL
- 2 = 2-DIMENSIONAL
- 3 = 3-DIMENSIONAL
- L = LATERALLY AVERAGED
- V = VERTICALLY AVERAGED
- 0 = ACTION IS NOT APPLICABLE TO THIS WATERBODY
- B = BRANCHING OR NETWORK

TABLE 6.3 REMEDIAL ACTIONS VS. MODEL MATRIX

| REMEDIAL ACTIONS | MINIMUM DIMENSIONS REQUIRED | IMPORTANT FACTORS | SELECTED MODELS | | | | | | | | |
|---|---|---|---|---|---|---|---|---|---|---|---|
| | | | DEM | FLESCOT | FETRA[1] | TOXIWASP | HSPF | EXAMS | SERATRA | TODAM | LARM |
| **REMOVAL** | | | | | | | | | | | |
| MECHANICAL DREDGING | 2D(L) 1D | P,S,C | P | P,S | P,S | A | A | A | A | A | P |
| EXCAVATION | 2D(L) 1D | P,S | P | A | A | A | A | A | A | A | P |
| HYDRAULIC DREDGING | 2D(L) 1D | P,S,C | P | P,S | P,S | A | A | A | A | A | P |
| SKIMMING | 2D(V) | P,C | - | P | - | A | C | C | C | - | P |
| BARRIERS/ DIVERSIONS | 2D(L) 1D(N) | P | A | A | A | A | A | A | A | - | - |
| DILUTION | 1D | P,C | A | P | A | A | A | A | A | A | P |
| **CONTAINMENT** | | | | | | | | | | | |
| COFFERDAMS | 2D(L) | P | A | A | A | A | A | A | A | - | P |
| SILT CURTAINS | 2D(L) | P,S | - | A | A | A | A | S | A | - | P |
| CAPPING | 2D(V) | S,C | - | S | S | A | A | A | C | - | - |
| BOOMS | 2D(V) | P | - | A | - | A | - | - | - | - | A |
| **TREATMENT** | | | | | | | | | | | |
| ON-SITE | 1D | P,C | P | P | P | A | A | A | A | A | P |
| IN-SITU | 2D(L) | S | - | A | A | A | S | - | A | - | - |

FOOTNOTES:

1. FETRA DOES EMPLOY FIRST-ORDER DECAY FOR A POLLUTANT, HOWEVER, THIS LUMPED PARAMETER MAY BE TOO SIMPLE FOR POLLUTANTS WITH MULTIPLE DEGRADATION PROCESSES.

LEGEND:

A = REPRESENTS ALL IMPORTANT PROCESSES (P,S, AND C).
P = REPRESENTS THE THE CRITICAL PHYSICAL PROCESSES.
C = REPRESENTS THE CRITICAL CHEMICAL/BIO-DEGRADATION PROCESSES.
S = REPRESENTS THE CRITICAL SEDIMENT-WATER INTERACTION.

B = BRANCHING OR NETWORK
L = LATERAL
V = VERTICAL
1D, 2D, 3D = REFERS TO DIMENSIONS REQUIRED

remedial action description (Section 3) for reference. Model evaluation criteria are based upon the environmental processes affected by the remedial action and the dimensionality needed to represent these processes. For simplicity and ease of use, these criteria have been stated as questions and organized into the following groups.

Physical processes (denoted by "P"):

- o  Can the model simulate inflow and outflows?

- o  Is the dimensionality sufficient to represent a change in the system boundaries and geometry due to the remedial action?

- o  Can dispersion be adequately simulated using empirical coefficients, or should it be calculated within the model equations to account for the effect of new barriers, or inflow/outflow?

- o  Can the removal of the waste source such as a contaminated sediment bed be simulated (i.e.: Are there adequate source/sink terms?)

Sediment/Water Interactions (denoted by "S"):

- o  Can partitioning between sorbed/desorbed phases of the pollutant be simulated?

- o  If the pollutant is sorbed or in particulate form, can sediment transport be simulated?

- o  Can the model simulate bed-water transfers, such as scouring, deposition, and diffusion over time?

Chemical/Biological Degradation (denoted by "C"):

- o  Does the model simulate the _important_ degradation processes?

The above groups will be represented by the letters P, S, and C, respectively. Appearance of a letter under a model corresponding to a specific remedial action indicates that the model can represent the critical environmental processes affected by the remedial action (or answering "yes" to the hypothetical questions posed within the specific group above). Important factors or groups are listed beside each remedial action for easy reference.

## 6.3 THE MODEL DEVELOPMENT AND APPLICATION PROCESS

The process of setting up a computer code so that it will simulate water and waste constituent movement at a specific site is called the "model development" process. It involves combining one's understanding of how a code represents individual processes with one's understanding of their actual occurrence in the field. The latter is based on available site data, information and past experience. Model application is the use of a developed and tested model to analyze target situations, in this case the performance of a potential remedial action. While a numerical code is often quite general, a developed model is specific to a particular site and, when applied, to a particular condition at that site. Figure 6.1, taken from Mercer and Faust (1981), represents one process for model development and application. Once the need for numerical modeling has been determined and appropriate models selected for each affected zone, the following steps may be taken.

1. The conceptual model of the site used to select model codes is further defined and quantified through the collection and analysis of site data. This conceptual model may also include approximate effects of potentially feasible remedial actions.

2. The conceptual model is then used to define the model structure required for the water body of concern, the types of outputs needed, and the required spatial and temporal resolution of model simulations.

3. The individual code is installed on an appropriate computer and the site model implemented by creating an appropriate model structure (i.e., element or grid size and orientations, boundary conditions, and sink and source node loctions).

4. Values for individual model parameters are estimated from field data and then verified by comparing model predictions with available site data (i.e., history matching). The model can then be calibrated on a different set of site data to identify the ranges of values for critical parameters. This process is based on the assumption that the ability to obtain complete sets of data is not limited by time or money constraints. If the data is incomplete, best engineering judgement of information from field sites with similar characteristics should be applied.

5. Adjustments to model parameters and localized model structure can then be made to represent the effects of

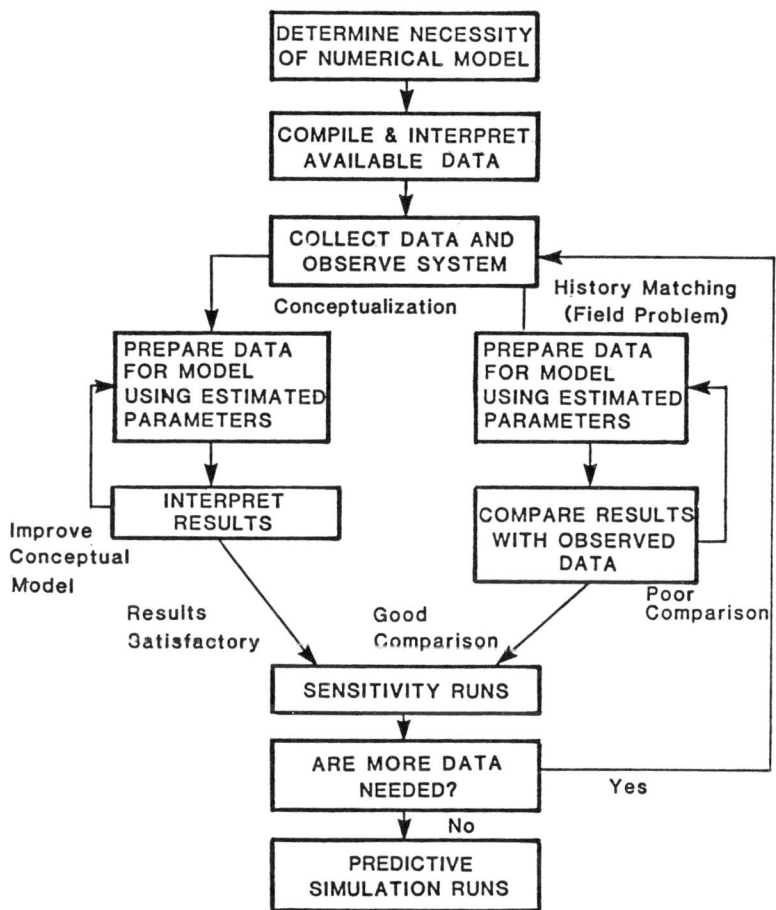

Figure 6.1  Model development and application process (from Mercer and Faust, 1981). Copyrighted by National Water Well Association.

alternative remedial actions on water and constituent movement. Procedures or adjusting model parameters to represent specific remedial action alternatives are discussed in detail in Section 5.

6. The verified and adjusted model can now be run to predict future conditions with and without remedial actions. Various combinations of actions can be explored. Where data uncertainties exist, sensitivity analyses can be used to estimate the range of outcomes.

Numerical models are potential tools for answering several important questions raised by the feasibility study process for evaluating remedial action alternatives, including:

1. existing exposure routes and levels of exposure for specific chemicals

2. future exposures if no action is taken

3. effects of alternative remedial actions on conditions at and near the site

4. future exposures during and after the implementation of alternative remedial actions

Most of these questions will need to be answered during the screening and analysis of alternatives. While screening will require, at most, analytical models, numerical models may find use in the alternatives analysis where complex site conditions exist or complex remedial actions are anticipated.

Site characterization involves data collection and evaluation efforts, (including the potential use of numerical models) required to specify chemical sources, chemical migration pathways, chemical fate, potential receptors, and human health and environmental effects. These efforts will be accomplished during the site investigation and analysis steps of the Remedial Investigation/Feasibility Study process.

# 7. Model Requirements for Surface Water Remedial Actions

## 7.1 OVERVIEW

This section provides modeling requirements for surface water remedial actions. Such requirements may apply to either analytical models (Level I) or numerical models (Level II). Model requirements refer to the type of model required (analytic or numerical) and any unique capabilities such as sediment transport; the model dimensionality and grid configuration; and parameter adjustments. For each remedial action, guidance is provided for the model adjustments required to simulate the environmental effects of that action. Most of these model adjustments involve parameters. As such, model parameter estimation guidance is also provided here to assist the user in deriving appropriate values for critical parameters. The model parameters that must be adjusted to simulate the effects of implementing different actions can vary. As Volume 1 notes, modeling requirements for all potential remedial actions must be considered early enough in the Feasibility Study/Remedial Investigation process to have an impact on the specific model(s) selected for use in remedial action evaluation. The remedial actions described in Section 3 were condensed into eight groups, according to their design objectives and conjunctive use with other actions in the same group. The actions are listed in Table 7.1.

Each remedial action scenario produces unique effects in the waterbody. Modeling requirements will be dictated by the spill/discharge mode, the degree of initial mixing or dilution and, to a lesser extent, the migration of contaminants through the waterbody to an exposure site. The processes governing contaminant transport and fate are different between the spill site and the exposure site. Spill site processes of importance include rate, duration, and type (i.e., point, nonpoint, pulse, continuous) of contaminant discharge, momentum and buoyancy of the contaminant flow, in-stream velocity distribution, and turbulent mixing. These processes are commonly incorporated in what is termed near-field models. Exposure or far-field models incorporate advection as the

TABLE 7.1  GROUPING OF REMEDIAL ACTIONS ACCORDING
           TO MODEL REQUIREMENTS

Dilution

Containment Measures

    Booms and partial barriers

    Cofferdams and full barriers

    Silt curtains

    Capping

Removal Measures

    Hydraulic and mechanical dredging

    Excavation

Treatment

primary process of contaminant transport, with degradation and sediment transport also being important. Therefore, model selection and application will be different for analysis of spill site processes and remedial actions than for evaluation of chemical transport and fate away from the source.

Unlike soil and ground-water remedial actions, surface water remedial actions have limited influence on the contaminant migration path, except in cases where flow is disrupted, as with barrier implementation. Consequently, remedial action modeling can often be confined to the immediate vicinity of the source. For some of the removal-type remedial actions a change in the source term for a far-field or exposure model is sufficient for representation. In these cases, selection of an appropriate model will be based on the complexity of the receiving waterbody; the remedial action should not affect model selection criteria.

The key questions for remedial action simulation are, then: must the source (remedial action site) be modeled, and once it is adequately represented, should remedial action impacts be input to a separate far-field model that will predict pre- and post-restorative concentration levels at an exposure site? As mentioned above, the source term can be empirically derived for some removal-type remedial actions, and can be input into an exposure model. If the remedial action's effects in the near-field cannot be simplified, the source and spill area must be modeled. The focus of this section, then, will be source modeling or representation, and will include near-field modeling requirements for the remedial actions that affect initial mixing and dilution processes. Far-field or migration modeling will not be discussed except where the remedial action affects advection. Modeling needs dictated by the waterbody characteristics will be addressed only as they affect remedial action modeling.

The following subsections detail modeling requirements for specific measures and provide parameter estimation guidance for those parameters that may be adjusted in order to represent the environmental effects of a remedial action. Prior to presenting modeling requirements for each group of measures, several key points need to be addressed.

1. Only those modeling requirements associated with a given group of remedial measures are discussed. Requirements associated with the use of numerical models for site characterization and assessment are not presented. Thus, the guidance presented herein is in addition to that needed to develop a model of the site.

2. Certain model parameter adjustments are highly site-

specific. Thus, it is difficult to provide guidance on parameter estimation.

3. Data on certain model parameters are, on the whole, quite sparse due to a lack of field data on the performance of most remedial measures. In many cases, only laboratory or pilot scale data or parameter values from previous modeling studies are available.

## 7.2 MODELING REQUIREMENTS

The modeling requirement for each group of measures are presented in terms of the following:

1. Model Type - Model type refers to the level of sophistication required in the selected model. Choices include numerical (or Level II, as referred to in Volume 1), analytical (or Level I), or empirical computation.

2. Dimensionality and Grid Configuration - Dimensionality refers to the directions (i.e., x, y, and z) of water and chemical movement that can be simulated, while grid configuration refers to the spatial discretization used to represent a site and the remedial action.

3. Parameter Adjustments - Parameter adjustments refer to the model inputs that must be modified to represent a remedial measure.

Table 7.2 summarizes the modeling requirements for each measure. The following discussion provides more detailed guidance.

### 7.2.1 Dilution

Dilution is the most simple remedial action to simulate, as well as being relatively simple to implement in the field. This type of measure was used to reduce aqueous concentrations of formalin in the Russian River (Ca.) following a spill in 1982.

#### Model Type

A source model is not required because changes in source concentrations and flow rates can be estimated directly. An exposure or far-field model can be used by applying new source

TABLE 7.2 MODELING REQUIREMENTS FOR REMEDIAL ACTIONS

| Remedial Actions | Model Type | Dimensionality/ Grid Configuration | Parameter Adjustment |
|---|---|---|---|
| Dilution | WB | WB | S/S |
| Containment Measures: | | | |
| Booms and partial barriers | 2P | 2D(x-z) | BC |
| Cofferdams and full barriers | WB | WB | S/S |
| Silt curtains | SD | 2D(x-z) | BC |
| Capping | WB | 2D(x) | S/S, SB |
| Removal Measures: | | | |
| Hydraulic and mechanical dredging | SD | 2D(x-z) | S/S, SB |
| Excavation | WB | WB | S/S |
| Treatment | WB | WB | S/S |

LEGEND:
- BC = Boundary conditions
- SB = Sediment bed parameters
- SD = Sediment transport
- S/S = Source on sink terms
- WB = Requirements are a function of the waterbody characteristics
- 1D(x) = One-dimensional, longitudinal direction
- 2D(x-z) = Two-dimensional, longitudinal and vertical directions
- 2P = Two-phase flow

NOTE: Grid configurations are generic in nature and are described in text.

concentration and flow rates.

### Dimensionality and Grid Configuration

The required model dimensionality and grid configuration will be a function of the velocity distributions and geometry of the waterbody. However, the exposure model must be able to represent a new input distribution for the source.

### Parameter Adjustment

The parameters that can be adjusted to represent dilution are the input concentration (or mass) or the source flow rate. The concentration can be reduced and be input for a longer period of time, thus assuring that no change in input mass is realized. If the contaminant is released into the receiving water via a waste stream, as with a point source, the source flow rate can be increased in order to dilute the incoming plume.

## 7.2.2 Containment: Booms and Partial Barriers

This group of measures is directed at controlling the spreading of light, immiscible contaminants on top of the water column. The use of skimmers is also included in this group, as skimming and booms are usually used in conjunction. A number of references provide information on deployment and configurations, such as Department of Transportation (1978); Petty et al., (1982); Fussell et al., (1981); and Huibregste et al., (1977).

### Model Type

The primary model requirement is the ability to simulate 2 phase flow (e.g., water and a floating, immiscible liquid). Only a very few numerical models have this capability. The selected model should incorporate turbulent mixing and shear between the two liquid layers, in order to adequately represent dissolution and mixing of soluble components of the pool into the water column. However, the control of pool spreading along with a specified dissolution rate may also be modeled on a gross level using empirical entrainment and containment calculations. One such model for pool spreading is in HACS (Hazard Assessment Computer System), a set of analytical models intended for use in rapid response situations of chemical spills. Raj and O'Farrel (1981) provide details on this model as well as the other water quality models used in HACS.

### Dimensionality and Grid Configuration

The recommended model dimensionality for most waterbodies is two-dimensional in the longitudinal and vertical planes. The two-dimensional (x-z) configuration allows resolution of the water column, which is important in controlling the spread of a selected layer or depth of water. A one-dimensional (x) model is sufficient for narrow rivers, where the pool spreads across the water surface laterally to both side boundaries. However, this dimensionality and configuration (well-mixed reaches) do not allow for the selective containment of the surface slick. Herbes et al., (1982) present such a model for transport of coal liquefaction product spills in rivers. A two-dimensional (x-z) configuration will allow variable grid spacing along the vertical plane. Most grid points should be specified around the boom or barrier in order to allow better resolution of the containment, and to reduce any numerical instability caused by a no flow boundary and high shear stress between the two liquid layers.

### Parameter Adjustment

The physical barrier or booms are represented by a no flow boundary within the grid. The removal of material via skimming must be approximated by reducing the solubility of the contaminant in water. This will serve to reduce or eliminate aqueous concentrations downstream, which is the purpose of containment and skimming.

### 7.2.3 Containment: Cofferdams and Full Barriers

This group of measures has only minor modeling requirements because the source is assumed to be completely isolated from the waterbody, including the period of implementation of the remedial action.

### Model Type

The model required, if any, will be dependent on the waterbody characteristics. If an exposure model is required because of waterbody complexity, the actions are represented by a change in the source term and possibly the boundary shape of the new shoreline.

### Dimensionality and Grid Configuration

Dimensionality and grid configuration will be a function of the waterbody characteristics.

Parameter Adjustment

The only parameter that needs to be adjusted is the source term, which is reduced according to the assumed efficiency of the barrier in isolating the waste, and the degree of dewatering of the spill site area.

### 7.2.4 Containment: Silt Curtains

Silt curtains are designed to reduce suspended sediments in the near-field water column resulting from dredging, excavation, and non-point sources. They usually force the turbid water to a lower elevation with minimal deposition and the suspended sediments resurface further down-stream. Since silt curtains are often used in conjunction with mechanical dredging, the user should refer to those modeling requirements as well when evaluating these actions.

Model Type

A numerical model with sediment and contaminant transport capabilities is required for simulation. The model should incorporate turbulent mixing and shear, and sediment scour and deposition processes.

Dimensionality and Grid Configuration

A minimum of 2 dimensions (longitudinal-vertical) is required to simulate the vertical distribution of sediments and allow better resolution of the trapping effect of the silt curtain. If the area to be contained is irregularly shaped, a 3D simulation may be required. It is important to compute the velocity distribution in the water column in order to simulate sediment scour and deposition accurately. The grid configuration along the vertical plane should reflect more points around the curtain and between the curtain and the bottom. This is done to represent the turbulent mixing and shear, and associated sediment deposition and transport in these locations in the water column. If a three-dimensional model is used, the boundaries should be set away from the curtains in order to mitigate any influences the artificial boundaries may have on the flow field.

Parameter Adjustment

No parameters need to be adjusted, as the curtain is simulated by no-flow grid points in the model. In this sense, water is not allowed to flow through the curtain as it would in the waterbody. Because the curtain impedes the flow of water and causes more turbulence and increased velocities around curtain edges, mixing-related parameters (dispersion coefficients) may also need adjustment.

### 7.2.5 Containment: Capping

The purpose of capping is to prevent desorption of contaminants and erosion of contaminanted sediments from the sediment bed. This type of action is limited in use because of the difficulty in locating and covering the total contaminated sediment beds.

Model Type

The model required will be dependent upon the waterbody geometry and flow field complexity. The action is represented by reducing the source term in an exposure model.

Dimensionality and Grid Configuration

A minimum of one-dimension (longitudinal) with a boundary layer profile calculation for sediment entrainment is required. Two-dimensional (x-z) models with sediment transport may provide better resolution of desorbed contaminant concentrations and sediment entrainment in the water column immediately above the cap.

Parameter Adjustment

The simplest method of simulating the cap is to reduce the contaminant mass per unit area of bed or the concentration in the sediment bed. The degree of reduction will depend on the percentage of contaminated bed that is assumed to be isolated in each segment or reach. This method was employed by Onishi (1979) when he simulated the effects of dredging (or removing) the Kepone-contaminated bed along a portion of the James River (see Section 7.2.6). The caps can also be simulated with more detail if the user wishes to examine potential erosion of the cap, exposure of the contaminated bed, or diffusion of the contaminant through the cap into the water column. Parameter adjustment could include: assign the contaminant concentration to deep burial in the lower portion of the bed; modify the sediment bed characteristics, such as bed shear strength, particle size, diameter, and density, to reflect the cap material (probably clay); and decrease the resuspension velocity and/or increase the settling velocity of the sediment particles. If a two-dimensional simulation is used, the bottom profile can also be adjusted to represent potentially increased velocities around the raised, capped areas. In this case, the depth of the cap should also be specified.

### 7.2.6 Removal: Hydraulic and Mechanical Dredging

Hydraulic and mechanical dredging constitute the most commonly employed remedial actions for restoration of contaminated surface waters. For some dredging scenarios, two sets of modeling requirements must be applied: one for the dredging period in order to examine potential adverse effects; and one for post-restoration, in order to examine concentration levels from residual contamination. Both sets of requirements are described seperately below. The first set of modeling requirements (i.e., during dredging) can be omitted for those cases where the dredging effects are considered to be completely isolated in the spill area. Examples of this would include scenarios when silt curtains which are 100% effective are used in conjunction, or when the spill site is isolated using a full barrier.

#### 7.2.6.1 During Dredging Operations

##### Model Type

To model the effects of the dredging operation, a numerical model with sediment transport capabilities and a vertical line source is required. Johnson (1981) evaluates a number of such models designed to simulate dredging and barge dumping activities. The selected model should incorporate sediment scour and deposition also. Schnoor et al., (1982) utilized such a model (Wechsler and Cogley, 1977) to simulate the suspended sediment concentrations resulting from open water disposal of dredged material on the Mississippi River.

##### Dimensionality and Grid Configuration

A minimum dimensionality of 2D (x-z) is recommended. However, most of the dredging models reviewed by Johnson (1981) are three-dimensional. The vertical dimension allows better resolution for the resuspension and deposition areas.

##### Parameter Adjustment

The parameters adjusted should reflect the effects of the increased turbulence induced by hydraulic dredging and increased supended sediment concentrations from both types of dredging. The flow source and sink terms associated with hydraulic dredging shoud be negligible compared to the in-stream flow, especially on a large river. Turbulent diffusion or mixing coefficients in the lateral and vertical directions should be increased. Gradation of the source sediment should be specified because it affects the transport of the material. A vertical sediment concentration distribution must be set for the line source. For the side bank disposal of dredged material, Schnoor et al., (1982) utilized the dredging rate of the barge and the channel depth, width, and velocity to determine input sediment concentration. More guidance on estimating suspended sediment loading is provided in Section 7.3.

#### 7.2.6.2 Post-Restoration

Model Type

A numerical far-field or exposure model with sediment transport capabilities can be used to evaluate post-restoration conditions. The model can utilize an empirical source term or predicted suspended and deposited sediment concentrations for initial conditions. The ability to simulate scour and deposition of dredged material is required.

Dimensionality and Grid Configuration

A minimum two-dimensional (x-z) sediment transport simulation is recommended, if vertical distribution of suspended sediments is important. The grid spacing should be closer along the bottom to represent large suspended sediment concentration fluctuations.

Parameter Adjustment

The chemical concentration in the bed must be adjusted in order to reflect the presence of deposited contaminated sediments. Bottom topography, in the form of sediment bed thickness, and water column depth may have to be adjusted for those areas of dredge related heavy scour or deposition. Suspended sediment concentrations predicted from the dredging model can be used for initial levels. If no dredging modeling is performed, the removal of contaminated bed by dredging, capping, or any isolation and complete removal methods may be simulated in a fashion similar to Onishi (1979). He used the two-dimensional (x-z) sediment transport model FETRA (Onishi et al., 1979) to locate areas along the river where contaminated sediment was being deposited. Ten locations for clean-up were simulated by removing the contaminated bed along selected reaches. Figure 7.1 illustrates the changes in Kepone concentrations from different clean-up areas. As evidenced in Figure 7.1, a 34.5 km length clean-up region reduced concentrations the most (55%), although a 22 km clean-up region was quite close in level of reduction (48%). This study did not evaluate the transport of contaminated sediments over a period of time due to dredging itself. However, given the size of the tidal river and location of deposition areas, such effects would have been local, and were

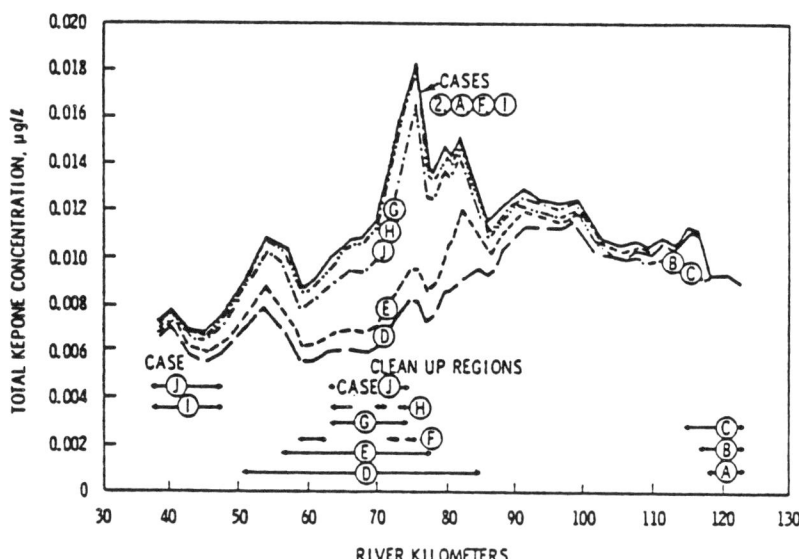

Figure 7.1 Reductions in total Kepone concentrations from different dredging scenarios (Onishi, 1979).

not important in terms of viewing Kepone loading into Chesapeake Bay.

### 7.2.7 Removal: Excavation

This action is usually used in conjunction with full barriers, as it can only be used on dry dewatered solids. Because the spill area is assumed to be completely isolated during implementation, no source area modeling is required.

Model Type

The model type for post-restoration will be dependent upon the waterbody.

Dimensionality and Grid Configuration

There will be no change in dimensionality or grid configuration; they are used dependent on waterbody characteristics.

Parameter Adjustment

The only parameter adjustment will be the change in source term. Before restoration, it is assumed that the source term will be from a contaminated sediment bed, in either concentration or mass per unit area form. Upon restoration, any contaminant leaving the spill area will be from dewatering operations. This source term may be represented for a finite period of time with a empirically derived aqueous concentration and flow rate.

### 7.2.8 Treatment

Modeling requirements for treatment actions will often be represented as reductions in the source term for the re-introduced waste water. Remedial action modeling is not required if the treatment action does not affect in-stream processes.

Model Type

Source area modeling is not required for either in-situ or on-site actions.

Dimensionality and Grid Configuration

Dimensionality and grid configuration requirements will be

dictated by the waterbody characteristics.

Parameter Adjustment

The only parameter adjustment is the reduction of a source term for the in-situ action, and the addition of sink and source terms for the on-site action. The new introduced source will be empirically reduced from in-stream concentrations.

## 7.3 PARAMETER ESTIMATION GUIDANCE

The parameters requiring adjustment to simulate the remedial measures described previously can be characterized by four groups: 1) source term parameters for contaminants and flow; 2) sediment-related parameters for bed and suspended sediments; 3) boundary conditions, including channel geometry; and 4) dispersion parameters. This section provides sources of data and techniques for estimation of model parameters.

The guidance presented herein is only meant to be used in support of, rather than in place of on-site field measurements, sampling and laboratory studies. To the extent possible, values for model parameters should be determined as part of the remedial investigation process. This process is meant to fill the data required to evaluate conceptual, remedial action alternatives (JRB Associates, 1983). Hopefully, this section can be used to more fully understand those data required for remedial action modeling and in the worst case aid in their estimation in absence of site-specific data.

Where available, data sources and estimation techniques pertinent to remedial action specific parameters are provided. Both are extremely limited, however. For this reason, more general data sources and estimation techniques are discussed to provide a basis for at least the initial determination of appropriate parameter values. Zison et al. (1978) is a good general source for transport and fate parameters and formulations.

### 7.3.1 Source Term Parameters

As evidenced throughout Section 7.2 on modeling requirements, the most common parameter adjustment for surface water remedial actions is the modification of source terms. The primary source discussed here is the introduction of a dissolved contaminant and flow into receiving water. Sources of contaminated sediments from a discharge or sediment bed are discussed in Section 7.3.2.

The simplest description for a concentration of a discharged contaminant into a waterbody involves mass, flow, and time in the following equation:

$$C_i = \frac{M}{V} = \frac{M}{Q} t \qquad (7.1)$$

where  $C_i$ = initial concentration
 $M$ = mass of contaminant
 $V$ = volume of released water
 $Q$ = flow of discharged water
 $t$ = time

This initial concentration is subsequently mixed or diluted in a mixing zone proximal to the discharge point. Complete and instantaneous mixing across the channel width is often assumed, except for wide rivers, estuaries, and lakes. Mills et al. (1982) provide a number of expressions to describe mixing zone geometry and concentrations. Reductions in sources terms from dilution will often be accomplished by increasing the source flow rate ($Q$). Reductions in the source term according to contaminant mass removal or isolation actions (i.e., excavation, installation of cofferdams) is reflected by lower mass or concentration inputs, whichever the exposure model requires. Actual reduction is determined empirically; the user may want to evaluate different source term reductions to reflect various clean-up efficiencies (i.e., 90%, 70%, 50%, 10%). Source terms may also be adjusted according to the period of release. The timing of the release during a spill can range from instantaneous to continuous. If we assume that clean-up is never 100% efficient, residual contamination may enter the water body for a finite period of time during and after restoration. The new source term is represented by a series of instantaneous releases at finite time intervals. Estimates of the removal mass and time over which discharge continues can be used to determine a new source concentration for an exposure model and the number of time steps it will be active.

### 7.3.2 Sediment Parameters

Many of the hazardous wastes that are discharged to receiving waters are transported as particulates or via contaminated sediments. Their ultimate sink can be the sediment bed, where the contaminant can desorb back into the water column or be resuspended with sediment particles as the result of scour.

Most of the remedial actions described previously are designed to isolate or remove contaminated sediments and solids. Two areas of parameter adjustments are important: 1) those affecting sediment transport, including scour and deposition; and 2) those affecting the contaminated sediment as a source, including contaminant mass, area of bed, and desorption and diffusion coefficients. Each group is described below.

### 7.3.2.1 Sediment Transport Parameters

The important parameters of sediment transport are the sediment diameter, specific gravity, settling (or fall) velocity, the critical shear stress associated with deposition of sediment, and the suspended sediment loading term. Sediment diameter and specific gravity along with shear velocity are the primary parameters used to describe sediment transport. These parameters are not often adjusted, except to represent new sediment material. Table 7.3 is a sediment grade scale with sizes of different materials. Table 7.4 lists specific weights (gravities) for sediments in representative waterbodies. If contaminant transport is associated with certain particle sizes, the model should have the ability to transport by grain size, not by total sediment load only.

The settling velocity strongly affects the rate at which sediments will be deposited on the bed. Net settling velocity may be calculated in the model using bed shear stresses or be a parameter input. In the latter case, it can be adjusted to represent different sediment types, such as clay, from a capping remedial action. Barnard (1978) developed an empirical relationship between settling velocity and particle diameter, presented in Figure 7.2. This relationship is based on the assumption that the particle has a certain specific gravity and shape factor at a specific water temperature. In estuary and sometimes in reservoir analysis, parameters that affect cohesive (silts and clays) sediment transport are chemical conditions (e.g., salinity, pH, valence of cations), concentration of suspended material, and mineral properties of the particles such as sodium adsorption ratio (SAR) and cation exchange capacity. This is paticularly important because the sorption/desorption activity of contaminants is usually associated with clays and silts.

The critical shear strength of the sediment bed will determine the degree of erosion that can occur. It is a function of the sediment type, in particular, the median sediment diameter in the bed. Figure 7.3 is a diagram that can be used for such a purpose.

Suspended sediment source terms, used to represent the effects

TABLE 7.3  SEDIMENTATION GRADE SCALE (from Vanoni, 1975) Copyrighted by the American Society of Civil Engineers

| Class Name | Size Range | | | Approximate Sieve Mesh Openings Per Inch | |
|---|---|---|---|---|---|
| | Millimeters | Microns | Inches | Tyler | United States Standard |
| Very large boulders | 4096-2048 | | 160-80 | | |
| Large boulders | 2048-1024 | | 80-40 | | |
| Medium boulders | 1024-512 | | 40-20 | | |
| Small boulders | 512-256 | | 20-10 | | |
| Large cobbles | 246-128 | | 10-5 | | |
| Small cobbles | 128-64 | | 5-2.5 | | |
| Very coarse gravel | 64-32 | | 2.5 -1.3 | | |
| Coarse gravel | 32-16 | | 1.3 -0.6 | | |
| Medium gravel | 16-8 | | 0.6 -0.3 | 2-1/2 | |
| Fine gravel | 8-4 | | 0.3 -0.16 | 5 | 5 |
| Very fine gravel | 4-2 | | 0.16-0.08 | 9 | 10 |
| Very coarse sand | 2-1 | 2.000-1.000 | 2000-1000 | 16 | 18 |
| Coarse sand | 1-1/2 | 1.000-0.500 | 1000-500 | 32 | 35 |
| Medium sand | 1/2-1/4 | 0.500-0.250 | 500-250 | 60 | 60 |
| Find sand | 1/4-1/8 | 0.250-0.125 | 250-125 | 115 | 120 |
| Very fine sand | 1/8-1/16 | 0.125-0.062 | 125-62 | 250 | 230 |
| Coarse silt | 1/16-1/32 | 0.062-0.031 | 62-31 | | |
| Medium silt | 1/32-1/64 | 0.031-0.016 | 31-16 | | |
| Fine silt | 1/64-1/128 | 0.016-0.008 | 16-8 | | |
| Very fine silt | 1/128-1/256 | 0.008-0.004 | 8-4 | | |
| Coarse clay | 1/256-1/512 | 0.004-0.0020 | 4-2 | | |
| Medium clay | 1/512-1/024 | 0.0020-0.0010 | 2-1 | | |
| Fine clay | 1/1024-1/2048 | 0.0010-0.0005 | 1-0.5 | | |
| Very fine clay | 1/2048-1/4096 | 0.0005-0.00024 | 0.5-0.24 | | |

TABLE 7.4 SPECIFIC WEIGHTS OF SEDIMENTS SHOWING EXTREME VARIATION (Vanoni, 1975) Copyrighted by the American Society of Civil Engineers

| Location (1) | Predominant class of sediment (2) | Specific weight, in pounds per cubic foot (3) |
|---|---|---|
| Lake Niedersonthofen, Bavaria, upper layer | marl[a] | 21.6 |
| Lake Niedersonthofen, 20 m depth | marl[a] | 89.6 |
| Lake Arthur, South Africa | clay | 38 |
| Iowa River at Iowa City, Iowa | silt | 52 |
| Missouri River near Kansas City, Mo. | silt | 74 |
| Lake Claremore, Oklahoma | silt | 54 |
| Lake McBride, Iowa | silt | 60 |
| Powder River, Wyoming | silt | 81 |
| Castlewood Reservoir, Colorado | sand | 92 |
| Cedar River near Cedar Valley, Iowa | sand | 109 |
| Lake Arthur, South Africa | sand | 100 |

[a] As used herein, marl is a mixture of calcium carbonate or dolomite and clay.

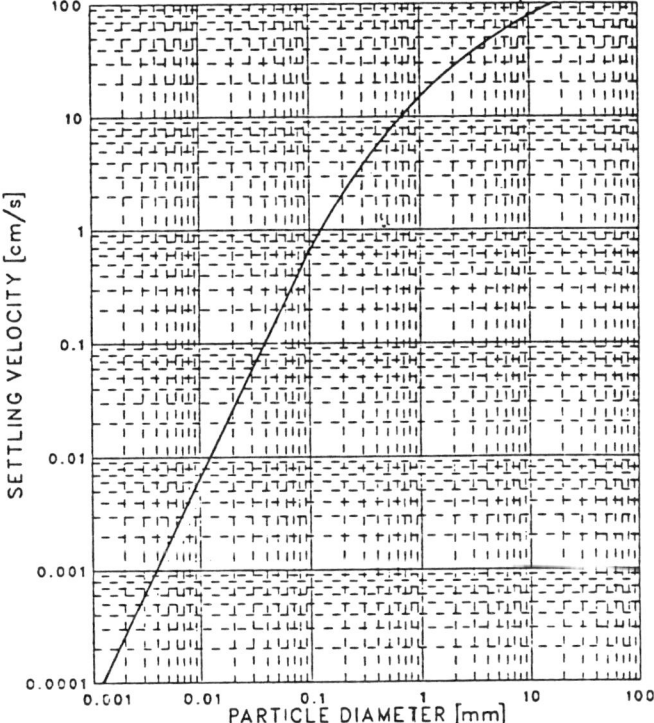

Figure 7.2  Particle size vs. settling velocity for suspended sediment (Barnard, 1978).

424 Modeling Remedial Actions at Uncontrolled Hazardous Waste Sites

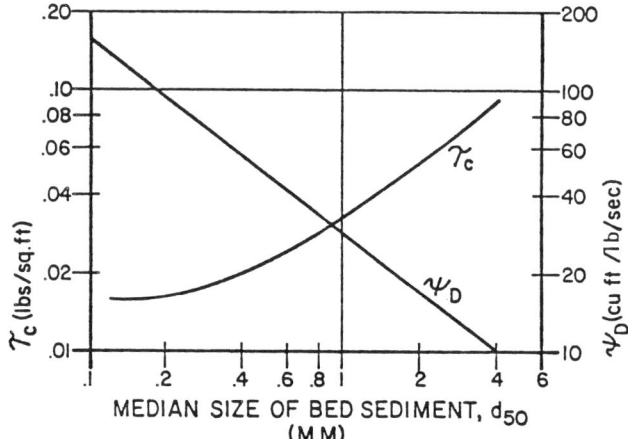

Figure 7.3  $\psi$ and $\tau_c$ for DuBoys relationship as functions of median size of bed sediment, where $\tau_c$ = critical shear stress and $\psi_D$ = coefficient depending on grain size (Task Committee on Preparation of Sedimentation Manual, 1971). Copyrighted by the American Society of Civil Engineers.

of a dredging operation, are often developed empirically based on the mode of dredging, type of sediment, and location of disposal. Schnoor et al. (1982) developed a suspended sediment source term for a sediment transport/dredging model (Wechsler and Cogley 1977) on the Mississippi River. Table 7.5 lists the factors they considered in order to obtain a suspended sediment concentration of 120 mg/l.

#### 7.3.2.2 Contaminated Sediment Bed Parameters--

These parameters are useful in deriving a source term to represent desorption and diffusion from a sediment bed, as well as to represent the residual sorbed contaminants that are subject to scour and transport with sediment. Desorption and diffusion parameters can be adjusted in a sediment transport model to represent a reduced contaminated bed size or concentration.

During the period of desorption (that is, after the contaminant has been spilled and some has advected through the waterbody with the rest settling in the bed in a sorbed phase) the average aqueous concentration can be described by the following equation:

$$C = \frac{X_o \sigma S}{K_p M_s} \qquad (7.2)$$

where
$X_o$ = concentration of pollutant in bed at time t=0
$\sigma$ = equivalent depth of water in sediment $M_s$, cm
$S$ = specific gravity
$K_p$ = partition coefficient
$M_s$ = mass of contaminated sediment per unit area of river bed, g/cm$^2$

If data are not available, $M_s$ and $\sigma$ can be estimated based on the depth of contaminated sediments and percent solids by weight values in Table 7.6.

In a sediment and contaminant transport model, the aqueous and sorbed contaminant concentration will be computed using the parameters above and flow parameters. The concentration ($X_o$) at t=0 can be reduced empirically to represent some removal action as can the mass of contaminated sediment per unit area of bed ($M_s$). The partition coefficient ($K_p$) does not need to be changed unless some treatment action is applied to the bed itself.

The contaminant can also diffuse back into the water column if there is a sufficient gradient. Ditoro et al. (1981) define an overall diffusive exchange coeficient ($K_L$) with the

TABLE 7.5 DETERMINATION OF A CONTINUOUS SUSPENDED SEDIMENT SOURCE TERM BY SCHNOOR ET AL.(1982)

| Parameter | Value | Source |
|---|---|---|
| Mean velocity | 0.5 m/s | Field measurement |
| Depth | 4.0 m | Field measurement |
| Source width | 20m | Observation |
| Dredge capacity | 1072 yd/hr | Communication with Army Corps of Engineers |
| Percent of Spoil that is solids | 85% | Best Engineering Judgement |
| Percent of solids that are actually entrained | >1% | Shallow depth near shore, most of solids were sand |

TABLE 7.6  MASS OF CONTAMINATED SEDIMENTS AND EQUIVALENT WATER DEPTH AS A FUNCTION OF DEPTH CONTAMINATION (Mills et al., 1982)

| Depth (mm) | Percent Solids by Weight | $M_s$ (g/cm²) | δ(mm) |
|---|---|---|---|
| 1 | 20 | 0.02 | 0.9 |
|   | 50 | 0.06 | 0.6 |
|   | 80 | 0.11 | 0.3 |
| 5 | 20 | 0.11 | 4.5 |
|   | 50 | 0.30 | 3.0 |
|   | 80 | 0.55 | 1.4 |
| 10 | 20 | 0.23 | 9.1 |
|    | 50 | 0.60 | 6.0 |
|    | 80 | 1.1 | 2.7 |
| 20 | 20 | 0.45 | 18. |
|    | 50 | 1.2 | 12. |
|    | 80 | 2.2 | 5.5 |
| 50 | 20 | 1.1 | 45. |
|    | 50 | 3.0 | 30. |
|    | 80 | 5.5 | 14. |
| 100 | 20 | 2.3 | 91. |
|     | 50 | 6.0 | 60. |
|     | 80 | 11.0 | 27. |

following equation:

$$K_L = \frac{D_2}{\Phi \sigma_z} \quad (7.3)$$

where
 $K_L$ = overall diffusive exchange coefficient, cm/day
 $D_2$ = interstitial water diffusion coefficient
 $\Phi$ = porosity of sediments in bed
 $\sigma_z$ = length or depth or the gradient in the bed, cm$^2$

$D_2$ can be estimated using the following equation from Manheim (1970):

$$D_z = D_{20} \Phi^2 \quad (7.4)$$

where  $D_{20}$ = molecular diffusion coefficient of the chemical

Lyman et al. (1982) provide a method to determine $D_{20}$ based on molecular weight. The diffusion of the chemical into the water column can then be adjusted in the model in order to represent the addition of new material such as a clay cap or another sediment type deposited on top of the bed, affecting the sediment porosity and gradient length, or depth of contamination.

### 7.3.3  Boundary Condition Parameters

Parameter adjustment for boundary conditions is very site specific, and will vary from case to case according to the channel geometry, the model dimensionality, and remedial action configuration.

Channel geometry changes as a result of the addition of a barrier can be simulated in two different ways, according to the model dimensionality. For example, the parameter adjustment in a one-dimensional (x) model to represent the barrier involves reducing the reach width for the spill site area. In a two-dimensional (x-y) simulation, the same barrier can be represented by applying no-flow conditions for the grid points or nodes along the barrier length.

Some remedial actions, such as silt curtains, require no-flow boundary conditions for certain grid points in the water column. In a two-dimensional (x-z) simulation, the no-flow points should be defined to a specific depth in the water column at some distance x which represents the curtain in the waterbody.

## 7.3.4 Dispersion Parameters

As described in Section 2, dispersion is the aggregate result of molecular diffusion, turbulent diffusion, and shear flow dispersion for each dimension (x, y, and z). Remedial actions such as barriers, cofferdams, silt curtains and dredging can cause an increase in dispersion, particularly turbulent diffusion, and cause the contaminant to spread more rapidly. Dispersion coefficents can be specified in numerical models to represent the change in mixing from remedial action implementation. Parameter adjustments for the longitudinal dispersion coefficient, transverse mixing coefficient, and vertical mixing coefficient are provided below. Fischer et al. (1979) is a good source for dispersion parameters.

### 7.3.4.1 Longitudinal Dispersion Coefficient ($K_x$)--

Several simple methods for evaluating $K_x$ are available in the literature. Compilations of available methods include Fischer et al. (1979) and Benedict (1978). The method Liu (1977) used (based on the work of Fischer, 1967) is presented here because it is relatively easy to calculate. Table 7.7 provides reported values K of representative channels.

The longitudinal dispersion coefficient $K_x$ is determined as follows:

$$K_x = \beta \frac{u_x^2 B^3}{u_* A} = \beta \frac{Q_B^2}{u_* D^3} \quad (7.5)$$

where (Liu, 1978):

$$\beta = 0.5 \left(\frac{u_*}{u_x}\right)^2$$

where  D = mean depth
B = mean width
$u_*$ = bed shear velocity
$u_x$ = mean stream velocity
A = cross sectional area
$Q_B$ = river discharge

This parameter may be adjusted locally when barriers cause a change in the depth, width, and cross-sectional area of the river.

TABLE 7.7  REPORTED VALUES FOR THE LONGITUDINAL MIXING
COEFFICIENTS FOR DIFFERENT CHANNELS
(Benedict, 1978)

| Channel | Depth (cm) | $K_x$ (M$^2$/sec) |
|---|---|---|
| Chicago Ship Canal | 807 | 3.0 |
| Sacramento River | 400 | 15. |
| River Derwent | 25 | 4.6 |
| South Platte River | 46 | 16.2 |
| Yuma Mesa Canal | 345 | 0.76 |
| Trapezoidal Lab Channel | 2.1-4.7 | 0.123-0.22 |
| Green-Duwamish River | 110 | 6.5-8.5 |
| Missouri River | 270 | 1500. |
| Clinch River | 58-210 | 14-47 |
| Copper Creek, VA | 48-85 | 9.5-21 |
| Powell River, TN | 85 | 9.5 |
| Sinuous Laboratory Channel | 2.7-7.0 | .51-3.1 |

### 7.3.4.2 Transverse Mixing Coefficient--

The transverse or lateral coefficient $K_y$ has been described by Elder (1959) as:

$$K_y = \sigma Du_*  \qquad (7.6)$$

where   $\sigma$ = coefficient
        D = depth
        $u_*$ = bed shear velocity

Transverse mixing becomes very important to contaminant dispersion when barriers are placed in rivers. The coefficient $\sigma$ can be adjusted to represent changes in channel geometry from such a cause. Yotsukura and Cobb (1972) reported values of $\sigma$ from 0.1 - 0.2 for straight channels, and 0.6 - 10 in the Missouri River. Fisher et al. (1979) recommend a value of 0.6. Table 7.8 lists $K_y$ values for represenative channels. In a two-dimensional (x-y) simulation with barriers that cause sharp bends in channel geometry, higher values of $\sigma$ should be used.

### 7.3.4.3 Vertical Mixing Coefficient--

Fischer et al. (1979) provide the following equation to determine the average vertical mixing coefficient $\overline{\varepsilon}_v$:

$$\overline{\varepsilon}_v - 0.067\ du^* \qquad (7.7)$$

where   $\overline{\varepsilon}_v$ = vertical mixing coefficient
        d = depth of an open channel flow
        $u^*$ = shear velocity at a wall boundary or channel bottom

This coefficient can be adjusted in two dimensional (x-z) or three dimensional models in order to change the value for $K_z$ for remedial actions that cause an increase in vertical mixing, such as hydraulic dredging and silt curtains.

TABLE 7.8 EXPERIMENTAL MEASUREMENTS OF TRANSVERSE MIXING IN OPEN CHANNELS WITH CURVES AND IRREGULAR SIDES (Fischer et al., 1979) Copyrighted by Academic Press

| Channel | Channel geometry | Channel width, W (m) | Mean depth of flow, d (m) | Mean velocity, $\bar{u}$ (m/s) | Shear velocity, $u^*$ (m/s) | Transverse mixing coefficient (m²/sec) |
|---|---|---|---|---|---|---|
| Missouri River near Blair, Nebraska | Meandering river | 200 | 2.7 | 1.75 | 0.074 | 0.12 |
| Laboratory | Smooth sides and bottom; 0.15 m long groins on both sides | 2.2 | 0.097 | 0.11 | - | - |
| | Smooth sides and bottom; 0.5 m long groins on both sides | 2.2 | 0.097 | 0.11 | - | - |
| Laboratory model of the IJssel River | Groins on sides and gentle curvature | 1.22 | 0.9 | 0.13 | 0.0078 | - |
| IJssel River | Groins on sides and gentle curvature | 69.5 | 4.0 | 0.96 | 0.075 | - |
| Mackenzie River from Fort Simpson to Norman Wells | Generally straight alignment or slight curvature; numerous island and sand bars | 1240 | 6.7 | 1.77 | 0.152 | 0.67 |
| Missouri River downstream of Cooper Nuclear station, Nebraska | Reach includes one 90° and one 180° bend | 210-270 | 4 | 5.4 | 0.08 | 1.1 |
| Potomac River; 29 km reach below the Dickerson Power Plant | Gently meandering river with up to 60° bends | 350 | 0.73-1.74 | 0.29-0.58 | 0.033-0.051 | - |

# References

Ambrose, R. 1982. Simulation Models for Assessing Eutrophication Problems in River Basins, U.S. EPA, Environmental Research Lab, Athens, GA. Presented at the International Workshop on the Comparison of Application of Mathematical Models for the Assessment of Changes in Water Quality in River Basins, both surface and ground waters. La Coruna, Spain.

Ambrose, R.B., S.I. Hill and L.A. Mulkey. 1983. Users Manual for the Chemical Transport and Fate Model TOXIWASP, Version 1, EPA 600/3-83-005, U.S. Environmental Protection Agency, Athens, GA.

Ambrose, R., and S. Roesch 1982. Dynamic Estuary Model Performance. ASCE, Vol. 108, No. EE1.

Barnard, W. 1978. Prediction and Control of Dredged Material Dispersion around Dredging and Open-water Pipeline Disposal Operations. U.S. Army Engineer Waterways Experiment Station. Vicksburg, MS.

Basta, D.J. and B.T. Bower, eds. 1982. Analyzing Natural Systems. Resources for the Future, Johns Hopkins University Press.

Benedict, B. 1978. Review of Toxic Spill Modeling, AD/A-073-22, U.S. Department of Commerce.

Burns, L., R. Lassiter and D. Cline. 1982. Documentation for the Exposure Assessment Modeling System (EXAMS), EPA 600/3-82-023, U.S. Environmental Protection Agency, Environmental Research Laboratory, Athens, GA.

Callahan, M.C., M.W. Slimak, N.H. Gabel, J.P. Map, C.F. Fowler, J.R. Freed, P. Jennings, R. L. Durfee, F.C. Whitmore, B. Maestri, W.R. Mabey, B.R. Holt and C. Gould. 1979. Water-Related Environmental Fate of 129 Priority Pollutants, EPA - 440/4-79-029ab, Volumes 1 and 2, U.S. Environmental Protection Agency, Washington, D.C.

Codell, R.B., K.T. Key and G. Whelan. 1982. A Collection of Mathematical Models for Dispersion in Surface Water and Groundwater. NUREG-0868, Division of Engineering, Office of Nuclear Realtor Regulation, U.S. Nuclear Regulatory Commission.

Colonna, G. 1982. Hazard Assessment Computer System (HACS) - Models Review and Validation. From Proceedings of the Control of Hazardous Material Spills Conference, April 19-22, Milwaukee, WI.

Csanady, G.T. 1973. Turbulent Diffusion in the Environment. D. Reidel Publ., Boston, MA.

Department of Transportation. 1978. CHRIS Response Methods Handbook. COMDTINST MI6465.14. U.S. Coast Guard, Washington, D.C.

Department of Transportation, U.S. Coast Guard. 1981. CHRIS Hazardous Assessment Handbook C6-446-3.

Di Toro et al., 1981. Analysis of Fate of Chemicals in Receiving Waters. CMA Project Env-7-W, Chemical Manufacturers Association, Washington, D.C.

Di Toro, D., J. Fitzpatrick and R. Thomann. 1982. Documentation for Water Quality Analysis Simulation Program (WASP) and Model Verification Program (MVP). Environmental Research Laboratory, U.S. Environmental Protection Agency, Duluth, MN. EPA Contract N1-68-01-3872.

Donigian, A.S. 1981. Water Quality Modeling in Relation to Watershed Hydrology, In: Modeling Components of Hydrologic Cycle, ed. V.P. Singh, Water Resources Publications, Litteton, CO.

Edinger, J. and E. Buchak. 1982. Description and Applications of the LARM Code (Laterally-Averaged Reservoir Model). By J. E. Edinger and Associates, Wayne, PA.

Elder, J.W. 1959. The Dispersion of Marked Fluid in Turbulent Shear Flow. Journal of Fluid Mechanics, Vol. 5, No. 4, pp. 544-560.

Environmental Protection Agency. 1982. The PCB Contamination Problem in Waukegan, Illinois. U.S. Environmental Protection Agency, Region V.

Environmental Protection Agency. 1983. Environmental Data Base and Model Directory. Two Volumes, EPA Information Clearinghouse (PM 211A), U.S. Environmental Protection Agency, Washington, D.C.

Fischer, H.B. 1967. The Mechanics of Dispersion in Natural Streams. Journal of the Hydraulics Division, American Society of Civil Engineers, Vol. 93, HY6, pp. 187-216.

Fischer, H.B., E.J. List, R.L.Y. Koh, J. Imberger and N.H. Brooks. 1979. Mixing in Inland and Coastal Waters. Academic Press, New York, NY.

Frick, W. E. 1981. A Theory and Users Guide for the Plume Model MERGE. Tetra Tech, Inc. Corvallis, OR.

Fussell, D.R., H. Godjen, P. Hayward, R.H. Lillie, A. Manlo and C. Panisi. 1981. Revised - Inland Oil Spill Clean-up Manual. Conservation of Clean Air and Water - Europe (Concawe), Report No. 7/81, Den Haag, Netherlands.

George, J.F. 1978. An Analysis of the Functional Capabilities and Performance of Silt Curtains. Technical Report No. D-78-39. Army Engineer Waterways Experiment Station, Vicksburg, MI.

Herbes, S.E., G.T. Yeh and G.R. Southworth. 1982. Transport Model for Coal-Liquefaction Product Spills in Rivers. Presented at the Third Annual Meeting of the Society of Environmental Toxicology and Chemistry, Arlington, VA.

Huibregste, K.R., R.C. Scholz, R.E. Wullschleger, J. H. Moser, E.R. Bollinger and C.A. Hansen. 1977. Manual for the Control of Hazardous Material Spills, Vol. 1: Spill Assessment and Water Treatment Techniques. EPA 600/2-77-227.

Hydrologic Engineering Center. 1977. HEC-6-Scour and Deposition in Rivers and Reservoirs, Users Manual. U.S. Army Corps of Engineers, Davis, CA. 723-62-62470.

JRB Associates. 1982. Handbook - Remedial Actions at Waste Disposal Sites, EPA 625/6-82-006, Environmental Protection Agency, Cincinnati, OH.

Johanson, R., J. Imhoff, H. Davis, J. Kittle and A. Donigian. 1981. User's Manual for the Hydrological Simulation Program - Fortran (HSPF): Release 7.0, EPA 600/6-82-0046, U.S. Environmental Protection Agency, Athens, GA.

Johnson, B. 1981. Discussion of Mathematical Models for Computing the Physical Fate of Solids Released in Open Water During Dredging or Dredged Material Disposal. In: Contaminants and Sediments, R. Baker, Ed. Chap. 15. Ann Arbor Science, Ann Arbor, MI.

Karickhoff, S., D. Brown and T. Scott. 1979. Sorption of Hydrophobic Pollutants on Natural Sediments. Wat. Res. Vol. 13, pp. 241-248.

Karickhoff, S. 1979. Semi-Empirical Estimation of Sorption on Hydrophobic Pollutants on Natural Sediments and Soils. Chemosphere. Vol. 10, No. 8, pp. 833-846.

Krenkel, P.A. and V. Novotny. 1980. Water Quality Management, Academic Press, New York, NY.

Liu, H. 1977. Predicting Dispersion Coefficient of Streams. Journal of the Environmental Division, American Society of Civil Engineers, Vol. 103, No. EE1, pp. 59-69.

Liu, H. 1978. Discussion of, Predicting Dispersion Coefficient of Streams. Journal of the Environmental Division, American Society of Civil Engineers, Vol. 104, No. EE4, pp. 825-828.

Liu, S. and J. Leenderste. 1972. A Three - Dimensional Model for Estuaries and Coastal Seas: Volume VI, Bristol Bay Simulations. R-2405-NOVA. Rand Corp., Santa Monica, CA.

Lyman, W.J., W.F. Reehl and D.H. Rosenblatt. 1982. Handbook of Chemical Property Estimation Methods--Environmental Behavior of Organic Compounds. McGraw-Hill Book Co., NY.

Maki, A., J. Dickson and J. Cairns, Eds. 1979. Biotransformation and Fate of Chemicals in the Aquatic Environment. Am. Soc. of Microbiology Press, Washington, D.C.

Manheim, F.T. 1970. The Diffusion of Ions in Unconsolidated Sediments. Earth Planet Sci. Letter, Vol. 9, pp. 307-309.

Mercer, J.W. and C.R. Faust. 1981. Ground-Water Modeling. National Water Well Association, Worthington, OH.

Mills, W., J. Dean, D. Porcella, S. Gherini, R. Hudson, W. Frick, G. Rupp and G. Bowie. 1982. Water Quality Assessment: A Screening Procedure for Toxic and Conventional Pollutants, EPA 600/6-82-004abc, Volumes 1,2.

Morrow, T., F. Dodge, E. Bowles and D.W. Astleford. 1981. Mixing and Dilution of Water Soluble Hazardous Chemical Released in Navigable Rivers. Am. Soc. Mech. Eng. 81-FE-17.

Mulkey, L., R. Ambrose and T. Barnwell. 1982. Aquatic Fate and Transport Modeling Techniques for Predicting Environmental Exposure to Organic and Other Toxicants -- A Comparative Study. U.S. Environmental Protection Agency, Environmental Res. Lab. Athens, GA. Presented at the International Workshop on the Comparison of Application of Mathematical Models for the Assessment of Changes in Water Quality in River Basins, both Surface and Ground Waters. La Coruna, Spain.

Neely, W., G. Blau and T. Alfrey. 1976. Mathematical Models Predict Concentration - Time Profiles Resulting From Chemical Spill in a River. Env. Sci. Tech., Vol. 10, No. 1.

Neely, W., D. Branson and G. Blau. 1974. Partition Coefficient to Measure Bioconcentration Potential of Organic Chemicals in Fish. Env. Sci. Tech.

O'Conner, D.J. 1967. The Temporal and Spatial Distribution of Dissolved Oxygen in Streams, Water Resources Research, Vol. 3, No. 1.

O'Connor, D. and K. Farley. 1981. Preliminary Analysis of Kepone Distribution in the James River. Environmental Engr. and Sci. Program, Manhattan College, Bronx, N.Y.

Onishi, Y. 1979. Mathematical Transport Modeling for Determination of Effectiveness of Kepone Clean-up Activities in the James River Estuary. Presented at the ASCE Hydraulics Division Conference.

Onishi, Y. 1981. Sediment - Contaminant Transport Model. ASCE Vol 107, No. HY9, J. Hydraulics Division.

Onishi, Y., E.M. Arnold and D.W. Mayer. 1979. Modified Finite Element Transport Model, FETRA, for Sediment and Radionuclide Migration in Open Coastal Waters, NUREG/CR-1026, PNL-3114, Pacific Northwest Laboratory, Richland, WA.

Onishi, Y. and S. Wise. 1979. User's Manual for the In-stream Sediment Contaminant Transport Model, SERATRA. U.S. Environmental Protection Agency, Athens, GA.

Onishi, Y., R.J. Serve, E.M. Arnold, C.E. Cowan and F.L. Thompson. 1981. Critical Review: Radionuclide Transport, Sediment Transport, and Water Quality Modeling; and Radionuclide Adsorption/Desorption Mechanisms. NUREG/CR-1322, Nuclear Regulatory Commission, Washington, D.C.

Onishi, Y., G. Whelan and R.L. Skaggs. 1982. Development of a Multimedia Radionuclide Exposure Assessment Methodology for Low-Level Waste Management. PNL-3370, Pacific Northwest Laboratory, Richland, WA.

Onishi, Y. and D.S. Trent. 1982. Mathematical Simulation of Sediment and Radionuclide Transport in Estuaries - Testing of Three-Dimensional Radionuclide Transport Modeling for the Hudson River Estuary, N.Y. NUREG/CR-2423, PNL-4109, Pacific Northwest Laboratory.

Orlob, G. 1971. Mathematical Modeling of Estuarial Systems. International Symposium on Mathematical Modeling Techniques.

Petty, J.E., W. Wakamiya, C.J. English, J.A. Strand and D.D. Mahlum. 1982. Assessment of Synfuel Spill Clean-up Options. PNL-4244, Prepared for the U.S. Department of Energy by the Pacific Northwest Laboratory, Battelle Memorial Institute.

Pritchard, D.W. 1969. Dispersion and Flushing of Pollutants in Estuaries. Journal of Hydraulics Division, ASCE, HY1, pp. 115-124.

Raj, P.K. and P.M. O'Farrel. 1977. Development of Additional Hazard Assessment Models. CG-D-36-77, U.S. Coast Guard, Dept. of Transportation, Washington, D.C.

Rausch, A.H., R.M. Kumar and L.J. Lynch. 1977. A Critical Technical Review of Six Additional Hazard Assessment Models. CG-D-54-77. U.S. Coast Guard, Dept. of Transportation, Washington, D.C.

Schnoor, J. 1981. Fate and Transport Modeling for Toxic Substances. Presented at the Pellston Conference on Modeling the Fate of Chemicals in the Aquatic Environment, Pellston, MI.

Schnoor, J.L., A.R. Giaquinta, C. Sato, C.P. Robinson and D.B. McDonald. 1982. Refinement and Verification of Predicted models of Suspended Sediment Dispersion and Desorption of Toxics from Dredged Sediments. IIHR Report No. 249, Iowa Institute of Hydraulic Research, The University of Iowa, Iowa City, IA.

Schnoor, J. and D.C. McAvoy. A Pesticide Transport and Bioconcentration Model, Submitted to Journal of American Society of Civil Engineers, Engineering Division. (In press).

Smith, R.C. and K.S. Baker. 1981. Optical Properties of the Clearest Natural Waters (200-800 NM). Applied Optics, Volume 20, No. 2, pp. 177-184.

Task Committee on Preparation of Sedimentation Manual, 1971. Sediment Discharge Formulas. Journal of the Hydraulics Division, ASCE. Vol. 97, No. HYU, Proc. Paper No. 7786.

Thibodeaux, L. 1979. <u>Chemodynamics: Environmental Movement of Chemicals in Air, Water, and Soil</u>. Wiley-Interscience, NY.

Thomann, R.V. 1972. Systems Analysis and Water Quality Management. McGraw-Hill, New York, N.Y.

Torguimsen, G. 1982. A Comprehensive Model for Oil Spill Simulation. NOAA, Modeling and Simulation Studies, Seattle, WA.

Tracor Inc., ed. 1971. Estuarine Modeling: An Assessment-Water Pollution Control Research Series, No. 16070 DZV 02/71, U.S. Environmental Protection Agency, Water Quality Office, Washington, D.C.

Vanoni, V.A., editor. 1975. <u>Sedimentation Engineering</u>. ASCE-Manuals and Reports on Engineering Practice - No. 54, New York, N.Y.

Wechsler, B.A. and D.R. Cogley. 1977. A Laboratory Study of the Turbidity Generation Potential of Sediments to be Dredged. Technical Report D-77-14, U.S. Army Engineer Waterway, Experiment Station, CE, Vicksburg, MS.

Wolfe, N.L. 1980. Determining the Role of Hydrolysis in the Fate of Organics in Natural Waters. In: <u>Dynamics, Exposure, and Hazard Assessment of Toxic Chemicals</u> (R. Haque, Ed.), Ann Arbor Science, Ann Arbor, MI.

Yotsukura, N. and E.D. Cobb. 1972. Transverse Diffusion of Solutes in Natural Streams. U.S. Geological Survey Professional Paper 582-C.

Zaccor, J. 1981. A Mobile Stream Diversion System for Hazardous Materials Spills Isolation. EPA - 600/S2-81-219.

Zison, S.W., W.B. Mills, D. Deimer and C.W. Chen. 1978. Rates, Constants, and Kinetics Formulations in Surface Water Quality Modeling. EPA 600/3-78-105, U.S. Environmental Protection Agency, Environmental Research Laboratory, Athens, GA.